MW00974794

Aquaculture Engineering

Odd-Ivar Lekang

Department of Mathematical Sciences and Technology,
Norwegian University of Life Sciences

Blackwell
Publishing

Blackwell Publishing editorial offices:
Blackwell Publishing Ltd, 9600 Garsington Road, Oxford OX4 2DQ, UK
Tel: +44 (0)1865 776868
Blackwell Publishing Professional, 2121 State Avenue, Ames, Iowa 50014-8300, USA
Tel: +1 515 292 0140
Blackwell Publishing Asia Pty Ltd, 550 Swanston Street, Carlton, Victoria 3053, Australia
Tel: +61 (0)3 8359 1011

First published 2007 by Blackwell Publishing Ltd

ISBN: 978-1-4051-2610-6

Library of Congress Cataloging-in-Publication Data
Lekang, Odd-Ivar. Aquaculture engineering / Odd-Ivar Lekang.
p. cm.
Includes bibliographical references and index.
ISBN: 978-1-4051-2610-6 (hardback : alk. paper)
1. Aquacultural engineering. I. Title.
SH137.L45 2006
639.8–dc22
2006019514

A catalogue record for this title is available from the British Library

For further information on Blackwell Publishing, visit our website:
www.blackwellpublishing.com

Contents

Preface

The aquaculture industry, which has been growing at a very high rate for many years now, is projected to continue growing at a rate higher than most other industries for the foreseeable future. This growth has mainly been driven by static catches from most fisheries and a decline in stocks of many major commercially caught fish species, combined with the ever increasing need for marine protein due to continuing population growth. An increased focus on the need for fish in the diet, due to mounting evidence of the health benefits of eating more fish, will also increase the demand.

There has been rapid development of technology in the aquaculture industry, particularly as used in intensive aquaculture where there is high production per m^3 farming volume. It is predicted that the expansion of the aquaculture industry will lead to further technical development with more, and cheaper, technology being available for use in the industry in the future years.

The aim with this book is to give a general overview of the technology used in the aquaculture industry. Individual chapters focus on water transfer, water treatment, production units and additional equipment used on aquaculture plants. Chapters where equipment is set into systems, such as land-based fish farms and cage farms, are also included. The book ends with a chapter which includes systematic methodology for planning a full aquaculture facility.

The book is based on material successfully used on BSc and MSc courses in intensive aquaculture given at the Norwegian University of Life Science (UMB) and refined over many years, the university having included courses in aquaculture since 1973. In 1990 a special Master's course was developed in aquaculture engineering (given in Norwegian), and from 2000 the university has also offered an English language international Master's programme in aquaculture (see details at www.umb.no).

During the author's compilation of material for use in this book, and also for earlier books covering similar fields (in Norwegian), many people have given useful advice. I would like especially to thank Svein Olav Fjæra and Tore Ensby. Further thanks also go to my colleagues at UMB: B.F. Eriksen, P.H. Heyerdal, T.K. Stevik, and from earlier, colleagues and students: V. Tapei. Mott, A. Skar, P.O. Skjervold, G. Skogesal and D. E. Thommassen.

Tore Ensby has drawn all the line illustrations contained in the book. All the photographs included in the book have been taken by the author.

O.I. Lekang
November 2006

1
Introduction

1.1 Aquaculture engineering

During the past few years there has been considerable growth in the global aquaculture industry. Many factors have made this growth possible. One is development within the field of aquaculture engineering, for example improvements in technology allowing reduced consumption of freshwater and development of re-use systems. Another is the development of offshore cages: sites that until a few years ago not were viable for aquaculture purposes can be used today with good results. The focus on economic efficiency and the fact the salaries are increasing have also resulted in the increased use of technology to reduce staff numbers.

The development of new aquaculture species would not have been possible without the contribution of the fisheries technologist. Even if some techniques can be transferred for the farming of new species, there will always be a need for technology to be developed and optimized for each species. An example of this is the development of production tanks for flatfish with a larger bottom surface area than those used for pelagic fish.

Aquaculture engineering covers a very large area of knowledge and involves many general engineering specialisms such as mechanical engineering, environmental engineering, materials technology, instrumentation, and monitoring, and building design and construction. The primary aim of aquaculture engineering is to utilize technical engineering knowledge and principles in aquaculture and biological production systems. The production of fish has little in common with the production of nails, but the same technology can be used in both production systems. It is therefore a challenge to bring together both technological and biological knowledge within the aquaculture field.

1.2 Classification of aquaculture

There are a number of ways to classify aquaculture facilities and production systems, based on the technology or the production system used.

'Extensive', 'intensive' and 'semi-intensive' aquaculture are common ways to classify aquaculture based on production per unit volume (m^3) or unit area (m^2) farmed. Extensive aquaculture involves production systems with low production per unit volume. The species being farmed are kept at a low density and there is minimal input of artificial substances and human intervention. A low level of technology and very low investment per unit volume farmed characterize this method. Pond farming without additional feeding, like some carp farming, is a typical example. Sea ranching and restocking of natural lakes may also be included in this type of farming.

In intensive farming, production per unit volume is much higher and more technology and artificial inputs must be used to achieve this. The investment costs per unit volume farmed will of course also be much higher. The maintenance of optimal growth conditions is necessary to achieve the growth potential of the species being farmed. Additional feeding, disease control methods and effective breeding systems also characterize this type of farming. The risk of disease outbreaks is higher than in extensive farming because the organism is continuously stressed for maximal performance.

Salmon farming is a typical example of intensive aquaculture.

It is also possible to combine the above production systems – this is called semi-intensive aquaculture. An example is intensive fry production combined with extensive ongrowing.

Another classification of an aquacultural system can be according to the life stage of the species produced on the farm, for instance eggs, fry, juvenile or ongrowing. Farms may also cover the complete production process, and this is called full production.

According to the type of farming technology used there are also a number of classifications based on the design and function of the production unit. This will of course be species and life-stage dependent. For fish the following classifications may be used: 1. Closed production units where the fish are kept in a enclosed production unit separated from the outside environment. 2. Open production units where the unit has permeable walls, such as nets and so the fish are partly affected by the surrounding environment. It is also possible to classify the farm based on where it is located: within the sea, in a tidal zone or on land.

Land-based farms may be classified by the type of water supply for the farm: water may be gravity-fed or pumped. In gravity systems the water source is at a higher altitude than the farm and the water can flow by gravity from the source to the farm. When pumping, the source can be at an equal or lower altitude compared to the farm. For tidal through-flow farms, water supply and exchange is achieved using the tide.

Farms can also be classified by how the water supplied to a farm is used. If the water is used once, flowing directly through, it is named a flow-through farm. If the water is used several times, with the outlet water being recycled, it is a water re-use or recirculating system.

1.3 The farm: technical components in a system

In a farm the various technical components included in a system can be roughly separated as follows:

- Production units
- Water transfer and treatment
- Additional equipment (feeding, handling and monitoring equipment)

To illustrate this, two examples are given: a land-based hatchery and a juvenile farm, and an ongrowing sea cage farm.

1.3.1 Land-based hatchery and juvenile production farm

Land-based farms normally utilize much more technical equipment than sea cage farms, especially intensive production farms with a number of tanks. The major components are as follows (Fig. 1.1):

- Water inlet and transfer
- Water treatment facilities
- Production units
- Feeding equipment
- Equipment for internal fish transport and size grading
- Equipment for transport of fish from the farm
- Equipment for waste and wastewater treatment
- Instrumentation and monitoring systems

Water inlet and transfer

The design of the inlet depends on the water source: is it seawater or freshwater (lakes, rivers), or is it surface water or groundwater? It is also quite common to have several water sources in use on the same farm. Further, it depends whether the water is fed by gravity or whether it has to be pumped, in which case a pumping station is required. Water is normally transferred in pipes, but open channels may also be used.

Water treatment facilities

Usually water is treated before it is sent in to the fish. Equipment for removal of particles prevents excessively high concentrations reaching the fish; additionally large micro-organisms may be removed by the filter. Water may also be disinfected to reduce the burden of micro-organisms, especially that used on eggs and small fry. Aeration may be necessary to increase the concentration of oxygen and to remove possible supersaturation of the gases nitrogen and carbon dioxide. If there is lack of

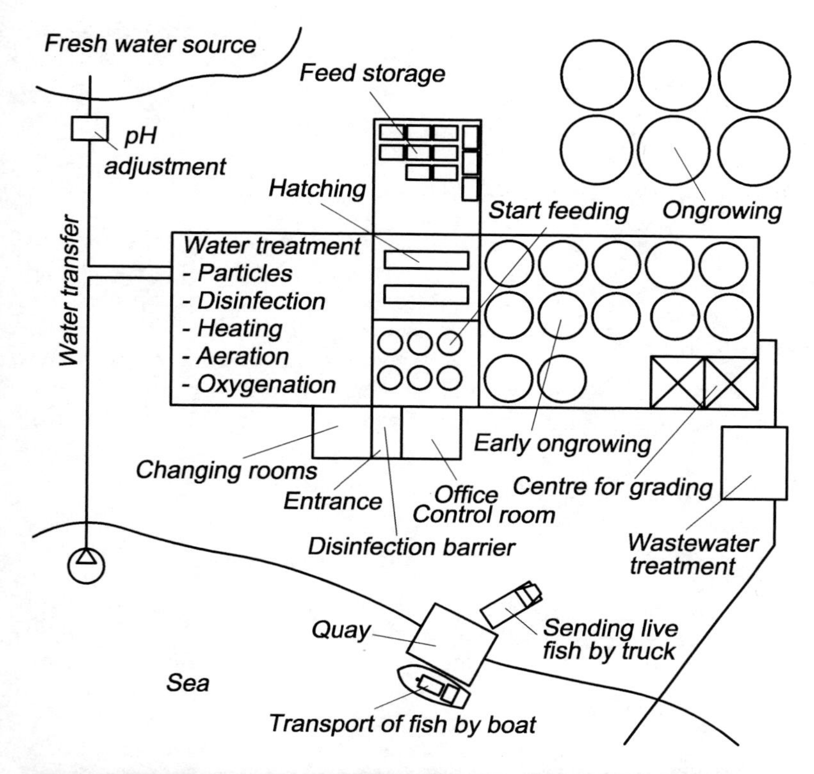

Figure 1.1 Example of major components in a land-based hatchery and juvenile production plant.

water or the pumping height is large pure oxygen gas may be added to the water. Another possibility if water supply is limited is to reuse the water, however, this will involve much water treatment. For optimal development and growth of the fish heating or cooling of the water may be necessary; in most cases this will involve a heat pump or a cold-storage plant. If the pH in a freshwater source is too low pH adjustment may be a part of the water treatment.

Production units

The production units necessary and their size and design will depend on the species being grown. In the hatchery there will either be tanks with upwelling water (fluidized eggs) or units where the eggs lie on the bottom or on a substrate. After hatching the fish are moved to some type of production tank. Usually there are smaller tanks for weening and larger tanks for further on-growing until sale. Weening start feeding tanks are normally under a roof, while on-growing tanks can also be outside.

Feeding equipment

Some type of feeding equipment is commonly used, especially for dry feed. Use of automatic feeders will reduce the manual work on the farm. Feeding at intervals throughout the day and night may also be possible; the fish will then always have access to food, which is important at the fry and juvenile stages.

Internal transport and size grading

Because of fish growth it is necessary to divide the group to avoid fish densities becoming too high. It is also common to size grade to avoid large size variations in one production unit; for some species this will also reduce the possibilities for cannibalism.

Transport of fish

When juvenile fish are to be transferred to an on-growing farm, there is a need for transport. Either a truck with water tanks or a boat with a well is normally used. The systems for loading may be an integral part of the farm construction.

Equipment for waste handling and wastewater treatment

Precautions must be taken to avoid pollution from fish farms. These include legal treatment of general waste. Dead fish must be treated and stored satisfactorily, for example, put in acid or frozen for later use. Dead fish containing trace of antibiotics or other medicine must be destroyed by legal means. Whether wastewater treatment is necessary will depend on conditions where the effluent water is discharged. Normally there will at least be a requirement to remove larger suspended particles.

Instrumentation and monitoring

In land-based fish farms, especially those dependent upon pumps, a monitoring system is essential because of the economic consequences if pumping stops and the water supply to the farm is interrupted. The oxygen concentration in the water will fall and may result in total fish mortality. Instruments are being increasingly used to control water quality, for instance, to ensure optimal production.

1.3.2 On-growing sea cage farm

Normally a sea cage farm can be run with rather less equipment than land-based farms, the major reason being that water transfer and water treatment (which is not actually possible) are not necessary because the water current ensures water supply and exchange. The components necessary are as follows (Fig. 1.2):

- Production units
- Feeding equipment
- Working boat
- Equipment for size grading
- Base station

Production units

Sea cages vary greatly in construction and size; the major difference is the ability to withstand waves, and special cages for offshore farming have been developed. It is also possible to have system cages comprising several cages, or individual cages. The cages may also be fitted with a gangway to the land. Sea cages also include a mooring system. To improve fish growth, a sub-surface lighting system may be used.

Feeding equipment

It is common to install some type of feeding system in the cages because of the large amounts of feed that are typically involved. Manual feeding may

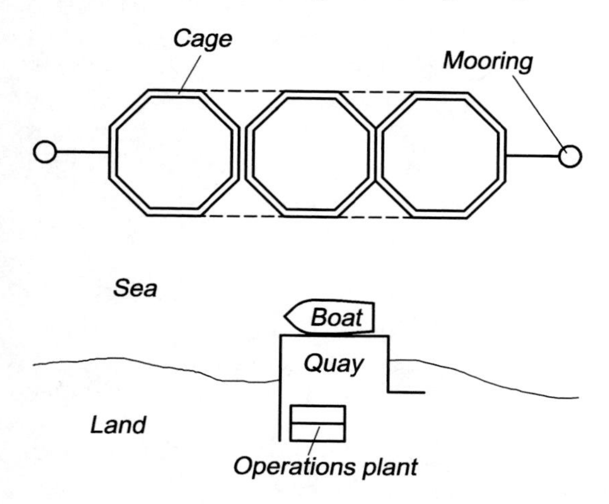

Figure 1.2 Example of major components in an on-growing sea cage farm.

also be used, but this involves hard physical labour for the operators.

Working boat

All sea cage farms need a boat; a large variety of boats are used. Major factors for selection are size of the farm, whether it is equipped with a gangway or not, and the distance from the land base to the cages. Faster and larger boats are normally required if the cages are far from land or in weather-exposed water.

Size grading

Equipment for size grading can be necessary if this is included in the production plan. It may, however, be possible to rent this as a service from subcontractors.

Base station

All cages farms will include a base station; this may either be land based, floating on a barge or both. The base station can include storage rooms, mess rooms, changing rooms and toilet, and equipment for treatment of dead fish. The storage room includes rooms and/or space for storage of feed; it may also include rooms for storage of nets and possibly storage of equipment for washing, maintaining and impregnating them. However, this is also a service that is commonly rented from subcontractors.

1.4 Future trends: increased importance of aquaculture engineering

Growth in the global aquaculture industry will certainly continue, with several factors contributing to this. The world's population continues to grow as will the need for marine protein. Traditional fisheries have limited opportunities to increase their catches if sustainable fishing is to be carried out. Eventually, therefore, increase in production must come from the aquaculture industry. In addition, the aquaculture industry can deliver aquatic products of good quality all year round, which represents a marketing advantage compared to traditional fishing. The increased focus on optimal human diets, including more fish than meat in the diet for large groups of the world's population, also requires more fish to be marketed.

This will give future challenges for aquaculture engineers. Most probably there will be an increased focus on intensive aquaculture with higher production per unit volume. Important challenges to housing the growth will be availability of freshwater resources and good sites for cage farming. Limited supplies of freshwater in the world mean that technology that can reduce water consumption per kilogram of fish produced will be important; this includes reliable, cost effective re-use technology. By employing re-use technology it will also be possible to maintain a continuous supply of high quality water independently of the quality of the incoming water. To have more accurate control over water quality will also be of major importance when establishing aquaculture with new species, especially during the fry production stage.

The trend to use more and more weather-exposed sites for cage farms will continue. Development of cages that can not only withstand adverse weather conditions but also be operated easily in bad weather, and where fish feeding and control can be performed, is important.

Rapid developments in electronics and monitoring will gradually become incorporated into the aquaculture industry. Intensive aquaculture will develop into a process industry where the control room will be the centre of operations and processes will be monitored by electronic instruments; robots will probably be used to replace some of today's manual functions. Nanotechnology will be included, for instance by using more and smaller sensors for more purposes. An example would be to include sensors in mooring lines and net bags to monitor tension and eventual breakage. Individual tagging of the fish will most probably also be a future possibility, which makes control of the welfare of the single individual possible; this can also be important regarding control of escaped fish.

1.5 This textbook

The aim with this book is to give a general basic review of the total area of aquaculture engineering. Based on the author's two previously published books on aquaculture engineering written in Norwegian.[1,2] Several of the illustrations are also based on illustrations in these books. The textbook is primarily intended for the introductory course in aquaculture engineering for the Bachelor and Master degrees in aquaculture at the Norwegian University of Life Science (UMB). Several other textbooks dealing with parts of the syllabus are available and referred to in later chapters. The same is the case with lecture notes from more advanced courses in aquaculture engineering at UMB.

The focus of the book is on intensive fish farming, where technology is and will become increasingly important. Most of it concerns fish farming, but several of the subjects are general and will have much interest for molluscan and crustacean shellfish farmers.

Starting with water transport, the book continues with an overview chapter on water quality and the need for and use of different water treatment units, which are described in the next few chapters. It continues with a chapter on production unit classification followed by chapters on the different production units. Chapters devoted to additional equipment such as that for feed handling and fish handling, instrumentation, monitoring and buildings follow. Chapters on planning of aquaculture facilities and their design and construction conclude the book.

References

1. Lekang, O.I., Fjæra, S.O. (1997) *Teknologi for akvakultur*. Landbruksforlaget (in Norwegian).
2. Lekang, O.I., Fjæra, S.O. (2002) *Teknisk utstyr til fiskeoppdrett*. Gan forlag (in Norwegian).

2
Water Transport

2.1 Introduction

All aquaculture facilities require a supply of water. It is important to have a reliable, good-quality water source and equipment to transfer water to and within the facility. The volume of water needed depends on the facility size, the species and the production system. In some cases can it be very large, up to several hundred m^3/min (Fig. 2.1). This is equivalent to the water supply to a quite large villages, considering that in Norway a normal value for the water supply per person is up 180 litres per day.

If something fails with the water supply or distribution system it may result in disaster for the aquaculture facilities. This also emphasizes the importance of good knowledge in this area. Correct design and construction of the water inlet system is an absolute requirement to avoid large unnecessary problems in the future. For instance, this may be apparent when the inlet system is too small and the water flow rate to the facility is lower than expected.

The science of the movement of water is called hydrodynamics, and in this chapter the important factors of this field are described with emphasis upon aquaculture. In addition, a description of the actual materials and parts for water transport are given: pipes, pipe parts (fittings) and pumps. Much more specific literature is available in all these fields (basic fluid mechanics,[1–3] pipes and pipe parts,[4–6] pumps[7–9]).

2.2 Pipe and pipe parts

2.2.1 Pipes

Pipe materials

In aquaculture the common way to transport water is through pipes; open channels are also used in some cases. Channels may be used for transport into the farm, for distribution inside the farm and for the outlet of water. They are normally built of concrete and are quite large; the water is transported with low velocity. Channels may also be excavated in earth, for example to supply the water to earth ponds. Advantages of open channels are their simple construction and the ease with which the water flow can be controlled visually; disadvantages are the requirement for a constant slope over the total length and there can be no pressure in an open channel. The greater exterior size compared to pipes, and the noise inside the building when water is flowing are other disadvantages.

Plastics, mainly thermoplastics, are the most commonly used materials for pipes. Thermoplastic pipes are delivered in many different qualities with different characteristics and properties (Table 2.1). A thermoplastic melts when the temperature get high enough.[10] Thermoplastic pipes can be divided into weldable (polyethylene; PE) and glueable (polyvinyl chloride; PVC) depending on the way the pipes are connected. The opposite of

Figure 2.1 The supply of water to a fish farm can be up to several hundred cubic metres per minute, as here for a land-based fish farm for growing of marked size Atlantic salmon.

Table 2.1 Typical characteristics of actual pipe materials.

Material	Temperature range (°C)	Common pressure classes (bar)	Common size range (mm)
PE	–40 to +60	3.2, 4, 6, 10 and 25	20–1600
PP	0 to +100	10 and 16	16–400
PVDF	–40 to +140	16	16–225
PVC-U	0 to +60	4, 6, 10, 16 and 25	6–400
PVC-C	0 to +80	16	16–225
ABS	–40 to +60	16	16–225

thermoplastic is hardening plastic, such as fibreglass which is made of different materials that are hardened; afterwards it is impossible to change its shape, even by heating. Fibreglass can be used in special critical pipes and pipe parts, but only in special cases (see below).

It is also important that materials used for pipes are non-toxic for fish.[11] Copper, much used in piping inside houses, is an example of a commonly used material that is not recommended for fish farming because of its toxicity. In the past steel, concrete or iron pipes were commonly used, but today these materials are seldom chosen because of their price, duration and laying costs.

PE pipes are of low weight, simple to handle, have high impact resistance and good abrasion resistance. Nevertheless, these pipes may be vulnerable to water hammer (see later) or vacuum effects. PE pipes are delivered in a wide variety of dimensions and pressure classes; they are normally black or grey but other colours are also used. Small diameter pipes may be delivered in coils, while larger sizes are straight, with lengths commonly between 3 and 6m. PE may be used for both inlet and outlet pipes. PE piping must be fused together for connection; if flanges are fused to the pipe fittings, pipes may be screwed together.

PVC is used in pipes and pipe parts inside the fish farm and also in outlet systems. This material is of low density and easy to handle. Pipe and parts are simple to join together with a special solvent cementing glue. A cleaning liquid dissolves the surface and makes gluing possible. There are a large variety of pipe sizes and pipe parts available. When using this kind of piping, attention must be given to the temperature: below 0°C this material becomes brittle and will break easily. PVC is also recommended for use at temperatures above 40°C. PVC is also vulnerable to water hammer. There are questions concerning the use of PVC materials because poisonous gases are emitted during burning of left-over material.

Fibreglass may be used in special cases, for example in very large pipes (usually over 1 m in diameter). The material is built in two or three layers: a layer of polyester that functions like a glue; a layer with a fibreglass mat that acts as reinforcement; and quartz or sand. The ratio between these components may vary with the pressure and stiffness needed for the pipe. A pipe is normally constructed with several layers of fibreglass and polyester. Fibreglass has the advantages that it tolerates low temperatures, is very durable and may be constructed so thick that it can tolerate water hammer and vacuum effects. The disadvantage is the low diversity of pipes and pipe parts available. For joining of parts, the only possibilities are to construct sockets on site using layers of polyester and fibreglass, or pipes equipped with flanges by the manufacturer can be screwed together with a gasket in between.

At present, materials such as polypropylene (PP), acrylonitrile–butadiene–styrene (ABS) and polyvinyl difluoride (PVDF) have also been introduced for use in the aquaculture industry, but to minor degree and for special purposes. They are also more expensive than PE and PVC.

Pressure class

Each pipe and pipe part must be thick enough to tolerate the pressure of water flowing through the system. To install the correct pipes it is therefore important to know the pressure of the water that will flow through them. The pressure (PN) class indicates the maximum pressure that the pipes and pipe parts can tolerate. The pressure class is given in bar (1 bar = 10 m water column (mH_2O) = 98 100 Pa); for instance, a PN4 pipe will tolerate 4 bar or a 40 m water column. This means that if the pressure inside the pipe exceeds 4 bar the pipe may split. In fish farming pressure classes PN4, PN6 and PN10 are commonly used. Pipes of different PN classes vary regarding wall thickness: higher pressure requires thicker pipe walls. Pipes of higher PN class will of course cost more, because more material is required to make them.

A complete inlet pipe from the source to the facility may be constructed with pipes of different PN classes. If, for instance, the water source to a fish farm is a lake located 100 m in height above the farm, a PN4 pipe can be used for the first 40 m drop, then a PN6 pipe for the following 20 m drop, and on the final 40 m drop a PN10 pipe is used.

Some problems related to pressure class are as follows:

Water hammer: Water hammer can occur, for instance, when a valve in a long pipe filled with much water is closed rapidly. This will generate high local pressure in the end of the pipe, close to the valve, from the moving mass of water inside the pipeline that needs some time to stop. The result is that the pipe can 'blow'. Rapid closing of valves must therefore be avoided. Water hammer may also occur with rapid starting and stopping of pumps. This can, however, be difficult to inhibit and it may be necessary to use special equipment to damp the water hammer effect. A tank with low pressure air may be added to the pipe system: if there is water hammer in the pipes the air in this tank will be compressed and this reduces the total hammer effect in the system.

Vacuum: A vacuum may be generated in a section of pipe, for example when it is laid at different heights (over a crest) and then functions as a siphon (Fig. 2.2). A vacuum may then occur on the top crest. It is recommended that such conditions be avoided, because the pipeline may become deformed and collapse because of the vacuum. Pipes are normally not certified for vacuum effects; however, if vacuum effects are possible, it is recommended that a pipe of higher pressure class is used in the part where the vacuum may occur. By using pipes with thicker walls, higher tolerance to vacuum effects is achieved; alternatively, a fibreglass pipe which tolerates a higher vacuum could be employed.

Classification of pipes

Pipe diameters are standardized. There are a number of sizes available for various applications in different industries. In aquaculture, pipes with the following external diameters (mm) are generally used 20, 25, 32, 40, 50, 63, 75, 90, 110, 125, 160, 180, 200, 225, 250, 280, 315, 355, 400, 450, 500, 560 and 630. The internal diameter that is used when

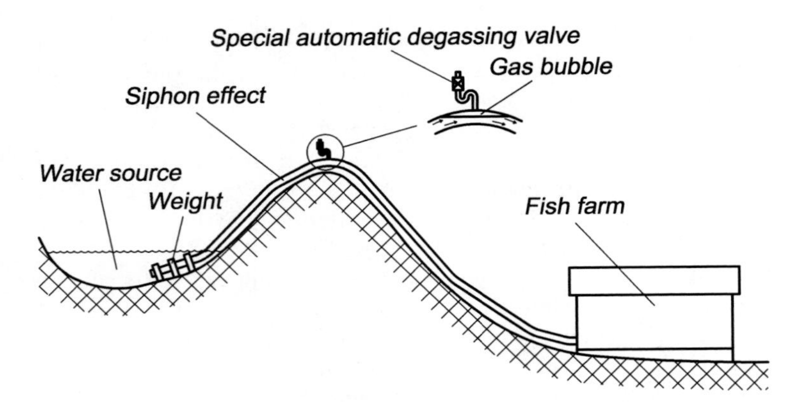

Figure 2.2 A vacuum may occur inside the pipe on the top crest causing deformation.

calculating the water velocity in the pipelines, is found by subtracting twice the wall thickness. Higher pressure class pipes have thicker walls than lower pressure class pipes.

All pipes and pipe parts must be marked clearly. For pipes the marking print on the pipe is normally every metre, and for pipe parts there is a mark on every part. The following is included in the standardized marking: pipe material, pressure class, external diameter, wall thickness, producer and the time when the pipe was produced. It is important to use standardized pipe parts when planning fish farms.

2.2.2 Valves

Valves are used to regulate the water flow rate and the flow direction. Many types of valve are used in aquaculture (Fig. 2.3). Which type to use must be chosen on the basis of the flow in the system and the specific needs of the farm. Several materials are used in valves, such as PVC, ABS, PP and PVDF, and the material chosen depends on where the valves will be used. Large valves may also be fabricated in stainless or acid proof steel.

Ball valves are low cost solutions used in aquaculture. The disadvantage is that they are not very precise when regulating the water flow. They are best used in an on/off manner, or for approximate regulation of the water flow. The design is simple and consists of a ball with an opened centre. When turning it will gradually open or close.

Valves constructed with a membrane pulled down by a piston for regulation of water flow are called diaphragm or membrane valves. These valves can regulate water flow very accurately. They cost considerably more than a ball valve, and the head loss through the valve is significantly higher.

Angle seat valves have a piston standing in an angled 'seat'. When the screw handle is turned the piston moves up or down. The opening is gradually reduced by pressing the piston down. This type of valve is also capable of accurate flow regulation, but is quite expensive and has also a higher head loss than a ball valve. For accurate flow regulation, for instance on single tanks, diaphragm valves or angle seat valves are recommended. When selecting these types of valves it is, however, important to be aware that the head loss can be over five times as high as with a ball valve.

Butterfly valves are usually located in large pipes (main pipeline or part pipelines) and regulate the water flow by opening or closing a throttle. Which is located inside a pipe part; by turning the throttle the passage for the water inside the pipe is changed. A slide valve or gate valve can be used for the same cases. This consists of a gate or slide that stands vertically in the water flow; the water flow is regulated by lifting or lowering the plate by a spindle. This valve type is also used in large diameter pipelines, but both butterfly valves and sluice valves are quite expensive, especially in large sizes. It is, however, better to use too many valves than too few. To have the facility to turn off the water flow at several

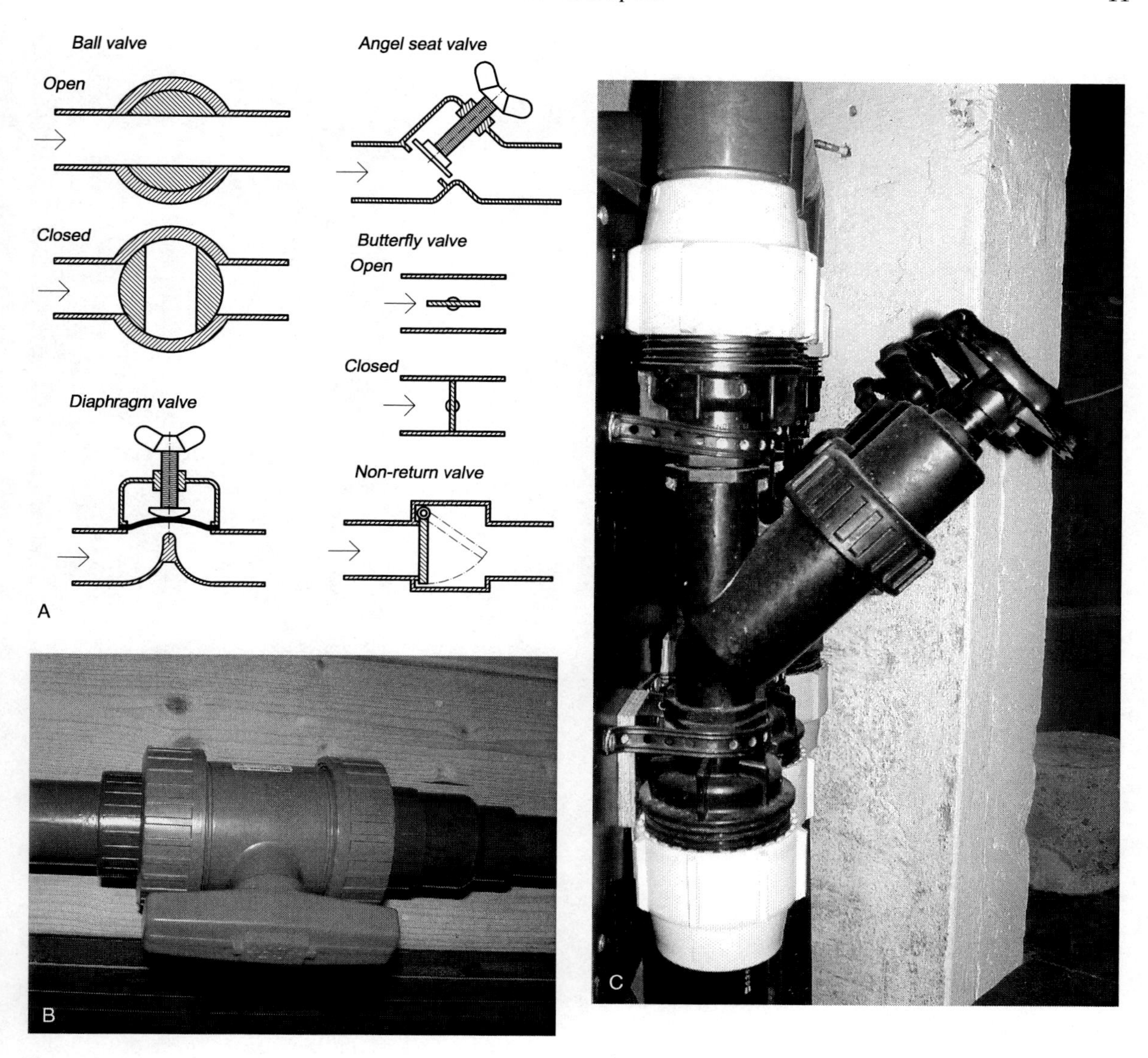

Figure 2.3 Valve types used on aquaculture facilities: (A) diagrams showing valve cross-sections; (B) ball valve; (C) angel seat valve; (D) diaphragm or membrane valve; (E) butterfly valve.

places in the farm, for instance for maintenance, will always be an advantage. These types of valves are, however, not recommended for precise regulation of water flow.

The check valve or 'non-return' valve is used to avoid the backflow of water; this means that the water can only flow in one direction in the pipe system. In many cases it is used in a pump outlet to avoid backflow of water when the pump stops. Normally the valve is comprises a plate or ball that closes when the water flow tends to go in the opposite direction.

Triple way valves may regulate the flow in two directions to create a bypass. There are also many other types of valves, for instance electrically or pneumatically operated valves which make it

Figure 2.3 *Continued.*

possible to regulate water flow functions automatically. In new and advanced fish farms such equipment is of increasing interest, especially when saving of water is necessary.

It is important to remember, however, that all valves create a head loss, the size of which depends on the type of valve being used; for example, diaphragm valves have a high head loss. This must considered when planning the farm and deciding which valve types to use. It is necessary to have enough pressure to ensure that the correct flow rate is maintained through the valves; if the head loss is too high the water flow into or inside the farm will be decreased.

2.2.3 Pipe parts – fittings

A large variety of pipe parts can be found, especially for PE and PVC pipes (Fig. 2.4). Various bends or elbows are normally used in aquaculture. T-pipes are also used to connect different pipes. Different conversion parts allow the connection of pipes or equipment with different diameters. Sockets, flanges or unions are used to connect pipes or pipe parts. Sometimes end-caps are used to close pipes that are out of use. A particularly useful part is the repair socket which allows connection of an additional pipe (a T-pipe) to a pipeline where the water in the installation flows continuously, which means that connections can be made to pipelines that are in use.

2.2.4 Pipe connections – jointing

The connection or jointing of pipes and pipe parts may be executed in various ways depending on the material used to make the pipe and the pipe part (Fig. 2.5). For PE, fusing (heating) is the only possible jointing method. This process may be carried out by a blunt heating mirror, or electrofusion may be used. When using a fusion mirror both the pipes to be joined are heated on the mirror to make them

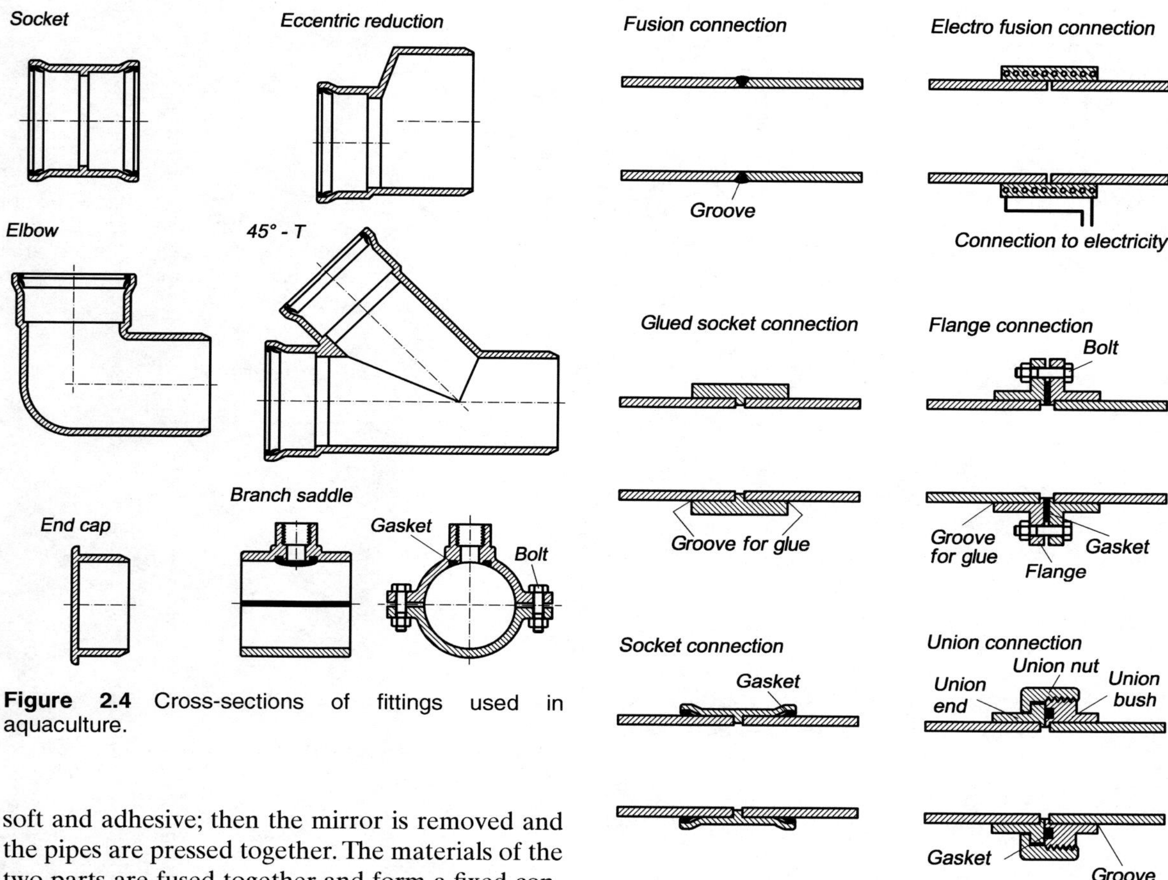

Figure 2.4 Cross-sections of fittings used in aquaculture.

Figure 2.5 Connection methods used for different pipe materials and in different places.

soft and adhesive; then the mirror is removed and the pipes are pressed together. The materials of the two parts are fused together and form a fixed connection. Resistance wire is an integral part of an electrofusion socket. When an electric current is passed the material around the wire will fuse, including the two pipe parts added to the socket, and then a fixed connection is established.

Pipes that are fused or glued together are permanently connected and cannot be separated. If there is a need to create non-permanent connections, it is possible to use flanges fused or glued to the separate pipe parts which are then screwed together. To obtain a completely watertight connection a gasket is placed between the parts before they are screwed together. A union is a very easy pipe connection to separate. It is always desirable to have some non-permanent connections because sometimes it is necessary to separate the pipeline for maintenance and exchange of equipment. The possibility of exchanging pipes and pipe parts in the water department of the fish farm must also be considered because of the need to allow for possible increase in farm production and also because of the constant requirement for modernization of the equipment.

It is common to use sliding sockets in the outlet pipe. This kind of connection system can only be used on unpressurized or very low pressure pipelines ($<0.2\,mH_2O$). If this type of connection is used on pressurized pipes they will easily slide apart.

2.2.5 Mooring of pipes

Pipes may carry large amounts of water at high velocities. This generates large forces that may cause movements of the pipe. In the worst case this

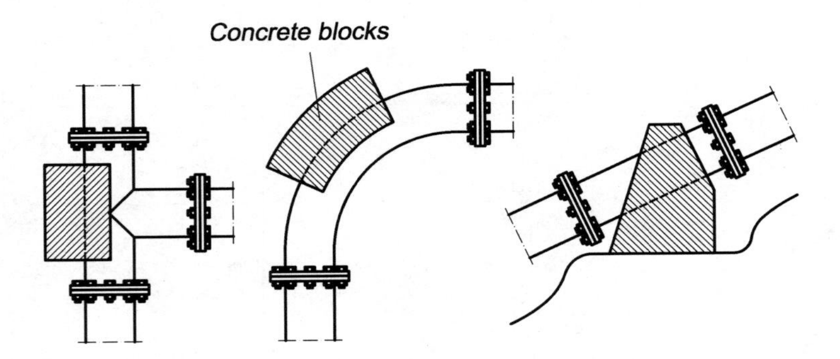

Figure 2.6 It is important that pipes are moored to avoid movement and possible breakages. On inlet pipes in the sea or in lakes specially designed block weights are used.

can damage the pipeline. For this reason a correct mooring system for the pipeline is of great importance (Fig. 2.6).

When there is a reduction in the pipe size, or when using T-pipes or elbows, there is an increase in the forces dependent on the velocity, and there will normally be a need for mooring to avoid movements and breakage of the pipes. Putting the pipes in a ditch will stabilize them and the ditch will function as a mooring for the pipe. In exposed places, however, it may also be necessary to have additional moorings; concrete blocks may, for instance, be used on elbows. This also shows the importance of having smooth pipe linings. In indoor facilities clamps are used to attach the pipes to the ceiling, walls or floor, and in this way moor and stabilize the pipes.

Inlet or outlet pipes placed underwater or subsurface in the sea or lakes require moorings. Specially designed concrete block weights are normally used to moor pipes to the ground (Fig. 2.6) to prevent them floating to the surface as a result of their buoyancy, especially when they are empty or only partly filled with water. The distance between the weights depends on the pipe type, diameter, weight size and expected water flow. When placing outlet pipes sub-surface, it is important to consider the weight of the pipe both when filled with water and filled with air, the buoyancy being much greater in the latter case which will also increase the requirements for weights. Usually pipe suppliers have their own mooring tables with recommended block weights and distance between them.

2.2.6 Ditches for pipes

The inlet and outlet pipes may be laid on the surface or in ditches (Fig. 2.7). It is generally cheaper to lay the pipes on the surface, but it is then more important to moor them, especially in connection with obstructions. Pipes on the surface may, however, inhibit transport and it can therefore be necessary to lay them in ditches. If the pipe is put in a ditch, care must be taken to avoid any damage when heavy traffic passes over the ditch. Therefore the ditch must be constructed and overfilled with gravel correctly. A ditch may be constructed in the following way: a layer of compressed crushed rock or gravel is laid as a base for the pipe in the ditch; then the pipe is laid with sufficient slope (>0.05%). Fine gravel is placed around and over the pipe to create good protection; this should only be hand compressed. Afterwards the ditch is filled with ordinary ditch material until ground level is reached. It is normal to overfill slightly because after a while the material will compress to normal terrain level.

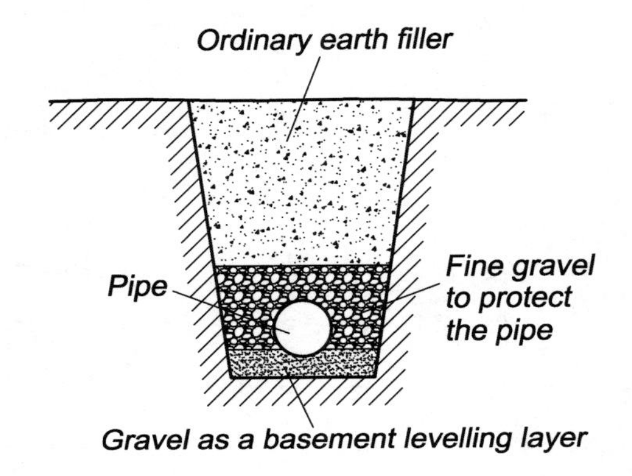

Figure 2.7 It is important that ditches for pipes are correctly designed.

The use of ditches will also improve the farm aesthetically because there are no pipes on the surface. Ditches also improve the possibilities for public traffic to use the area.

2.3 Water flow and head loss in channels and pipe systems

2.3.1 Water flow

The amount of water that flows through a pipe or in an open channel depends on the water velocity and the cross-sectional area of the pipe or the channel where the water is flowing. The following equation may be used for pipes and channels; it is also called the continuity equation:

$$Q = VA$$

where:

Q = water flow (l/min, l/s, m^3/s)
V = water velocity (m/s)
A = cross-sectional area of where the water is flowing. For full pipes the cross-sectional area will be the interior cross section of the pipe.

The above equation can be used as a basis for construction of a chart. If two of the sizes are known the last can be read from the chart and no calculation is necessary. Often the head loss is also included in the chart (see below).

Example
The water flow to a farm is 1000 l/min (0.0167 m^3/s). The acceptable velocity in the pipeline is set at 1.5 m/s. Find the necessary pipe dimensions if one pipe is to be used.

$$A = Q/V$$

$$A = 0.0167/1.5 = 0.011\,\text{m}^2$$

Now

$$A = \pi r^2$$

where:
r is the internal radius of the pipe and rearranging gives

$$r = \sqrt{\frac{A}{\pi}}$$

Therefore

$$r = \sqrt{\frac{0.011}{\pi}} = 0.059\text{m} = 59\text{mm}$$

The internal diameter in the pipe must therefore be 2 × 59 = 118 mm. Standard dimension pipes are available with an exterior diameter of 125 mm; a PN6 pipe with a wall thickness of 6 mm (supplier information) therefore has an internal diameter of 113 mm. This is actually slightly too small, but as the next stardard exterior dimension is 160 mm, it is best to choose the 125 mm pipe. This will result in the water velocity being slightly higher.

For an open channel the flow velocity depends on the slope, the hydraulic radius and the Manning coefficient. The Manning equation is used to calculate the flow velocity:

$$V = \frac{R^{\frac{2}{3}} S^{\frac{1}{2}}}{n}$$

where:

V = average flow velocity in the channel
R = hydraulic radius
S = channel slope
n = Manning coefficient.

The hydraulic radius is the ratio between the cross-sectional area where the water is flowing and the wetted perimeter, which is the length of the wetted surface of the channel measured normal to the flow.

$$R = \frac{\text{cross-sectional area}}{\text{wetted perimeter}}$$

To achieve water transport through the channel it must be inclined. The slope is defined as the ratio between the difference in elevation between two points in the channel and the horizontal distance between the same two points.

Example
The horizontal distance between two points A and B is 500 m. Point A is 34 m above sea level and point B

is 12 m above sea level. Calculate the slope (S) of the channel.

$$S = (34\,\text{m} - 12\,\text{m})/500\,\text{m} = 0.044 = 4.4\,\text{cm/m}$$

This means that for each metre of elevation the horizontal distance is 22.7 m.

To ensure drainage, it is recommended that the slope is more than 0.0013, while self-cleaning is ensured with slopes in the range 0.005–0.010.[11]

The Manning coefficient is determined by experiment, some actual values being about 0.015 for concrete-lined channels and 0.013 for plastic, while unlined channels made of straight and uniform earth have a value of 0.023 and those made of rock 0.025.[11]

Based on the flow velocity and the cross-sectional area, the flow may rate be calculated with the continuity equation which also can be expressed as:

$$Q = VA = \frac{R^{\frac{2}{3}} S^{\frac{1}{2}}}{n} A$$

where:

Q = water flow
A = cross-sectional area where the water is flowing
V = average flow velocity in the channel
R = hydraulic radius
S = slope of the channel
n = Manning coefficient.

2.3.2 Head loss in pipelines

All transport of water through a pipe or a channel between two points results in an energy loss (head loss). This is caused by friction between the water molecules and the surroundings. In all pipe parts where there is a change in the water direction (bends) or narrow passage (valves) additional friction will occur; this will also increase the head loss.

Inside a pipe there is a velocity gradient, with the highest water velocity in middle of the pipe and the lowest close to the pipe walls because friction is highest against and close to the wall. In addition to friction loss against the wall there will be friction between the water molecules because their velocities are not equal.

The amount of energy in water is constant (Bernoulli equation) if during passage no energy is supplied to or extracted from it. When friction occurs, the energy in the water is transformed into another form of energy, normally heat. This is very difficult to perceive with the large amounts of water that are common in aquaculture, since much energy is required to heat the water (see Chapter 7). However, in a thin pipe with a large amount of water passing through at very high pressure, it is possible to observe heating of the water.

As a result of frictional losses when flowing through a pipeline, the energy of the water must be higher at the beginning (inlet) than at the end (outlet); energy lines can be used to illustrate this. If the water is pumped, the pump pressure must overcome these frictional energy losses in addition to the pump height.

The energy loss (h_m) due to friction through a pipeline may be calculated using the Darcy–Weisbach equation:

$$h_m = fLV^2/2gd$$

where:

f = friction coefficient
L = length of pipeline
d = diameter of pipeline (wet)
V = water velocity
g = acceleration due to gravity.

The friction coefficient depends on the pipe surface; this is normally called the roughness of the pipe. Relative roughness (r) is defined as the relation between the absolute roughness (e) and the diameter (d) of the pipe, $r = e/d$ (Fig. 2.8). High

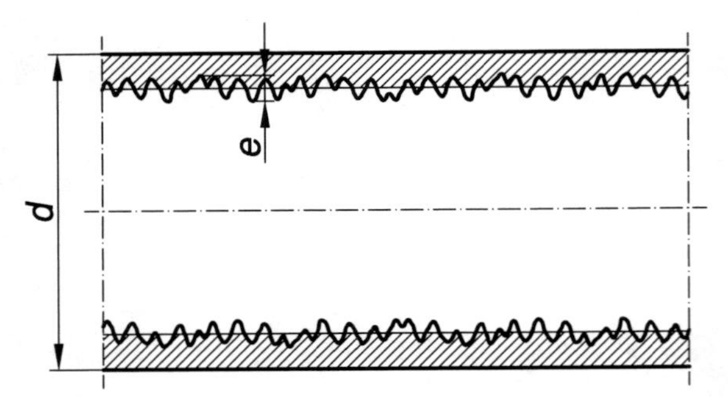

Figure 2.8 Relative roughness describes the relation between the absolute roughness (e) and the pipe diameter (d).

relative roughness gives high friction. The amount of friction depends on the pipe material, the connection method and the age of the pipe. For example, a new plastic pipe will have a lower friction coefficient than an old pipe. The fouling that occurs in pipes that have been in use for some time will increase the roughness of the pipe. The f value for the pipe is given by the manufacturer and for PE or PVC pipes normally ranges from 0.025 to 0.035. For new pipes the value is lower, but when doing calculations values for old pipes must be used.

The friction coefficient also depends on flow type. The flow pattern can be divided into laminar and turbulent. The frictional losses are much higher with turbulent flow. This will always be the case in pipes used in aquaculture, because the water velocity is so high. Laminar flow may occur in open channels with low water velocity. The Reynolds number $\overline{R_e}$ is a non-dimensional number used to describe the flow conditions. If $\overline{R_e}$ is less than 2000 the flow is laminar; when it is above 4000 the flow is turbulent. Between these $\overline{R_e}$ values the flow is unstable and both turbulent and laminar conditions may occur. $\overline{R_e}$ can be calculated from the following equation:

$$\overline{R_e} = \frac{Vd}{\nu}$$

where:

V = average water velocity
d = internal pipe diameter
ν = kinematic viscosity.

Kinematic viscosity is the absolute viscosity divided by the density of the liquid; the unit is m^2/s (formerly the stoke was used: $1\,St = 16^{-4}\,m^2/s$). The kinematic viscosity tells us something about how easily the liquid flows: for instance, oil will flow out slowly when drops are allowed to fall onto a horizontal plate, while water will be distributed much faster. The kinematic viscosity of water decreases with temperature; for example, is it reduced from $1.79 \times 10^{-6}\,m^2/s$ at 0°C to $1.00 \times 10^{-6}\,m^2/s$ at 20°C.[11] Salinity will also increase the kinematic viscosity of water: with a salinity of 3.5% it is $1.83\,m^2/s$ at 0°C and $1.05\,m^2/s$ at 20°C.

Example
The average velocity of fresh water in a pipe of internal diameter 123.8 mm is 1.5 m/s (0.1238 m). The temperature is 20°C. Calculate the Reynolds number.

$$\overline{R_e} = \frac{Vd}{\nu} = \frac{1.5 \times 0.1238}{1 \times 10^{-6}} = 185700$$

This clearly illustrates that the water flow in the pipe is in the turbulent area.

By calculating the Reynolds number and the relative roughness of the pipe, the friction coefficient f, can be found from the Moody diagram (Fig. 2.9).

Computer programs and special diagrams (Fig. 2.10) are available for calculating the head loss caused by friction inside a pipe. It is important to be aware that these diagrams are specific for given pipes because the f- value of the actual pipe is used to construct them.

Example
Calculate the head loss in an old PE pipe with internal diameter of 110 mm (0.11 m). The length of the pipe is 500 m and the velocity in the pipe is 1.5 m/s; the friction coefficient is 0.030.

$$h_m = fLV^2/2gd = 0.030 \times 500 \times 1.5^2/2 \times 9.81 \times 0.11 = 11.47\,m$$

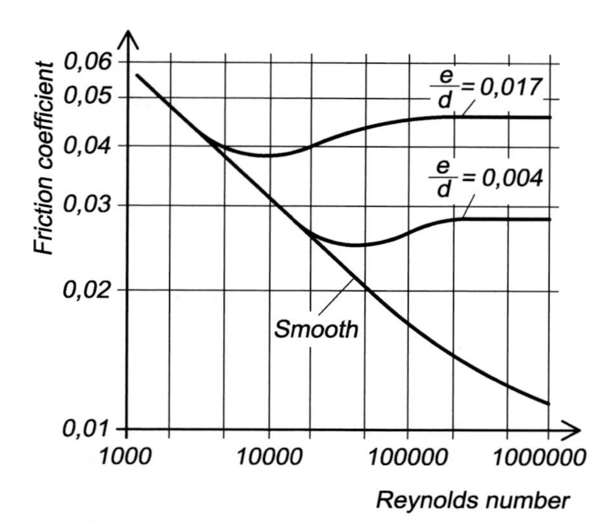

Figure 2.9 Principle of the Moody diagram showing the relation between relative roughness and Reynolds number.

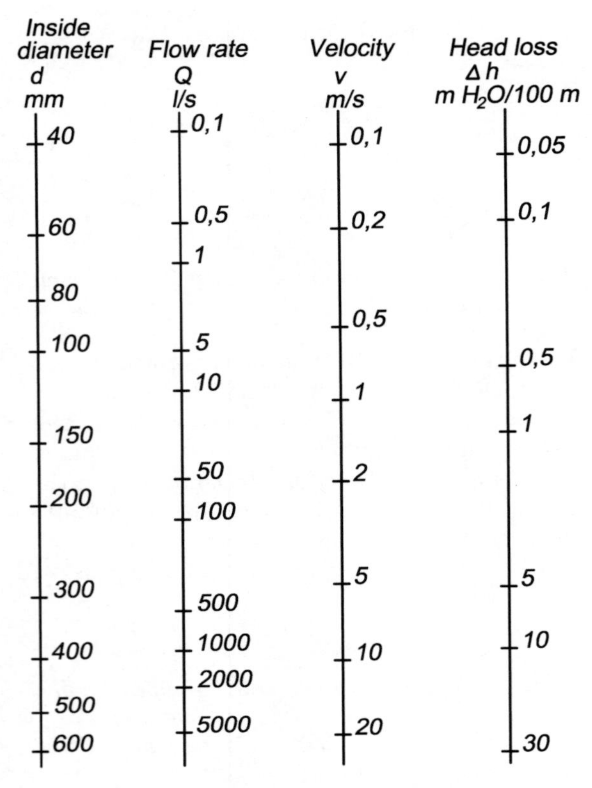

Figure 2.10 Diagram showing the relation between internal diameter, water flow (1000 l = 1 m^3), water velocity and head loss for a pipe with a known *f* value. (Reproduced with permission from Helgeland Holdings.)

This means that the head loss in the water flowing through the pipeline is 11.47 m. If the water is to flow in the pipe, the intake pressure must be at least 11.47 mH_2O in addition to atmospheric pressure; if it is less, the flow rate through the pipe will be reduced.

2.3.3 Head loss in single parts (fittings)

In addition to the head loss in the pipe there is energy loss due to friction in pipe parts (fittings) because any obstructions in the pipe which create extra turbulence will increase the head loss. Additional turbulence occurs in the inlet and outlet of the pipe, in valves, bends, reductions, connections, etc. The head loss can be calculated from the equation

$$H_t = kV^2/2g$$

where:

H_t = head loss in the single part
k = resistance coefficient for the pipe part
V = water velocity
g = acceleration due to gravity.

Example
The water must flow through a 90° elbow: either two 45° elbows or one 90° elbow with k *values of 0.26 and 0.9, respectively, can be used to achieve this. The flow velocity is set to 1.5 m/s. Calculate the head loss in the two cases.*

For the two 45° elbows

$$\begin{aligned} H_t &= \Sigma kV^2/2g \\ &= 2 \times 0.26 \times 1.5^2/2 \times 9.81 \\ &= 0.06\,\text{m} \end{aligned}$$

For the 90° elbow

$$\begin{aligned} H_t &= kV^2/2g \\ &= 0.9 \times 1.5^2/2 \times 9.81 \\ &= 0.10\,\text{m} \end{aligned}$$

As this example illustrates, there is a great advantage in using two 45° bends rather than one 90° bend to reduce the head loss. This will apply, for instance, for the outlet pipe from a fish tank.

The *k* values for different parts may be found from special tables (e.g. Table 2.2). They are also found in catalogues published by suppliers of fittings.

When constructing the pipe system the head loss that results from fittings in the pipeline must be considered in addition to the head loss in the pipe itself. The resistance of every single part must be added, so the sum of every single resistance plus the head loss in the pipeline gives the total head loss.

When designing the inlet pipe to a fish farm, it is important to use smooth bends to reduce the total head loss in the pipeline.

2.4 Pumps

Pumps are mechanical devices that add energy to fluids by transforming mechanical energy (normally from electric motors) to potential and/or kinetic energy of the fluid. Increase in potential energy is illustrated by the lifting of water to an elevated tank, while the increase in velocity and hence

Table 2.2 Typical resistance coefficients, *k*, for different fittings. Values of *k* will vary with the producer of the fitting.

Fitting	*k*	Comments
Pipe entrance in a basin or in a river	0.05–1	Lowest value with rounded inlet pipes; highest with pipes with sharp edges. The value is increased when the pipe goes into the basin.
Contraction of pipes	0.05–0.5	Lowest value with conical contraction and small alteration in the diameter.
Expansion of pipes	0.05–1	Lowest value with conical expansion and small alteration in the diameter.
Elbow		Increasing with reduction in the pipe diameter. Long, smooth bends will have reduced *k* values. Smaller angles will have lower *k* values.
90 degree	0.5–1	
45 degree	0.1–0.3	
T-pipe (divided flow)	0.3–1.8	Depends on the proportion divided from the main flow; increased with increased part flow.
T-pipe (connection flow)	0.1–0.8	Depends on how much water is supplied via part flow in the T-pipe; increases with increased size.
Valves		Values are highly dependent on the specification and the producer of the valve. Values shown are for fully open valves.
Ball valve	0.1	
Angle seat valve	1.3	
Diaphragm valve	4	
Check valve	2	
Gate valve	0.2	

the flow rate through a pipeline by pumping increases the kinetic energy of the water. Pumps are commonly used in aquaculture systems, usually to increase the system pressure and thereby force the water to move against an energy gradient. In most aquaculture situations pumps are used to lift water from one level to another. Water will flow only when energy is available to create a flow, i.e. there is a positive energy gradient. In hydraulic systems pumps are used to create the pressure which is normally high. This allows the fluid to do work, such as turn shafts or extend a hydraulic cylinder against a load. Oil is commonly used in such systems, but those using water are also available.

Pumps are fairly efficient machines for transferring energy to water, provided they are correctly selected for the job. The key requirement when selecting a pump is that there shall be a close correlation between the system requirements and the maximum operating efficiency of the pump; sub-optimal pump selection may result in significantly increased operating and maintenance costs and/or result is system failure.

2.4.1 Types of pump

There are several types of pump based on different principles (Fig. 2.11). The type of pump chosen depends on a number of factors, including the amount of fluid to be pumped and its characteristics, and the head.

A major pump type is the displacement pump in which liquid is displaced from one area to another. An example is the piston pump: when the piston moves up and down it creates, respectively, a vacuum and pressure, and in this way the liquid is transported; back-flow valves must be included. Gear pumps and screw pumps are other types of displacement pump. The pumps may break if the outlet is blocked.

The ejector pump is based on another principle. Here a part flow under high pressure is used to draw a main stream with much higher water flow but lower pressure. By pumping water into a specially designed narrow passage a vacuum effect will occur and create a drag on the main stream. The design of the ejector is most important. This principle is used in pumps for fish transport, for example.

In air-lift pumps, air is supplied inside an open pipe standing partly below the surface and partly filled with water. The air bubbles will then drag the water towards the surface and in this way a water flow is created inside the pipe. This principle may be used to pump water, add air (aeration) and for fish transport.

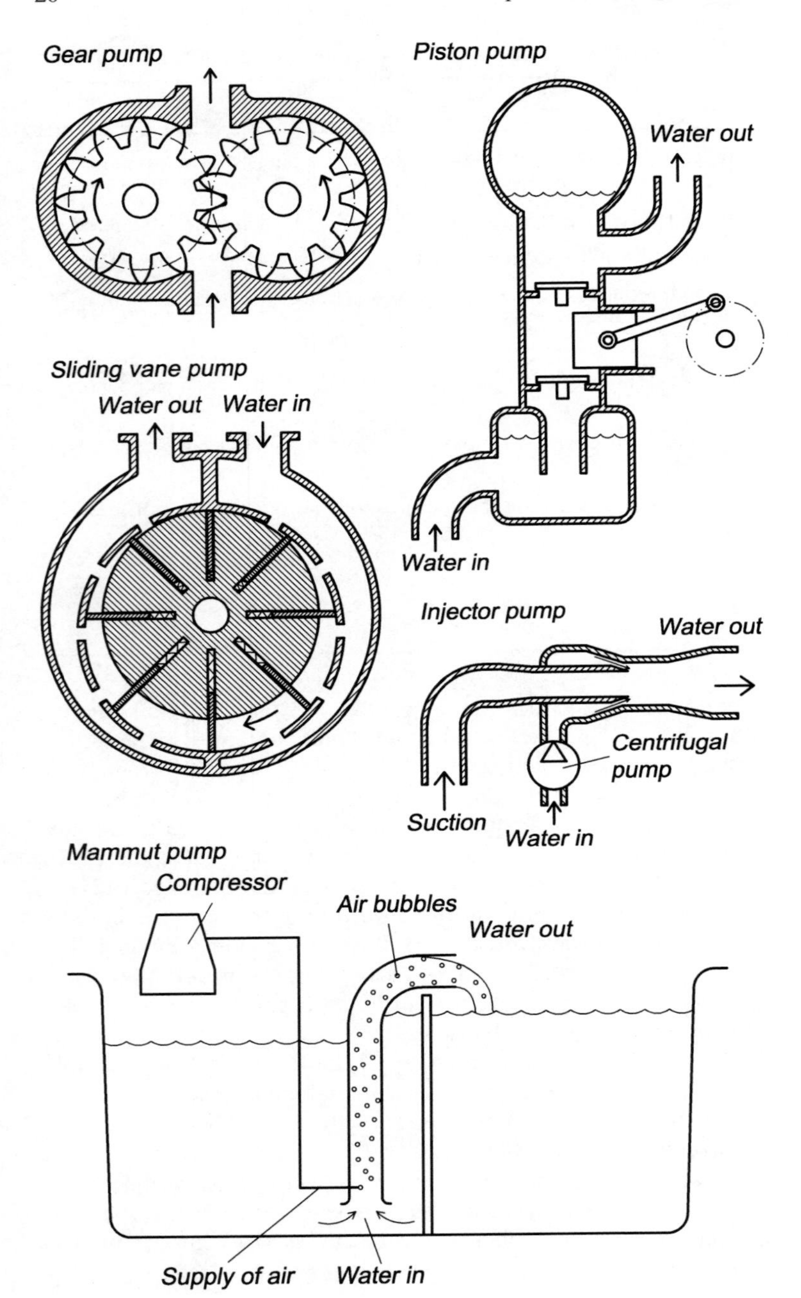

Figure 2.11 Diagrams to show the principles of different pump types.

An endless screw pump is based on another principle; among other uses, such pumps are employed for sludge.

For aquaculture facilities there is a need to pump a large amount of water with a relatively small lifting height. Centrifugal pumps or propeller pumps are the most suitable and most commonly used. Centrifugal and propeller pumps are described in Section 2.4.4.

2.4.2 Some definitions

Pump height

When water is lifted from one level to another the height difference is called the static lifting height. The lift height can either equal the pressure head (in the case of a submerged pump) or it can be a combination of vacuum and pressure head depending on where the pump is placed. In addition to this, the pump has to overcome the head loss caused by the friction in the pipe on both sides of the pump. If a manometer that measures the pressure is connected to the pump outlet the measured pressure is the sum of the pressure head (static head) and head loss. When the water passes through the pump it needs to have certain velocity in order to flow; this is called the velocity head. The total pressure head is obtained by summing the manometric height and velocity head. The actual pressure head at the end of the pipe, in addition to the difference in level must also be considered, for example when the pump is required to deliver water to a pressurized tank.

To collect water from a lower level, a vacuum head must be overcome by removing air from the inlet pipe to create a vacuum inside it. The pressure of the atmosphere will force the surrounding water up the inlet pipe. For this reason the vacuum head must not exceed atmospheric pressure which is normally 1013 mbar (10.3 mH_2O), although it depends to some extent on the weather (low pressure, high pressure) and the altitude. When a pipe is completely emptied of air, the water will therefore be forced up the pipe to a height of 10.3 m. The actual suction head of a pump is, however, lower than 10.3 mH_2O, because of losses such as from the velocity head in the inlet pipe. To make a centrifugal pump self-suctioning it must include a mechanism, for instance a specially designed impeller, to remove the air from the inlet pipe.

Cavitation

If the vacuum into the pump is too high, the water may boil and vaporize. That the temperature of vaporization pressure dependent can easily be illustrated with a pressure boiler, where the boiling temperature of the water is increased; similarly, below normal atmospheric pressure, the boiling temperature will decrease (see below). When a mixture of liquid and gas goes through a pump the boiling point will increase because the pressure around the water molecules increases from vacuum upwards. In changing from gas to liquid the bubbles undergo violent compression (implosion) and collapse creating very high local shock, i.e. a sharp rise and fall in the local pressure; the phenomenon is called cavitation. If this happens in connections to a pump or in the impeller, small metal parts can be dislodged. Multiple indentations or dimples in the material can result. The same may occur on boat propellers where worm like holes may be observed in the material of the propeller.

Cavitation reduces the effectiveness of pumps and will also shorten pump life. A characteristic 'hammer' noise is produced inside the pump when it cavitates. Cavitation may also occur if there are leakages in the pipe or pipe connection on the suction side of the pump. If air leaks in here (known as 'false air'), it will create air bubbles that enter the pump chamber with the water where they implode.

Cavitation can happen if the suction head is too high. When the pressure around the water molecules drops, the water will boil at a lower temperature (i.e. the boiling point of water is reduced). For example, if the pressure drops from atmospheric (10.3 mH_2O) to 1 mH_2O, water will boil at 46°C. This phenomenon can be observed when boiling water at high altitude, for instance in the Himalayas. Here the water will boil below 100°C because the atmospheric pressure is less than 10.3 mH_2O. Atmospheric pressure also depends on the weather. The safe static suction head will also decrease with surface water temperature from 10.4 mH_2O at 10°C to 7.1 mH_2O at 21.1°C.[12]

Net positive suction head (NPSH)

If the pump is not self suctioning, the water level must be higher than the level of the pump. This means that the impeller needs a certain pressure to function optimally. The net positive suction head (NPSH) gives the absolute lowest pressure the water must have when flowing into the pump chamber, or (more easily) the actual height of water over the impeller. If the water pressure is lower than the NPSH the pump will cavitate. NPSH depends on the water flow and increases with increasing flow; it can be described as follows:

$$\text{NPSH} = h_b - h_v - h_f + h_h$$

where:

h_b = barometric pressure
h_v = vapour pressure of the liquid at the operating temperature
h_f = frictional losses due to fluid moving through the inlet pipe including bends
h_h = pressure head on pumping inlet (negative if it is a static lift on the suction side of the pump).

Example
A pump is to be chosen for a land-based fish farm. With an actual discharge (Q) and head (H), the necessary NPSH can be read from the pump performance curves to be 4 mH_2O*. The fish farm is situated close to the sea and the barometric pressure (*h_b*) is measured to be 10.3* mH_2O*. The maximum temperature during summer time is 30°C which corresponds to a vapour pressure (*h_v*) of 4.25* N/m^2 *equal to 0.44* mH_2O*. The friction loss in the inlet pipe including loss in fittings (*h_f*) with the actual water velocity is 1.5* mH_2O*. The static suction lift (*h_h*) is 2 m.*
The NPSH can then be calculated as follows:

$$\begin{aligned}\text{NPSH} &= h_b - h_v - h_f + h_h \\ &= 10.3 - 0.44 - 1.5 + (-2) \\ &= 6.36\,\text{mH}_2\text{O}\end{aligned}$$

This is higher than the NPSH value of 4 mH_2O *that the pump requires, which means that there will not be any problems regarding NPSH when using the pump.*

The NPSH requirements of a specific pump are given in the pump diagram (see section 2.4.5). This value must be higher than the value calculated from the above equation. Remember that NPSH is given in units of pressure (mH_2O, bar or pascal).

2.4.3 Pumping of water requires energy

Energy is required to pump water from one level to another. Energy consumption is usually expressed as power (P), which is energy supplied per unit time. P is measured in joules per second; 1 J/s = 1 watt (W).

The following equation can be used to calculate the energy requirement for pumping:

$$P = \rho g h Q$$

where:

ρ = density of water (kg/m^3)
g = acceleration due to gravity (m/s^2)
Q = water flow rate (m^3/s)
h = height that the water is pumped.

Example
Calculate the energy required to lift 1000 l/min of water by 5 m and 15 m (including the friction head). The density of water is 1025 kg/m^3*, the flow rate is 0.016* m^3/s *and acceleration due to gravity is 9.81* m/s^2*.*

Case 1: 5 m lift

$$\begin{aligned}P &= \rho g h Q \\ &= 1025\,\text{kg/m}^3 \times 9.81\,\text{m/s}^2 \times 5\,\text{m} \times 0.016\,\text{m}^3/\text{s} \\ &= 804.4\,\text{J/s} \\ &= 804.4\,\text{W}.\end{aligned}$$

Case 2: 15 m lift

$$\begin{aligned}P &= \rho g h Q \\ &= 1025\,\text{kg/m}^3 \times 9.81\,\text{m/s}^2 \times 15\,\text{m} \times 0.016\,\text{m}^3/\text{s} \\ &= 2413.3\,\text{W} \\ &= 2.4\,\text{kW}.\end{aligned}$$

This illustrates that by tripling the pump height the energy requirement is also tripled.

The power supplied from the pump to the water is called the water effect and is the sum of the velocity head, head loss and static head. As a result of energy losses in the pump, more power is supplied to the pump than the pump supplies to the water. The efficiency of the pump, is given by the equation

$$\eta = P_D/P_S$$

where:

η = efficiency
P_D = energy delivered from the pump to the water
P_S = energy supplied to the pump.

Example
A pump has an efficiency of 75% and consumes 5 kW of electric power. How much of this power is used to pump the water?

$$\eta = P_D/P_S$$

$$\begin{aligned}P_D &= \eta P_S \\ &= 0.75 \times 5 \\ &= 3.75\,\text{kW}\end{aligned}$$

Energy losses occur at several places in the pump. This results in efficiencies below unity (Fig. 2.12). Losses occur in the pump motor (m), transmission (t) and in the impeller (p), the sum of which gives the pump efficiency. Low pump efficiency results in the creation of heat, because energy cannot disappear; for example, when using submerged pumps this energy will be transferred to the water, which is heated. In water re-use systems this heating can be noticeable. The total efficiency of a pump η_A, can be calculated as follows:

$$\eta_A = \eta_p + \eta_t + \eta_m$$

Efficiency may also be defined as hydraulic efficiency η_H (loss when the water flows through a pump), volumetric efficiency η_v (leakage of water between suction and pressure side of the pump, for example in centrifugal pumps) and mechanical efficiency η_m (losses in the motor and transmission).

In fish farming the usual efficiency of well-suited pumps is around 0.7. Efficiency normally varies between 0.4 and 0.8.

Pump costs

To calculate the cost of pumping, the following equation can be used:

$$\text{Pumping cost} = Pd\text{EP}$$

where:

P = power (kW)
d = duration of pumping (h)
EP = electricity price per kWh (kilowatt per hour).

Example
A centrifugal pump with a 5 kW power supply runs continuously to supply a fish farm with water. Calculate the yearly electricity cost of running the pump with an electricity price of 0.1 €.

$$\begin{aligned}\text{Yearly pumping cost} &= \text{P}d\text{E}P\\ &= 5\,\text{kW} \times 24\,\text{h} \times 365\text{ days} \times 0.1\,\text{€/kWh}\\ &= 4380\,\text{€}.\end{aligned}$$

2.4.4 Centrifugal and propeller pumps

Centrifugal pumps

Centrifugal pumps account for the majority of those used by aquaculture enterprises. A centrifugal pump contains three major units: the power unit, the pump shaft and the impeller (Fig. 2.13). The power unit, an electric motor, causes the pump shaft to rotate and with it the attached impeller. Around the rotating shaft where the impeller and motor are fixed there are seals that prevents leakage via the shaft into the motor.

If the inlet and the pump cavity are filled with water (primed), the rotation of the impeller causes movement of water molecules which will be accelerated outwards towards the periphery of the impeller. This will reduce the pressure in the centre of the impeller and new water is drawn into the eye of the impeller via the pump inlet. Vanes on the impeller may direct the flow of water and help to transfer the energy from the impeller to the water.

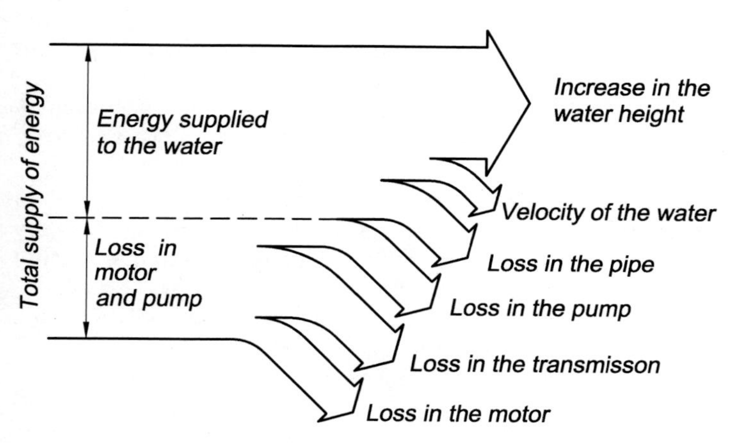

Figure 2.12 Only part of the electric energy fed to the pump is used to transport the water.

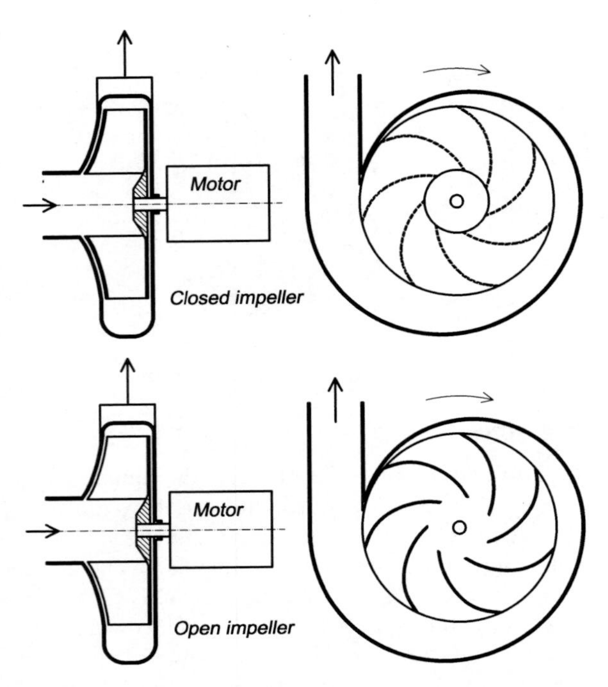

Figure 2.13 A centrifugal pump contains three major units: the power unit (motor), the pump shaft and the impeller which can be open, closed or semi-closed.

The rotation of the impeller imparts high velocity to water molecules at the periphery. When this water leaves the impeller, the velocity is rapidly reduced and the dynamic head is converted to static head. How the pressure changes from the inlet to the outlet on a centrifugal pump depends on the characteristics of the pump. The outlet pressure is a function of the inlet pressure and pump characteristics, mainly the design of the impeller.

The impeller may be open, semi-closed or enclosed; all have different characteristics (Fig. 2.13). The choice depends on the amount of particulate matter in the water. If there are many particles (sludge like), an open or semi-closed impeller is normally used; if there are few particles or high pressure is required, an enclosed impeller is used. The enclosed impeller has the highest efficiency, but tolerates the lowest amount of suspended solids in the water. In aquaculture facilities all types of impeller are used.

Propeller pump

Propeller pumps are of simple construction with a propeller rotating inside a pipe (Fig. 2.14). The principle is the same as a propeller on a boat, but instead of moving the boat, the propeller is fixed so the water moves instead. Propeller pumps have the advantage that they can deliver large amounts of water at low pressure (normally less than 10 mH_2O). The reason for the low pressure is leakage that occurs between the two sides of the propeller (head and suck). On some pumps the flow rate can easily be varied by adjusting the angles of the propeller vanes. The propeller is normally installed in a vertical pipe, but it is also possible to place it in a horizontal pipe to create a flow.

Dry placed or submerged pumps

Centrifugal pumps can be dry placed, either above water level or in a dry well below water level. They can also be installed in the water as submerged pumps (Fig. 2.15).

A dry placed pump consists of a pump chamber with an impeller, a transfer shaft and an electric motor. On dry placed pumps the motor is cooled by a fan and on submerged pumps by water. Between the pump chamber and the motor there is a seal to prevent water leaking into the motor. Dry placed pumps may also be made self-suctioning by use of a specially designed impeller. Dry placed pumps are normally bolted to a rack that is fixed to the floor.

In a submerged pump the motor and the pump chamber are usually built together and encapsulated in one unit which is lowered into the water. As for dry placed pumps, it is important to have a good seal between the pump chamber and the motor. The whole must be watertight, to prevent water from entering the motor. The motor is cooled by the surrounding water and the encapsulating part can be equipped with cooling ribs to facilitate this.

Both dry placed and submerged pumps are commonly used in aquaculture. The advantage with dry placed pumps is that maintenance is simple because it is easy to access the pump and pump parts. A disadvantage could be that artificial cooling of the motor is necessary, needing a fan. If there are leakages in the inlet pipeline false air may be drawn in,

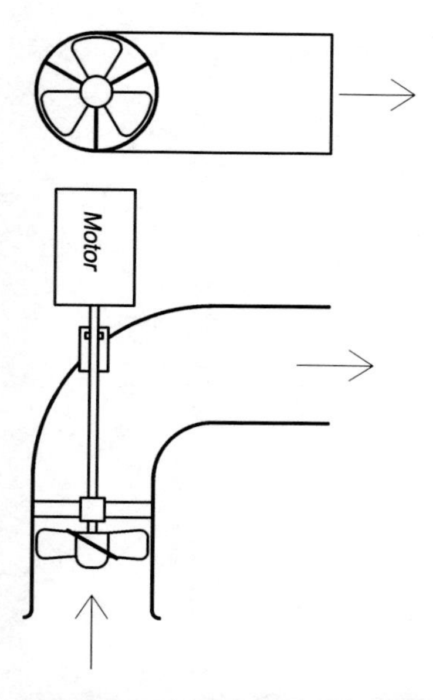

Figure 2.14 A propeller pump comprises a fixed propeller rotating inside a pipe. This results in transport of water through the pipe. The photograph shows water delivery from submerged propeller pumps.

causing cavitation in the pump and supersaturation of nitrogen (see Chapter 8), especially when using dry placed pumps standing above the water surface.

2.4.5 Pump performance curves and working point for centrifugal pumps

Characteristics curves

The characteristics curves (pump diagram) are used to describe the performance of centrifugal pumps (Fig. 2.16). The most important is the head versus capacity curve which shows the connection between the water flow (discharge) and the head. Depending on its construction the same centrifugal pump can either be used to deliver a large water flow with a low head or a smaller water flow with a larger head. Adding a valve on the pump outlet and gradually closing it will decrease the water flow but increase the head, because the cross-sectional area through which the water is forced is reduced. As the valve is closed more pressure is needed to discharge

Figure 2.15 Centrifugal pumps are either (A) submerged (pump shown ready to lower into the water) or (B) dry placed.

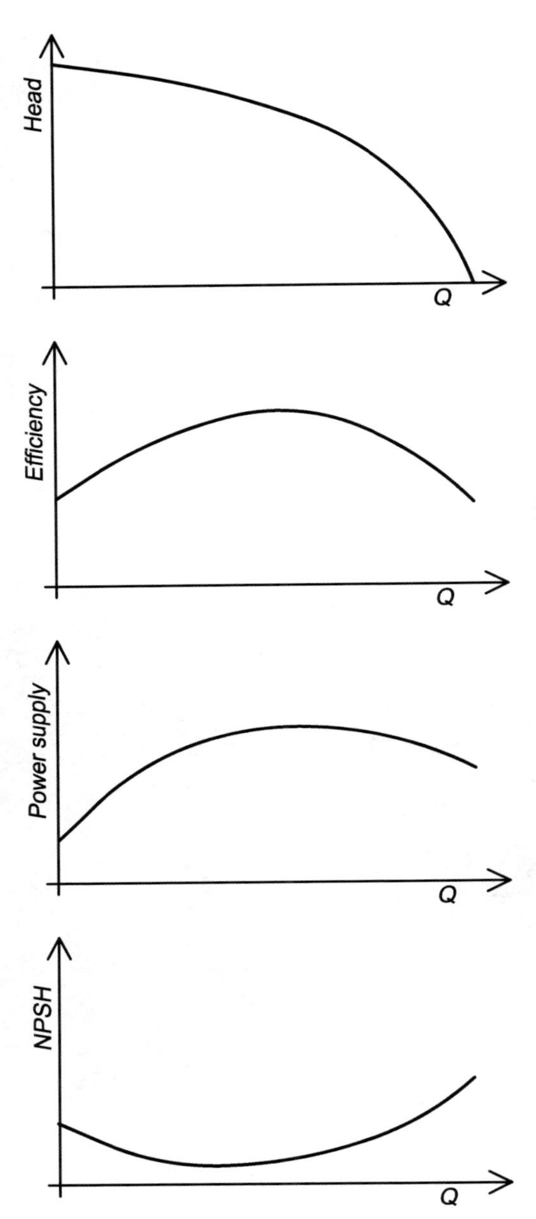

Figure 2.16 Pump performance (characteristics) curves for centrifugal pumps: head, efficiency, power supplied and net positive suction head are plotted against water flow rate (discharge).

the water; again this increases the pump head. A pump characteristics curve can be constructed by measuring the water flow that the pump delivers with different heads: if pump behaviour is ideal, the plot should be a straight line. This seldom happens because of the reduction in pump efficiency at the limits of performance, so the plot is curved. If large amounts of water are being pumped, the friction loss through the pump becomes large. In addition to the frictional losses there is so called impact loss resulting from the impact of the water molecules hitting the impeller and the inlet and outlet parts of the pump chamber. The highest losses occur with large heads and large flows at the ends of the pump characteristics curve, close to the *x*- and *y*-axes. This is to be expected, because a pump has the highest efficiency at its construction point in the middle of the characteristics curve. Normally this point is distinctly marked.

A pump's characteristics will of course depend on design and size of the chosen impeller, each of which will have individual characteristics. A centrifugal pump can normally be delivered with different impellers, and it is quite easy to change them. In many cases several pump characteristics are given in the same diagram, one for each diameter of the impeller.

The pump diagram may also show a plot of pump efficiency versus discharge. A pump that works as closely as possible to the maximum efficiency (construction point) should be chosen; operation away from the construction point will result in decreased efficiency. Pump efficiency may also be given in the so-called shell or muscle diagrams in which curves and/or circles represent percentage efficiency.

The power requirements are also given in the pump diagram. Here the necessary power that must be supplied to the pump is plotted versus different discharges and heads.

The NPSH curve gives information about the necessary 'over-pressure' required by the pump which, as described earlier, increases with increasing water flow. It is important to be aware that the curve showing the NPSH is constructed in relation to atmospheric pressure, i.e. 10.3 mH_2O. The supplied (inlet) pressure is the sum of the air pressure and water pressure to the pump impeller. If this is lower than the NPSH requirements shown on the curve, the pump will cavitate. There will not normally be problems with NPSH if there is positive water pressure into the pump. Problems are more likely to occur when there is static suction lift into the pump.

Pipeline characteristics

When delivering water through a pipeline, the pressure and/or head that the pump has to deliver depends on the flow rate to be pumped through the pipe. If the pump is not running, the only resistance is the fixed static head; however, when the pump starts to work there is in addition a resistance head (manometric head) caused by friction losses in the pipe.

The manometric head increases with the flow rate through the pipe. If the pump does not lift the water to a higher level, but only overcomes friction in the pipes, there is no static head, but only a resistance head. To find the total friction, the characteristics of the total pipe system, including pipe length, pipe material, number of bends, etc., must be known. With this information the total head loss through the system with different water flows can be calculated. The results may be plotted to show the connection between the total head and the water flow. This curve is called the pipeline characteristic curve. It starts at zero if there is no start head and increases exponentially with the water flow (Fig. 2.17).

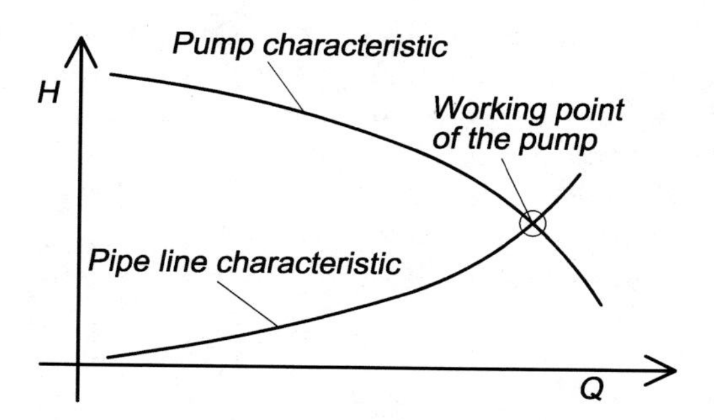

Figure 2.17 By drawing the pipeline characteristic curve and the pump characteristic curve on the same diagram the working point of the pump is found at the intersection.

Working point

To find the working point for a pump with given pipeline characteristics, the pipeline and pump characteristic curves can be drawn on the same diagram (Fig. 2.17). The actual working point of the pump will be at the intersection of the two curves. This point shows the water flow that the pump will deliver through the pipeline. To achieve the best possible efficiency it is important that this point is as close to the construction point of the pump as possible. If not, another pump should be chosen.

2.4.6 Change of water flow or pressure

When choosing a pump it is important that the pump works as closely as possible to its construction point. What may then be done if the chosen pump does not fulfil this condition and cannot be changed?

Change of impeller (pump wheel)

Most of the centrifugal pumps can be supplied with different impellers. By selecting an impeller of different shape and diameter it is therefore possible to change the pump characteristics (Fig. 2.18). In this way it is possible to find an impeller that is better adjusted to the working point of the pump, and better efficiency can be achieved.

The following connections can be used to show the pump performance when reducing or increasing the diameter of the impeller:

$$Q_2/Q_1 = d_2/d_1$$

$$H_2/H_1 = (d_2/d_1)^2$$

$$P_2/P_1 = (d_2/d_1)^3$$

where:

d_2 = diameter of smallest impeller
d_1 = diameter of largest impeller
Q = flow rate
H = head
P = power requirement.

Example
The diameter of an impeller to a centrifugal pump is increased from 515 mm (d_1) to 555 mm (d_2). The flow rate in the first case was 700 l/s (Q_1) and the head is 9 m (H_1). Calculate the new flow rate (Q_2) and the new head (H_2).

$$Q_2/Q_1 = d_2/d_1$$

$$Q_2 = 1.08 \times 700 = 754\ \text{l/s}$$

$$H_2/H_1 = (d_2/d_1)^2$$

$$H_2 = 9 \times 1.16 = 10.5\ \text{m}$$

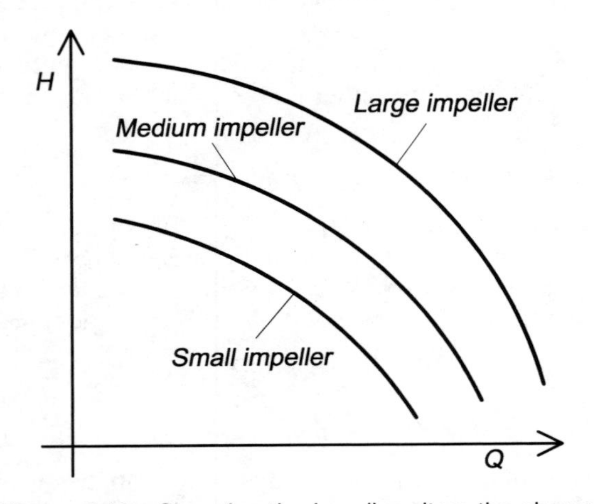

Figure 2.18 Changing the impeller alters the characteristics of the pump.

These equations are most reliable with a range of 20% increase/decrease in diameter, because large changes of the diameter of the impeller may alter pump geometry.[12]

Connection of pumps

Pumps can be connected in series (one after another) or in parallel (beside each other). In this way the water flow and water pressure can be changed (Fig. 2.19). When pumps are connected in series, they are placed one after another in the same pipeline, so increasing the head while maintaining the same water flow. The inlet pressure of the second pump is the outlet pressure from the first pump so the pressure head is increased.

When pumps are connected in parallel, the main pipeline is divided into two sub-pipelines. One pump is placed in each sub-pipeline; these are then connected to the main pipeline. Parallel connection doubles the water flow, while the head remains constant.

High pressure pumps

High pressure pumps are special types of centrifugal pumps that utilize a system of pumps connected

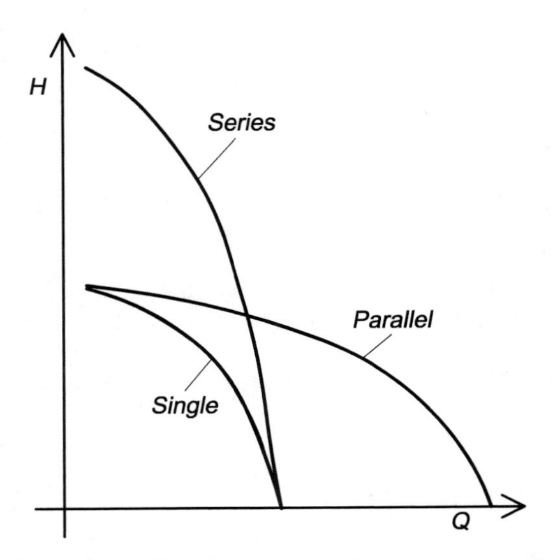

Figure 2.19 The flow rate and pressure may be changed by connecting pumps in series (one after another in the same pipeline) or in parallel (beside each other in separate sub-pipelines).

in series (Fig. 2.20). In a high-pressure (multi-stage) pump the impellers are connected in series, and can share the same motor and shaft. The motor and shaft are of course larger than if a single impeller is used. When the water leaves the first impeller it is collected and directed into the centre of the next impeller and so on, and the head gradually increases depending on the number of impellers. This is possible through the design of the pump chamber. High-pressure pumps are used for instance for pumping groundwater from great depths or for other purposes where there is either a large head to be overcome or if a high water pressure is required.

2.4.7 Regulation of flow from selected pumps

In many cases there is a requirement to adjust the water flow from specific pumps, for instance to reduce it. It is possible to use several methods for this purpose and these are briefly described below.

Adjustment of RPM

One method for regulating the water flow from a pump is to change the speed (RPM) of the motor and hence the impeller. When the speed is diminished both the water flow and the head are reduced. The usual type of electric motor employed for pumps is an asynchronous motor. Its speed may be changed by coupling in and out extra sets of poles in steps by a switch, which will change the RPM of the motor, also in steps. The following connection is given between the RPM (n), water flow (Q), head (H) and the power (P):

$$Q_2/Q_1 = n_2/n_1$$

$$H_2/H_1 = (n_2/n_1)^2$$

$$P_2/P_1 = (n_2/n_1)^3$$

From these equations it can be seen that the water flow increases in direct proportion to the RPM, the head increases in proportion to the square of the RPM, and the power increases in proportion to the cube of the RPM.

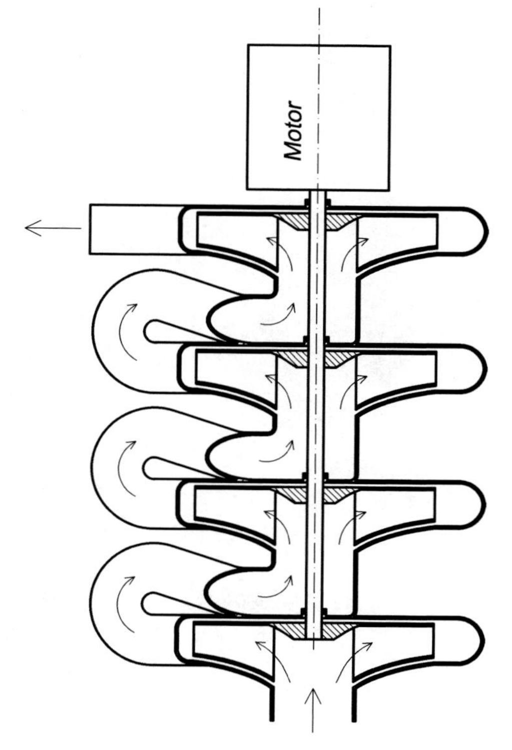

Figure 2.20 In a high-pressure (multistage) pump the impellers are connected in series and have a common motor and shaft.

Continuous adjustment of the RPM in pumps is, however, difficult in practice. The motor on a centrifugal pump normally uses alternating current and the continuous adjustment of a.c. motors is quite complicated and expensive. The usual way to adjust the motor speed is to transform the frequency of the current (Fig. 2.21). If, for instance, the normal frequency of the electricity is 50 Hz, reducing it to 40 Hz will reduce the speed of the motor, which will reduce the water flow out of the pump but maintain the head. Reducing the a.c. frequency can affect the pump motor; if it is reduced too much, the motor will stop and may be destroyed. The same thing can happen if the frequency is increased to above 50 Hz.

The disadvantage of this system is that frequency regulators are quite expensive and some energy loss occurs during frequency transformation. Large improvement have, however, been made in the technology in this field during the past few years, and the cost has been considerably reduced.

Figure 2.21 A frequency transformer changes the frequency of the current and may therefore be used to regulate the pump speed.

Example
A pump delivers 300 l/s. The speed of the pump motor must be changed so that the pump only delivers 200 l/s. The original speed of the pump is 2800 RPM/min. What is the new speed? The power supply is 30 kW at the start. What is the new power supply?
The new RPM is going to be:

$$Q_2/Q_1 = n_2/n_1$$

$$n_2 = Q_2 n_1/Q_1$$
$$= 200 \times 2800/300$$
$$= 1867\,\text{RPM}$$

The new power supply will be:

$$P_2/P_1 = (n_2/n_1)^3$$

$$P_2 = P_1(n_2/n_1)^3$$
$$= 30(200/300)^3$$
$$= 8.9\,\text{kW}$$

Throttling

By placing a throttle valve on the pressure side, it is possible to close the outlet of the pump (throttle) (Fig. 2.22). In this way an artificial head is created and hence a reduction in the water flow, according to the pump characteristics. Higher throttling results in a move to the left on the pump curve. To throttle on the pressure side is only possible for centrifugal pumps, because the water is inhibited from going to the peripheries of the impeller. Throt-

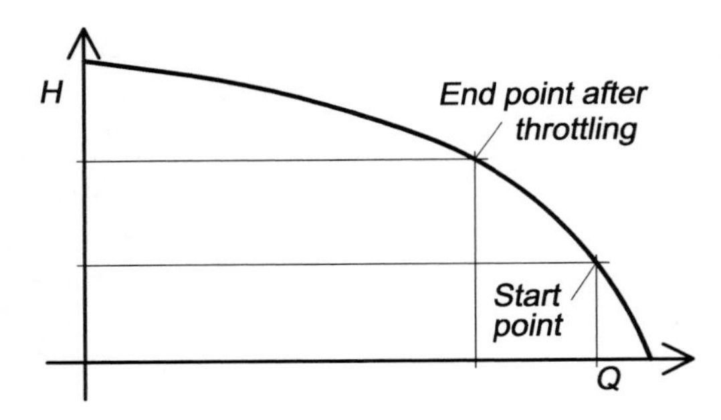

Figure 2.22 Throttling on the pressure side may be used to reduce the water flow from a centrifugal pump, but is not a good solution.

tling on the pressure side on pump types other than centrifugal and propeller can result in pump breakage. Throttling centrifugal and propeller pumps is not very satisfactory as there is significant energy loss because the throttling valve creates an artificial head. The power consumption is the same when the water flow is reduced, which easily can be seen from the left shift on the characteristics curve. To throttle a centrifugal pump on the suction inlet must be avoided as this easily creates cavitation conditions in the pump.

Several pumps

If there are large water requirements and the discharge varies, it is an advantage to use several pumps of different sizes. In this way it is possible to couple pumps in and out and by doing this vary the total flow rate (parallel connection) and at the same time achieve an overall high efficiency for all the running pumps. To use several pumps will also improve the reliability because one pump can stop without halting the total water flow to a farm.

References

1. Finnemore, J., Franzini, J.B. (2001) *Fluid mechanics with engineering applications.* Mc-Graw Hill.
2. Kundu, P.K., Cohen, I.M. (2001) *Fluid mechanics.* Academic Press.
3. Munson, B.R., Young, D.F., Okiishi, T.H. (2005) *Fundamentals of fluid mechanics.* John Wiley & Sons.
4. Nayyar, M.L. (1999) *Piping handbook.* McGraw-Hill Professional.
5. Frankel, M. (2001) *Facility piping systems.* McGraw-Hill Professional.
6. Willoughby, D. (2001) *Plastic piping handbook.* McGraw-Hill Professional.
7. Karassik, I.J., Messina, J.P., Cooper, P., Heald, C.C. (2000) *Pump handbook.* Mc-Graw Hill.
8. Rishel, J.B. (2002) *Water pumps and pumping systems.* McGraw-Hill.
9. Sanks, R.L., Tchobanoglous, G., Bosserman, B.E., Jones, G.M. (1998) *Pumping station design.* Butterworth-Heinemann.
10. ASM International (1988) Engineering plastics. In: *Engineered materials handbook, vol 2.* ASM International.
11. Huguenin, J.E., Colt, J. (2002) *Design and operating guide for aquaculture seawater systems.* Elsevier Science.
12. Lawsons, T.B. (2002) *Fundamentals of aquacultural engineering.* Kluwer Academic Publishers.

3
Water Quality and Water Treatment: an Introduction

3.1 Increased focus on water quality

As the aquaculture industry becomes ever more intensive, the focus on water quality in the rearing units will also increase. Higher production densities will also increase the requirements for optimal water quality, because of the degeneration in quality when the water flows through the production unit.

The importance of water quality is independent of the type of rearing unit and location of the production facilities. If using open production units in the sea such as sea cages, it will of course be more difficult to treat the water to improve quality, even though this is starting to happen (see Chapter 8 concerning oxygenation in open sea cages). On land-based fish farms with control of both the inlet and outlet, water treatment to improve the water quality will be much easier to perform. The increase in aquaculture production based on water re-use systems will focus attention on water treatment to improve quality.

During the past few years, the focus on the environmental impacts of the aquaculture industry have increased.[1–5] In future, more stringent requirements will be set to reduce these impacts. It is possible to reduce the discharge from the facility by optimal production management, and also by treating the outlet water from closed production facilities. It will also be important that production is adapted to site capacity, and of course that it is sustainable.[6,7]

3.2 Inlet water

While fish will grow in water of sub-optimal quality, their growth rate will not be maximized. With high investments costs involved in the running of rearing units it is of course vital that the production per unit of farming volume is as high as possible. Fish can live well in the wild even if water quality is sub-optimal. However, the food supply is usually limited and the growth rate will be much lower than that possible under optimal conditions.

There is often some kind of stress response involved when there is an outbreak of disease. Disease can be latent in the stock but can become a problem if the fish are exposed to some kind of stressor, for instance sub-optimal water quality. In fish farms, where fish are grown as quickly as possible, they are already under stress and so disease outbreaks are more likely due to sub-optimal environmental conditions. It has been shown that by catching wild fish and holding them under farming conditions at a high stocking density, it is very easy to cause a disease outbreak, even if the water quality is equal to that found at the wild site where the fish were caught.

Many experiments have been carried out for the species farmed today, and there are quite good data for recommended water quality.[8,9] However, as the amount of new water added per kilogram of fish is continuously decreasing, research is focused on accurately documenting lowest acceptable levels of nutrients, etc. to maintain optimal growth. Norwegian salmon smolt production can be taken as an example. In 1985 the average size of the fish was 40 g while today is it over 100 g, as the result of improved feed and increased individual growth rate. Most of the sites used have limited fresh water resources and therefore the amount of new water supplied per kilogram of fish has been reduced, mainly by adding pure oxygen: in 1985 the licence requirement for new water was 0.38 l/kg fish/min, while today in dry periods it is down to 0.1–0.2 l/kg fish/min.[10]

Water quality requirements depend upon the species. The same is the case regarding requirements for the various life stages; the early stages normally have the highest requirements for optimal water quality. Even if the quality requirements vary, it is better to have good quality water than bad, if this is possible.

Optimal water temperature is species specific and so general advice is impossible. Species can be defined generally as warm water species (>20°C) and cold water species (<20°C). Some species temperatures prefer below 10°C. If the water temperature falls below 0°C freezing will be a problem. The oxygen content of the water enclosure (see full saturation, p. 119) is reduced with increasing temperature. At 5°C the available oxygen content is 12.8 mg/l while at 25°C is it reduced to 8.2 mg/l.

The water ought to be fully saturated or supersaturated with oxygen gas. It is very important that the oxygen content of the rearing water is high enough: for instance, 7 mg/l (70% saturation) is the typical value for the outlet water in salmonid farming, 30% having been consumed by the fish.

Of the other gases dissolved in the water, the concentrations of nitrogen (N_2) and carbon dioxide (CO_2) must not be too high. The nitrogen gas concentration should be below 100.5% saturation. For carbon dioxide, levels are not only dependent upon the inlet concentration, they also increase in the tank as a result of fish metabolism which releases CO_2 into the water; the outlet concentration must not therefore be too high.

Water pH must neither be too low nor too high (the latter is seldom the case). This applies to freshwater, since seawater has stable pH values of 7.5–8.2. Sufficient alkalinity in the water helps to control fluctuations in pH.

Too many particles in the inlet water may have negative effects on the fish, for instance by clogging their gills. Fish faeces will increase the particulate content of the water, the outlet concentration of which must not be too high.

Ammonia may be a problem in production unit outlet water because of waste products from fish metabolism, but only with very limited supply and exchange of water. With normal water sources there are no problems with ammonia concentration in the inlet water.

The concentration of metal ions in the inlet water may be of levels that are toxic to fish; low pH may increase this toxicity. Problem metals include aluminium, copper, iron, zinc and cadmium.

Micro-organsims, including parasites, bacteria, viruses and fungi, may be present in the inlet water at concentrations unfavourable for aquaculture.

Interactions between several water quality parameters, for instance between pH and metals, may also pose quite a challenge. To fully understand water treatments effects, a good knowledge of basic water chemistry is an advantage; this topic is not covered in this book but extensive literature is available, for example refs 11-13.

3.3 Outlet water

All outlet water discharged from aquaculture facilities can present environmental problems which create an imbalance in the ecosystem in the recipient water body. This is especially important when the outlet water is discharged into freshwater. Freshwater recipient water bodies are of limited volume, whereas the sea represents an infinitely large water body; therefore the consequences of a discharge into a freshwater recipient, such as a small lake, will be much greater for both open and closed production units.

The effluent discharged from a fish farm can contain three classes of pollutant (Fig. 3.1):

- Nutrients and organic matter
- Micro-organisms
- Escaped fish

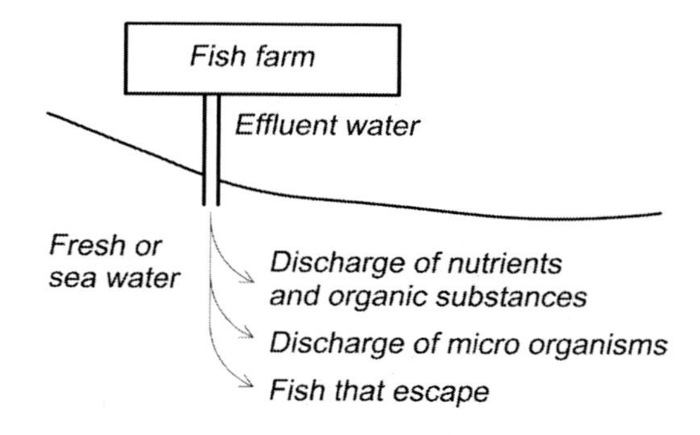

Figure 3.1 Every fish farm, whether it is on land or in the sea, will discharge effluent.

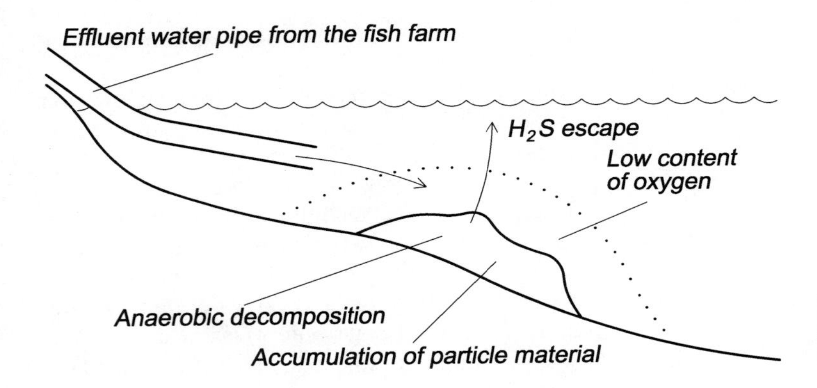

Figure 3.2 Local anaerobic decomposition may occur in the recipient water body if the point discharge is excessive.

The quantity of nutrients and organic matter discharged depends on the amount of feed used and farm management practices. The nutrient content in the feed must be optimal for the fish, and as much as possible must be available to and taken up by the fish. The amount of feed supplied must be optimal in relation to appetite, so that feed loss is avoided.

Discharge of nutrients to the recipient water body will result in increased algal growth leading to eutrophication and imbalance in the recipient ecosystem. Discharge of too much organic matter to the recipient water body may result in lack of oxygen during the night as a result of decomposition. Local accumulation of fish faeces in the recipient water body may cause anaerobic decomposition, possibly accompanied by release of hydrogen sulphide (H_2S) which is toxic for small animals and fish (Fig. 3.2). This shows the importance of having an adequate water current at the point where the discharge is released.

The larger concentration of biological material in a restricted volume compared to the case in natural conditions means that possibilities for disease outbreaks in a fish farm are greater than in the wild. The effluent water may therefore also contain a higher concentration of pathogenic micro-organisms such as parasites, bacteria, viruses and fungi, that can cause disease in the fish population. This may again have significant consequences for the wild fish in the recipient water body. If there are possibilities for 'short circuiting', the fish farm may function as a facility for increasing the concentration of pathogens (Fig. 3.3). For land-based fish farms this can be illustrated with pathogens that are ejected in the effluent water from the farm and are taken in again with the inlet water. This also shows the importance of not having these pipes too close to each other. The best arrangement is to have them in different water bodies, or to treat either the inlet or the outlet water. Fish that migrate can also be a host for transporting micro-organisms up rivers, meaning that short circuiting also can occur here. Migration obstructions in the river may be a solution.

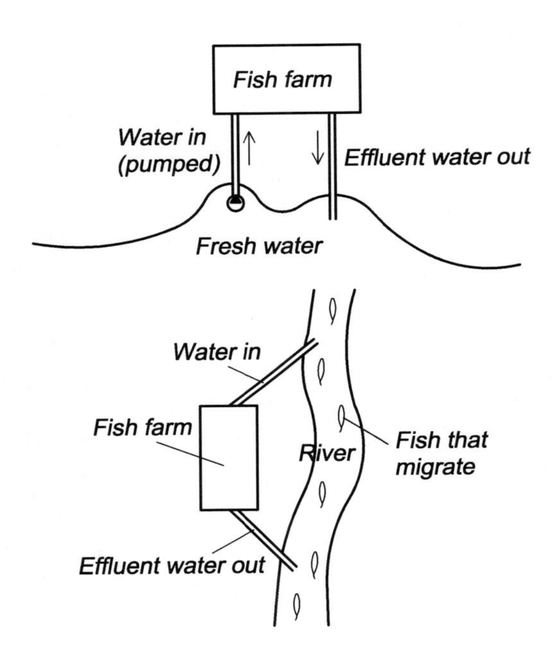

Figure 3.3 Short circuiting between the outlet and inlet water to the fish farm must be avoided to prevent the concentration of pathogenic micro-organisms increasing in the recipient water body.

Moving aquatic organisms from one farm to another may, via the water used for transport, bring new micro-organisms to the recipient farm. The result can be an outbreak of disease on the fish in the recipient farm. Some wild stocks may, through natural evolution, have developed natural immunity against some pathogenic micro-organisms while others may represent a large threat. An example of this is the salmon parasite *Gyrodactylus salaris* where the Atlantic salmon in a few rivers have developed immunity, whereas in most rivers the stocks have no immunity. When moving fish containing such micro-organisms between rivers, or between farms with outlets to different recipients, the consequences can therefore be fatal. Treatment of the effluent water is absolutely necessary in such cases.

Escape of fish or other aquatic organisms from farm conditions may also present environmental problems, except where local stock that not has gone through to a breeding programme is used on the farm. If the stocks have gone to a centralized breeding programme where mixing of local stocks from different districts occurs, or much more seriously, if genetic manipulations have been performed, there may be significant consequences attending escapes. The fish farming industry has benefitted greatly from national breeding programmes; for example, growth rates and feed utilization have been improved.[14]

What can happen when several fish escape from a farm? One possibility is that they can establish their own stock in the recipient water body. This may result in competition for feed and habitat with the naturally occurring stocks. Another possibility is that they can interbreed with the local stocks and create unwanted hybrids. Even if no interbreeding occurs the escaped fish can destroy the breeding grounds or breeding nests of the wild stocks. Of course, none of this is wanted and therefore is it very important to try to avoid escape of fish and other aquatic organisms from farm conditions.

3.4 Water treatment

All treatment of water leads to a change in the water quality, and it is improvement that is wanted. Regardless of the incoming water quality, it will always be possible to obtain a quality good enough for growing aquaculture products. The problem is, however, the cost; all water treatment operations involve expenditure. A major advantage for a good farming site is therefore to have good quality incoming water with low treatment requirements.

Several processes may be needed to adjust the water quality. The inlet water to land-based farms is aerated; pH adjustment and particle removal are also required. Heating and chilling are normally used to create optimal growth conditions. In some cases disinfection is needed to reduce the burden of micro-organisms, especially in fry production (Fig. 3.4).

The outlet water from the fish farm may also be treated to avoid affecting the water quality of the

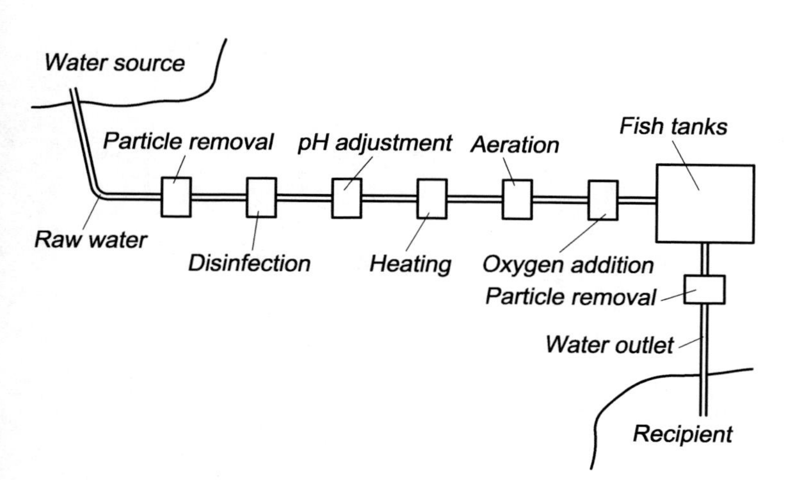

Figure 3.4 Water treatment processes.

recipient water body. If the recipient water body is highly eutrophicated the outlet water must be treated, and this is also the case if there are valuable wild stocks in the receiving water. Treatment of outlet water will, however, increase production costs, and is done only when necessary. Often government regulations will dictate the level of treatment required. Sites that require less water treatment are therefore favoured.

Treatment of the outlet water is normally restricted to removal of part of the suspended solids. From a cost perspective, is it normally impossible to remove dissolved substances such as nutrients and small micro-organisms in a flow-through farm with relatively high fish densities, because of the size of the water flow; if re-use systems are employed, such treatments may be included. However, re-use systems have much higher investment costs than flow-through farms.

The treatment requirements of a complete system are described in Chapters 4–10: the following six chapters will focus on the methods most commonly used in aquaculture for water treatment, concluding with a chapter on water re-use systems.

References

1. Midlen, A., Redding, T.A. (1998) *Environmental management for aquaculture*. Chapman & Hall.
2. Black, K.D. (2001) *Environmental impacts of aquaculture*. Sheffield Academic Press.
3. Read, P., Fernandes, T. (2003) Management of environmental impacts of marine aquaculture in Europe. *Aquaculture*, 226: 139–163.
4. Boyd, C.E., McNevin, A.A., Clay, J., Johnson, H.M. (2005) Certification issues for some common aquaculture species. *Reviews of Fisheries Science*, 13: 231–279.
5. Pillay, T.V.R. (2005) *Aquaculture and the environment*. Blackwell Publishing.
6. Ervik, A., Hansen, P.A., Aure, J., Stigebrandt, A., Johannessen, P., Jahnsen, T. (1997) Regulating the local environmental impact of intensive marine fish farming 1. The concept of the MOM system (Modelling–Ongrowing fish farms–Monitoring). *Aquaculture*, 158: 85–94.
7. NS 9410. (2000) *Environmental monitoring of marine fish farms*. Norwegian Standardization Association.
8. Alabaster, J.S., Lloyd, R. (1980) *Water quality criteria for freshwater fish*. Butterworth.
9. Poxton, M. (2003) Water quality. In: *Aquaculture, farming aquatic animals and plants* (eds J.S., Lucas, P.C. Southgate). Fishing News Books, Blackwell Publishing.
10. Bergheim, A. (1999) Redusert vannforbruk og påvirkning av vannkvaliteten ved settefiskanlegg. *Kurs i vannkvalitet for settefiskprodusenter*. Arrangør Hydrogass (in Norwegian).
11. Snoeyink, V.L., Jenkins, D. (1980) *Water chemistry*. John Wiley & Sons.
12. Stum, W., Morganm, J.J. (1996). *Aquatic chemistry*. John Wiley & Sons.
13. Benjamin, M.M. (2001) *Water chemistry*. McGraw-Hill Science.
14. Gjedrem, T. (2005) *Selection and breeding programs in aquaculture*. Springer-Verlag.

4
Adjustment of pH

4.1 Introduction

On some sites the freshwater pH is too low to achieve optimal growth for fish or shellfish. At other sites the buffering capacity of the water is low, and it is difficult to avoid pH fluctuation in the water. This again results in negative effects on growth. Sites where acid rain is a problem are particularly exposed to this. Further, in re-use systems (Chapter 10) biological filters are used to remove ammonia and this causes a drop in pH that must be corrected to maintain optimal growing conditions.

4.2 Definitions

pH is the measure of acidity or alkalinity in a solution. It is presented on a number scale between 1 and 14, where 7 represents neutrality and the lower numbers indicate increasing acidity and higher numbers increasing alkalinity. Each unit of change represents a tenfold change in acidity or alkalinity. What is measured is the negative logarithm of the effective hydrogen-ion concentration or hydrogen-ion activity in gram equivalents per litre of the solution.

$$pH = -\log[H^+]$$

If substances are added to the water they may act as acids, bases or be neutral. Acids give free hydrogen ions (H^+) and bases free (hydroxyl ions) (OH^-).

The alkalinity of the water is a measure of its capacity to neutralize acids, meaning its ability to keep the pH constant. If the alkalinity of the water is low fluctuation in pH occurs easily. The carbonate system normally represents the major part of the alkalinity in aquaculture systems, together with hydroxides (OH^-). In the carbonate system compounds are related to each other via different equilibria:

$$CO_2 + H_2O \leftrightarrow H_2CO_3$$
$$H_2CO_3 \leftrightarrow HCO_3^- + H^+$$
$$HCO_3^- \leftrightarrow CO_3^{2-} + H^+$$

where:

H_2CO_3 = carbonic acid; HCO_3^- = bicarbonate ion; CO_3^{2-} = carbonate ion; CO_2 = carbon dioxide.

Water with free bicarbonate will take up H^+ ions. The amount of each ion in the water is pH-related: in water of low pH there is excess carbon dioxide/carbonic acid; in water of pH 7 there is excess bicarbonate, and in water of high pH excess carbonate ion. The units for alkalinity are either mg/l as $CaCO_3$ or milliequivalents per litre (meq/l) where 1 meq/l is equal to 50 mg/l $CaCO_3$.

A buffer is defined as a substance capable of neutralizing both acids and bases in a solution and thereby maintaining the original acidity or alkalinity of the solution and the resistance to pH changes when adding moderate amounts of base or acids.

Hardness is sometimes confused with alkalinity, mainly because it can be expressed using the same units. Hardness is, however, a term for the sum of all metal ions in the water. This is dominated by the bivalent cations of calcium (Ca^{2+}) and magnesium (Mg^{2+}), but manganese (Mn^{2+}), iron (Fe^{2+}), sodium (Na^+) and potassium (K^+) may also be important. Because calcium and magnesium are included, hard water may also have high alkalinity, but this is not necessarily always so. If, for instance, sodium and potassium are responsible for the alkalinity, the hardness can be low. The following water classification system can be used: soft water, hardness less than 50 mg $CaCO_3$/l; moderately hard water 50–150 mg $CaCO_3$/l; hard water 150–300 mg $CaCO_3$/l; very hard water, above 300 mg $CaCO_3$/l.[1] Conductivity may be used as a unit to determine hardness, measured as microsiemens per cm (μS/cm).

4.3 Problems with low pH

pH that is too high or too low will have negative effects on the fish.[2–5] Low pH can cause damage to the gills, skin and eyes. Higher H^+ concentrations will also increase the permeability of the gills, leading to leakage of Na^+ and Cl^- which creates osmotic problems. First, the effects can be registered as a reduction in growth; too low a pH will kill the fish. In natural populations, the pH may vary from 5 to 9, but for aquaculture facilities it is recommended to be in the range 6.5–9.[2,3] Problems with metals in the water are best avoided; tolerance may vary with fish species and life stage, with newly hatched fry being especially sensitive. For crayfish, for example, the pH and alkalinity must be high because they utilize Ca^{2+} in the water for shell synthesis.

The solubility of metal ions in the water will increase with reduction in pH. There have been particular problems with the concentrations of aluminium (Al^{+++}) in fish farming; this metal leaches from the soil or bedrock in the catchment area. The toxicity of the complexes of Al or Al precipitates varies. A drop in pH will change the existing Al complexes to more toxic ones, meaning that fatalities can occur even if the pH itself does not represent any danger. The most stable and non-toxic forms of Al are in the pH range 6.5–6.8. Calcium will ameliorate problems with aluminium because it protects the gills from aluminium and also from acidity.[6] Toxic effects also depend on temperature, because rate of reaction increases with rise in temperature. Normally is it therefore not the pH that is dangerous, but the combination of low pH and metal ions.

Some of the aluminium or other metal complexes that are toxic are unstable and will only persist for a short period. This may occur when mixing water with different qualities and characteristics, and a mixed zone of water with different qualities is achieved. If this reaches the fish tanks, fatalities may result. It is really important to be aware of this in intensive fish farming, where water from sources of different quality and temperature is mixed just before entering the fish tank. For instance, a brief drop in pH may occur in a water source due to fall of acid rain or ice melting in the catchment area; this may create a mixed zone in the water source (river or lake).[7] Where single Al complexes coagulate and create larger, more toxic Al complexes.

In mixing zones is aluminium entering the gills causes osmotic stress.[6] When it is taken up, it can cause damage to the nervous system and block enzymatic reactions. As a defence mechanism, mucus may be secreted and oxygen uptake is thereby reduced, which may result in death. If there is a possibility of problems with metals ions, choice of the correct adjustment agent is important.

Water alkalinity for intensive fish farming and pond aquaculture is recommended to be above 40 mg/l $CaCO_3$ to stabilize the pH and protect health and physical quality.[3] Alkalinities in the range 100–200 mg/l $CaCO_3$ will, however, give several additional advantages, including making a stable water source for biofilters in a re-use circuit, adding buffer capacity to avoid pH fluctuation in ponds, and reducing the toxicity of heavy metals.[8] In ponds there will be a fluctuation in pH during the day and night due to biological processes. During the night the algae (phytoplankton) and fish will release CO_2 so the pH will drop if the alkalinity is low; during the daytime the algae will consume CO_2 by photosynthesis faster than the fish release it; therefore the pH will increase. The pH can vary from 5 to 10.[9]

4.4 pH of different water sources

The pH will of course vary with the water source. Seawater normally has pH value between 7.8 and 8.3 and has a good buffering capacity due to the available free bicarbonate. There is normally no need for pH adjustment. The only exception being when re-using water to a great extent.

For fresh surface water, whether river or lake, the pH will be highly dependent on the ground characteristics and whether the catchment area is exposed to acid rain. Normal pH values are between 4 and 8.5. Groundwater has a more stable pH, but the buffering capacity can be reduced and values between 5.5 and 8.5 are common. If there is limestone rock in the catchment area the pH will be high, while the pH of water coming from marshy areas can be low. The alkalinity of the different sources can vary from below 10 mg/l $CaCO_3$ in soft freshwater to several hundred of mg/l $CaCO_3$ in seawater and hard freshwater.[10]

It is important to be aware that there can be fluctuations in the pH of freshwater during the year. Floods during spring may result in dramatic reductions in pH, especially when snow is melting in the catchment area where acid has accumulated in the snow. Also dry weather may affect the pH of the water.

4.5 pH adjustment

The need for pH adjustment or neutralization of the water depends on the water source, the species farmed and the production system (e.g. water re-use system).

The principle used to adjust pH in acid water is to remove the free H^+ ions. Methods must therefore attract and bind the H^+ ions and will therefore involve basic solutions. The pH is usually regulated to between 6.5 and 7 either by adding hydroxides (OH^-) or carbonate compounds. Examples of the hydroxide group include lye, sodium hydroxide ($NaOH$), calcium hydroxide or slaked lime ($Ca(OH)_2$) and magnesium hydroxide ($Mg(OH)_2$). Carbonate compounds include different forms of lime (calcium carbonate ($CaCO_3$) and quick lime (CaO)), in addition to dolomite ($CaMg(CO_3)_2$, magnesium carbonate ($MgCO_3$), sodium carbonate (Na_2CO_3) and sodium bicarbonate (Na_2HCO_3). Where there are problems with aluminium silica lye has also been used with advantage to reduce the acute toxicity.[11,12] Silica (SiO_2) will attract labile aluminium and prevent the occurrence of long Al chains and creation of mixed zones.

If carbonate compounds are used there will, in addition to the increase in pH, be an increase in the buffering capacity of the water, i.e. an increase in the alkalinity of the water, which will then be more stable against pH drops. If adding lye complexes this effect will be minor but the pH will increase; it can, however, be simple to overdose with lye so the pH gets too high.

The pH must be adjusted before the water reaches the fish tanks. Normally there will be a need for some retention time so that the adjustment agents can function and to prevent unstable toxic metal complexes reaching the fish tank (mixed zone problems). Treatment at the start of the fish farm inlet pipe or using a holding tank to retain the water for some time after adding the pH treatment can solve this agent problem. The solubility and rate of solution will also depend on the chemical chosen; for example, sodium bicarbonate will react quickly, while dolomite reacts more slowly.

It is also important to achieve good mixing of the pH treatment agent with the incoming water; some kind of mixing equipment is quite normal and can, for instance, be a mixer, use of air bubbles or by addition of the chemical before pumping. The form of the pH agent (liquid, meal, powder or larger grain or rock) will dictate the requirements for the mixing equipment.

In a fish tank there will be an increase in the CO_2 concentration resulting from the fish metabolism. If the fish density is high compared to the water supply and pure oxygen is supplied in addition, the reduction in pH may be noticeable. When adding pH regulating agents to the water, care must be taken because the water will also contain ammonia that is toxic for the fish, the amount of which is pH dependent (see Chapter 9); decreasing the pH will reduce the amount of dissolved ammonia. If the pH is adjusted without doing anything to reduce the concentration of ammonia, fatalities may result.

4.6 Examples of methods for pH adjustment

4.6.1 Lime

Lime of various forms is a good substance for increasing the pH of acid water. When adding lime (calcium carbonate ($CaCO_3$)) to water with low pH the following process will take place, as mentioned earlier:

$CaCO_3 \Rightarrow Ca^{2+} + CO_3^{2-}$ (calcium carbonate is dissolved in water)

$CO_3^{2-} + H^+ \Leftrightarrow HCO_3^-$ (carbonate ion attracts H^+ from the water)

$HCO_3^- + H^+ \Leftrightarrow H_2CO_3$ (bicarbonate attracts H^+ from the water)

These reactions are part of the carbonate system, which is the most important contributor to the good buffering capacity of the water. What can be seen from the chemical equations is that when adding $CaCO_3$ to water, CO_3^{2-} ions will react with free H^+ and reduce the amount, so increasing the pH.

Lime may be used in different forms, such as limestone or as limestone powder. Limestone powder ($CaCO_3$), with a particle size of less than 0.005 mm, may also be dissolved in water and create lime slurry (approximately 75% dry matter), which is much used for pH regulation. Lime slurry may be added in concentrated form to the water or as diluted (5–10 times in a mixing basin) lime slurry.

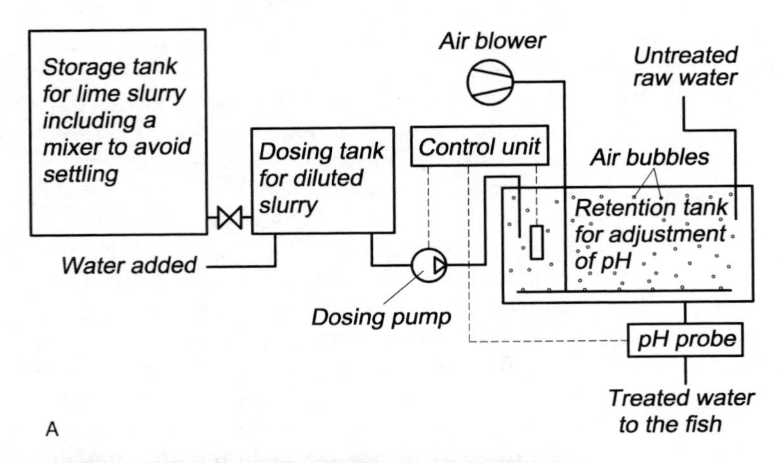

Figure 4.1 (A) A lime slurry plant for regulation of pH in a fish farm; (B) lime slurry tank.

The container for storing the lime slurry must have an efficient mixing system to avoid settling out of the lime particles. Lime slurry is added with a dosage pump to a mixing basin where mixing of the slurry with the water where pH is to be regulated takes place (Fig. 4.1). Automatic pH control can be achieved by having a pH sensor in the mixing basin, the signals from which can be used to control the dosage pump and the amount of slurry added. In this way any variation in the quality of the incoming water can be registered and adjusted before the water reaches the fish tanks.

To ensure good mixing of the slurry with the water, a diffuser creating air bubbles can be used in the mixing basin. Use of lime slurry may increase the particles content (turbidity) of the water. This must be considered when using lime slurry on water for fry production.

A shell-sand filter or limestone filter represents a simpler system where the water has to pass through a layer of crushed limestone (particle size 1–3 mm) or shell sand.[13] As the water passes through the filter there will be an increase in pH. After using the filter for a period of time there will be a gradual drop in pH, because the amount of CO_3^{2-} is reduced. To avoid these drops in pH several filters can be connected in parallel. This makes it possible to refill, clean or maintain a separate filter without stopping the whole system. When using such systems, a part flow of the water to be treated is sent over the limestone filter. The pH increases, and the water is sent back to the main water flow into the fish tanks. Care must, however, be taken to avoid possible mixing zones. If CO_2 gas is added just before the limestone filter, the CO_2 will increase the dissolution of the limestone. In this way, it is possible to automatically control the pH by controlling the addition of CO_2 gas before the filter. The reaction process in this filter can be described with the following equations:

$$CaCO_3 + H_2O + CO_2 \Rightarrow Ca^{2+} + 2HCO_3^-$$

$$HCO_3^- + H^+ \Leftrightarrow H_2CO_3$$

A shell-sand well is a very simple and low cost system for regulating the pH (Fig. 4.2). The incoming water is forced to flow up through a layer of shell-sand and in so doing the pH of the water will increase. This method requires manual refilling with shell sand to keep the pH stable.

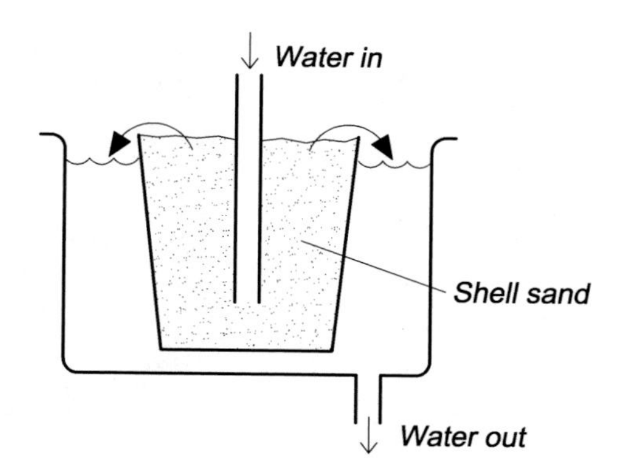

Figure 4.2 A shell-sand well for regulation of pH.

4.6.2 Seawater

Seawater has a high buffering capacity and contains free carbonate ions (CO_3^{2-}) and/or bicarbonate ions (HCO_3^-) which, similarly to limestone, will take up H^+ ions and increase the pH. Addition of 2–4% seawater to freshwater will increase the pH and the buffering capacity; the conductivity of the water will also increase. Since measuring the conductivity is quite simple, this method can be used for controlling the addition of seawater. Also, the amount of seawater may be fixed manually depending on how much freshwater is used.

When using this method for pH regulation, it is important to consider that the seawater contains pathogenic micro-organisms, poisonous algae and other substances that may be harmful to the fish. A solution is to pump the seawater from large depths, where the amount of algae and micro-organisms is lower. Seawater pumped from groundwater wells may also be used. This water normally has a very low content of micro-organisms. It is generally recommended that seawater is disinfected before using it for neutralizing freshwater. The operator must also be aware of mixed zone problems.

4.6.3 Lye or hydroxides

Different types of lye, such as sodium hydroxide (NaOH), may be used for pH regulation, but special care must be taken before use in fish farming because it is quite easy to overdose, especially when the water quality varies. The result can be water having a pH that is too high, which is also toxic for the fish. Sodium hydroxide is strongly corrosive on

Figure 4.3 Adjustment of pH using liquid sodium hydroxide (lye): (A) the tank and dosing pump for adding lye to the inlet water flow (note the small plastic tube going up to the pipe); (B) equipment for monitoring the pH of the water flow and for regulating the addition of lye.

metals and is not particularly friendly to work with. In solid form it is a white crystalloid non-odorous substance that is easily dissolved in water; the mixing process produces heat and steam.

Plants for adding of dilute sodium hydroxide solution (3%) to the inlet water work on the same principle as slurry plants (Fig. 4.3). A dosage pump is necessary for adding the sodium hydroxide solution into the mixing tank. The mixing tank for sodium hydroxide and water may be slightly smaller than for lime slurry because it is easier to mix.

The reaction process when using sodium hydroxide is:

$$NaOH \Rightarrow Na^+ + OH^-$$

$$OH^- + H^+ \Leftrightarrow H_2O$$

In water of low pH, the hydroxyl ions (OH^-) will react with the free hydrogen ions (H^+) and increase the pH of the water. The use of sodium hydroxide will, as seen from the chemical equation, only neutralize the water and not increase the buffering capacity; this is also a reason for preferring other methods. In water of low alkalinity either lye or hydroxides may be used.

References

1. Hammer, M.J. (1996) *Water and wastewater technology*. Prentice Hall.
2. Randall, D. (1991) The impact of variations in water pH on fish. In: *Aquaculture and water quality* (eds D.E. Brune, J.R. Thomasso). World Aquaculture Society, Louisiana State University.
3. Wedemeyer, G.A. (1996) *Physiology of fish in intensive fish culture systems*. Chapman & Hall.
4. Willougby, S. (1999). *Manual of salmonid farming*. Fishing News Books, Blackwell Science.
5. Poxton, M. (2003) Water quality. In: *Aquaculture, farming aquatic animals and plants* (eds J.S. Lucas, P.C. Southgate). Fishing News Books, Blackwell Publishing.
6. Rosseland, B.O. (1999) Vannkvalitetens betydning for fiskehelse. In: *Fiskehelse og fiskesykdommer* (ed. T. Poppe). Universitetsforlaget (in Norwegian).
7. Krogelund, F., Teien, H-C., Rosseland, B.O., Salbu, B. (2001) Time and pH-dependent detoxification of aluminium in mixing zones between acid and non-acid rivers. *Water, Air and Soil Pollution*. 130: 905–910.
8. Wedemeyer, G.A. (2000) pH. In: *Encyclopedia of aquaculture* (ed. R.R. Stickney). John Wiley & Sons.
9. Boyd, C.E., Tucker, C.S. (1998) Pond aquaculture water quality management. Kluwer Academic Publishers.

10. Wedemeyer, G.A. (2000) Alkalinity. In *Encyclopedia of aquaculture* (ed. R.R. Stickney). John Wiley & Sons.
11. Åtland, Å., Hektoen, H., Håvardstun, J., Kroglund, F., Lydersen, F., Rosseland, B.O. (1997) *Forsøk med dosering av silikatlut ved syrtveit fiskeanlegg*. Niva rapport 3625 (in Norwegian).
12. Camilleri, C., Markich, S.J., Noller, B.N., Turley, C.J., Parker, G., van Dam, R.A. (2003) Silica reduces the toxicity of aluminium to a tropical freshwater fish (*Mogurda mogurda*). *Chemosphere*, 50: 355–364.
13. Lekang, O.I. Stevik, A.M., Bomo, A.M., Stevik, T.K., Herland, H. (1999) *Bruk av skjellsand til regulering av pH og alkalitet i småskala resirkuleringsanlegg for fiskeoppdrett*. ITF rapport 102, Universitet for Miljø og Biovitenskap (in Norwegian).

5 Removal of Particles

5.1 Introduction

Removal of particles from a water flow is called water treatment or water purification. In aquaculture, removal of particles from a water flow is necessary in several places: for the inlet water to the farm; for the outlet water from the farm; or if the water is re-used. The inlet water is treated to avoid too high a concentration of particles reaching the fish. High concentrations will have a negative effect on growth and may increase mortality.[1,2] Some parasites in the water are also of a size that makes it possible to remove them with a particle filter.[3] They can therefore be removed from the water before it reaches the farm, or if used on the outlet water a filter could remove them from the water flow before it reaches the recipient water body. Another reason for removal of particles from the inlet water is that the function of other water treatment equipment can be affected negatively by the particle content (see Chapter 6). An example here is the disinfection plant where a low particle content is required. In the outlet water, particle removal is done to reduce the effect of the outlet water on water conditions in the recipient body.[4,5] For re-use systems particle removal is particularly important to avoid accumulation of particles in the system and reduction in fish growth.

The aim of using a filter to remove particles is to extract a certain proportion of particles from the water flow, not all. How much is removed depends on the design and function of the filter. The biggest particles are the easiest to remove, regardless of the chosen method. Before the water flow reaches the filter unit where the particles are removed, it must be treated as gently as possible to avoid breaking the particles and reducing their size, so increasing the size of the filter necessary for extraction.[2] Gentle handling of the particles includes using low water velocity and having few bends, valves, etc., in the system that create extra turbulence where the particles are flowing. For the same reason, the filter ought to be placed as close to the source as possible; for inlet water this means as close to the water source and for outlet water as close to the production unit as possible. It is also important to have a sufficient flow to prevent particles settling in the pipes, and leakage of nutrients.

The particles in the water are of various different forms and numbers. Several methods and definitions are used to define the particle content of the water. Total suspended solids (TSS) is defined as the amount of particles stopped by a special fibreglass filter with a pore size of 0.45 μm. Total solids (TS) represents the total amount of particles in the water; this quantity can also be expressed as total dry matter (DM).

Particles can also be classified according to size.[2] Those smaller than 0.001 μm are classified as soluble, 0.001–1 μm as colloidal, 1–100 μm as supercolloidal and larger than 100 μm as settleable. Some nutrients may be totally dissolved in the water, which means that they cannot be re-moved with a particle filter but with other filter types. Examples here are biofilters (Chapter 9) that remove dissolved substances, such as NH_4^+ or NO_3^-.

When removing particles from the water flow there will also be a reduction in the discharge of nutrients because some are included in the particles. There will also be a reduction in the number of micro-organisms because some are attached to the surface of the particles.

To give some ideas of the different particle sizes, the following can be used for illustration: cocoa and talcum powder 5–10 μm, hair straw 50–70 μm, table salt 90–110 μm. The lower limit for easily identifying single particles is around 40 μm.

5.2 Characterization of the water

Good characterization of the water to be filtered is necessary so that the correct filter can be chosen. The characteristics of the inlet water will vary from site to site, whether it is lake water, river water, groundwater or seawater. Before choosing a filter it is therefore necessary to take samples to be able to characterize the water.

The volume of wastewater coming from fish farms is normally much higher and the concentration of the discharged substances much lower than those entering a municipal wastewater treatment plant; they are, however, and comparable to those in the water discharged from municipal wastewater treatment plants, i.e. water that has been purified.[6,7] Requirements for the design and construction of wastewater plants for fish farming are therefore different to those used to treat muni-cipal wastewater. Hence, the purification equipment and technology used in municipal wastewater treatment cannot be transferred directly to fish farming conditions, even if the basic principles are the same.

The composition of the outlet water from a fish farm depends upon a number of factors, including species, growth rate, feed composition and utilization, feed conversion rate and water amount (see for example, ref. 8). The first step in reducing the discharge from the fish farm, without using any filter at all, is therefore to have an optimal feed that is fully utilized and consumed by the aquatic organisms. This also includes optimal management of the farm, having correct water quality and quantity, and feeding in an optimal way.[4]

Experiments have shown that the predominant particle size in the outlet water from fish farming is less than 30–40 μm.[9,10] The large number of small particles account for only a limited part of the total volume of discharged particles. Since the volume of particles is much more important than the number of particles when talking about the load on the recipient water body, it is of great importance to remove the few large particles. However, in re-use systems the small particles will normally dominate, since it is easy to remove the larger particles. This can also been seen with water that goes brown in high re-use systems, because the small particles remain in the water.

The density of faeces from fish farming varies. Reported densities are above 1, from 1.005 to 1.2, which means that the faeces will settle in water.[9,11] A study of intact faeces from rainbow trout showed an average sinking velocity of the faeces of 1–2.5 m/min depending on fish size.[12]

5.3 Methods for particle removal in fish farming

Several principle and methods are used to remove particles from a water flow.[13–18] These can be classified as follows:

- Mechanical filtration, also called straining or micro screens
- Depth filtration, also called sand filtration or just filtration
- Settling

It is also possible to use other methods to remove particles, such as flotation, membrane filtration and ozonation. These methods are normally utilized for removing smaller particles and are usually too expensive to use in aquaculture facilities, perhaps with the exception of water supplied to small fry, such as for some marine species.

Regardless of the method chosen, method it is important to remember that all filter systems will cause a head loss. This can be quite high, for instance when using pressurized filters on the inlet water, with the loss depending on the principle employed. Acceptable head loss is therefore an important criterion when selecting an appropriate filter.

5.3.1 Mechanical filters and micro screens

A mechanical filter is an obstruction that is set into the water flow to collect the particles and larger objects and allow the water to pass through. The principle of a mechanical filter is to separate particles from water in a straining plane, either a screen or a bar rack. Particles bigger than the aperture in the screen or the distance between the bars in the rack will be stopped. The simplest type of mechanical filters comprise a static screen, a grating or perforated plate, or a bar rack that is set down into the water flow. The screen, which has apertures or meshes of defined size, will stop particles larger than the aperture/mesh size moving with the water flow; they are caught on the surface of the screen (Fig. 5.1). After a while the screen will gradually become blocked; the head loss will increase until the screen is completely blocked with particles which prevent any water passing through it. This results in an overflow. The same will be the case with a bar rack. Typically bar racks are used to remove larger particles and objects (>15 mm), while screens can also be used on smaller objects (>6 μm).[16]

When a screen is used, the particles have to be removed from the surface to avoid blockage. One way to remove them is to manually take the screen up from the water and clean it. This method is very labour intensive and is only used in special cases where the pore size is very large compared to the major particle size in the water. Examples are removal leaves from the water in the autumn, and stopping other large floating objects from entering the inlet pipe. A major aim when constructing a mechanical filter or screen to be used for removing smaller particles, is therefore to find ways of preventing blockage, which means being self-cleaning. The bar rack can be made self-cleaning by using a scraper mechanism, but this, as mentioned, is a device for removing larger particles.

It is important that the screen surface is smooth so that the particles are not crushed. The screen can be made of perforated plates when the apertures are large (mm or cm). When a screen is used to filter inlet water or outlet water in fish farming, a screen cloth of metal or plastic threads woven to the wanted mesh size is employed. It is important that the screen surface is easy to keep clean.

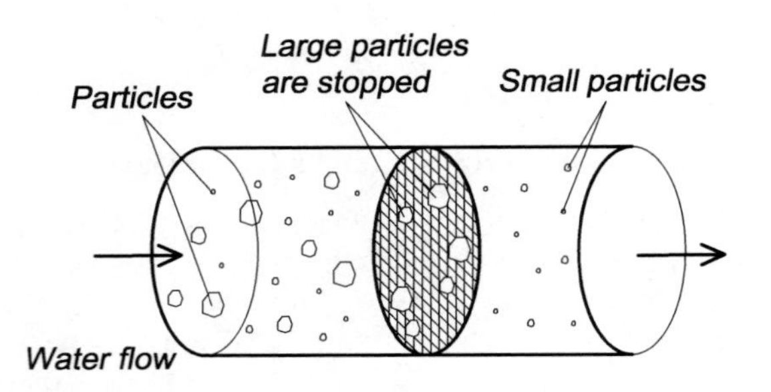

Figure 5.1 A static screen.

Several methods can be used to make the screen self-cleaning. A basic separation can be done by back-flushing, vacuuming or mechanical vibration of the filter cloth. If mechanical vibration is used, the filter cloth will shake and the trapped particles will fall off by gravity. If such equipment is used, the filter cloth must be installed at an angle to the horizontal plane; this method is, however, not commonly employed in aquaculture. The simplest method of cleaning the filter is to back-flush the screen (Fig. 5.2).

When using back-flushing or vacuuming it is desirable to have as great an area of the new cleaned screen as possible in the water. The screen is used until it gets blocked, when it is removed for cleaning and new cleaned screen cloth is substituted. This process must not be so rapid that the efficiency is reduced or mechanical breakdown occurs. One common set-up used in aquaculture is a rotary screen that rotates partly above and partly below the water surface. The meshes in the screen are cleaned by back-flushing either with air or water when the screen is above the water surface and where the back-flushing water that contains the particles removed from the screen can be collected. Straining or micro screening has been shown to be the most effective cleaning method per unit surface area; in aquaculture it is especially effective for removal of relatively big particles, with the head loss for the water flow through the screen also being quite low.

Rotary screen construction can be classified on the basis of how the screen rotates; various systems are available, of which the most common are (Fig. 5.3):

- Axial rotating screen
- Radial rotating screen (drum)
- Rotating belt
- Horizontally rotating disc.

An axial rotating screen is placed vertically and stands normal to the direction of water flow. One common type based on this principle is the disc filter. Which can comprise one or several vertically installed filter plates, with a gradual

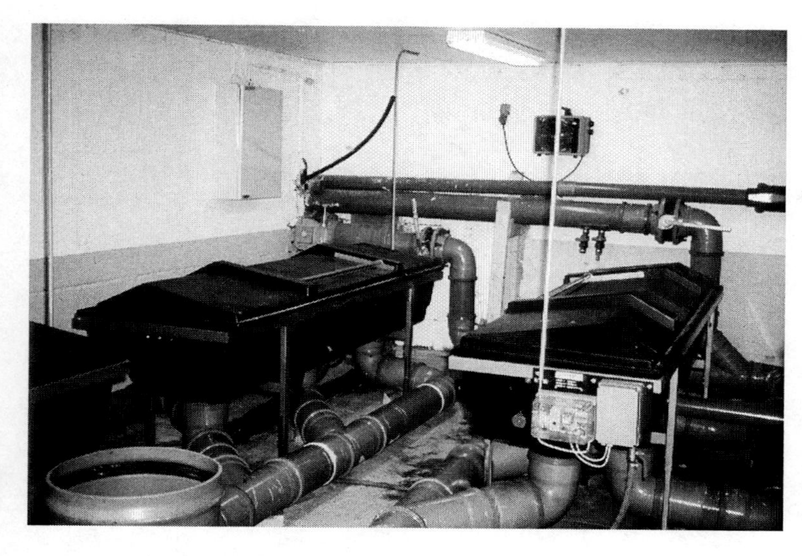

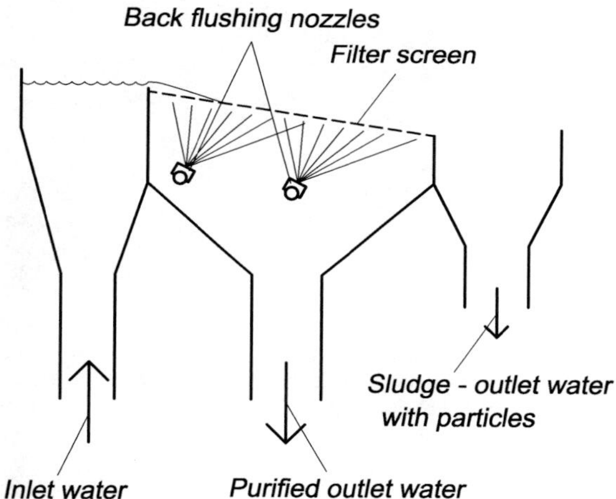

Figure 5.2 A system where the static screen is back-flushed to avoid blockage. The slide shows a typical static screen while the sketch shows one principle of back-flushing.

reduction in the mesh size of the screens. This means that the largest particles are removed on the first screen and smaller ones on subsequent screens.

In a radial rotating screen, the flow is radial towards or away from the axis of rotation. A rotating drum is a typical filter using this construction. Water to be purified flows into the drum, which comprises a straining cloth of appropriate mesh size fixed on a frame. The water has to pass through the drum, which means that it must go out normal (radial) to the main flow direction. The particles will be trapped in the straining cloth when flowing through the drum.

Rotary screens, whether axial or radial, are constructed so that the screen operates when only partially submerged in the water flow that is to be filtered. Back-flushing of the screen, which is used to clear the mesh and remove trapped particles, is done when the screen is rotating above the water surface. High-pressure water from nozzles is directed on to the screen so the particles are dislodged in the same direction that they entered the screen. The back-flush water containing the particles is collected and represents the sludge water from the filter in which the concentration of particles is high. Back-flushing of the straining cloth can be continuous or step-wise: the latter method is

Figure 5.3 Different types of rotating filter with automatic back-flushing of the straining cloth. (A) Overhead view of a disc filter, where axially rotating screens are vertically placed normal to the water flow direction; here two screens with different mesh sizes are shown. (B) A rotating belt filter where the water passes through the belt while the particles are transported to the surface. (C) Close-up of a drum filter with straining cloth. (D) Nozzles used for back-flushing the straining cloth.

most usually used, but choice depends on the load on the filter and the mesh size. The water level in front of the screen can be used to control the rotation and back-flushing. When the screen is clogged, the head loss through the straining cloth will increase together with the water level in front of the straining cloth. This increase in water level can be used to start the rotation of the screen and the back-flushing of the straining cloth. It is, however, important to remember that a rotary screen with back-flushing will produce significant quantities of back-flush water. It is also possible to back-flush with hot water to remove the layer of fat that can be created on the screen surface; this is done only from time to time, not on every back-flush, because of the increased costs.

In a rotating belt filter the straining cloth takes the form of a belt stretched out by rollers at both ends. One of the rollers is motorized and causes the belt to rotate partly above and partly below the water surface, so back-flushing and removal of particles is possible on the part of the belt that is above the water surface. Instead of using water for back-flushing, air may be used. Air at high pressure is blown on the straining cloth from nozzles and dislodges the particles from the mesh in the opposite direction together with some water; this device is known as an air knife. When using air instead of water, the particle concentration (TS) in the sludge water is increased, but is a more cost effective method.

Another type of filter is the horizontally rotating disc standing above the water surface the water to be filtered comes in from above and trickles through the straining cloth. Particles larger than the mesh size of the cloth will be trapped on the cloth surface and must be removed, possibly by vacuum to avoid flooding. The concentration of particles in the water will then be quite high, i.e. the water has high TS value.

The flow rate through the screen is determined by the mesh size of the screen, head loss across the screen, desired purification efficiency, amount and characteristics of particulate material in the inlet water, and the cross-sectional area of the screen. Mesh size selection is based on the maximum particle size that can be allowed in the effluent. Head loss depends on the percentage hole area in the straining cloth, amount and characteristics of particulate material in the inlet water, efficiency of back-flushing, screen rotation speed, and flow rate. Typically the volume of back-flushing water is about 0.2–2% of the bulk flow.[7]

The choice of mesh size depends on the conditions and where the filter is to be used. In inlet water the mesh size may be as small as 20 μm because some parasites will also be removed. On outlet water that is going to a recipient water body, a mesh size between 90 and 100 μm is commonly used. In re-use systems, mesh sizes down to 30 μm are used. Reduction of mesh size will increase the need for new screen cloth exponentially, especially if it is reduced below 60 μm.[19]

5.3.2 Depth filtration – granular medium filters

Depth filtration, also called sand filtration, or just filtration, means removal of particles when water is forced to flow through a layer of a material with particles (granular filter medium) of various sizes and depths, this filtration layer can be of sand or another granular material, depending on the purpose for which the filter is to be used. Because sand is a commonly used the filter is often called a sand filter. The layer is not dense but contains a number of channels and holes created between the particles that constitute the filter medium (Fig. 5.4).

When the water that contains particles goes through the filter medium, particles larger than a certain size will be trapped by several mechanisms.[16] They may be too large to go through channels (straining); they may settle both unaltered and due to flocculation and adhesion, and may be adsorbed by chemical and physical forces that attract them to the filter mass.

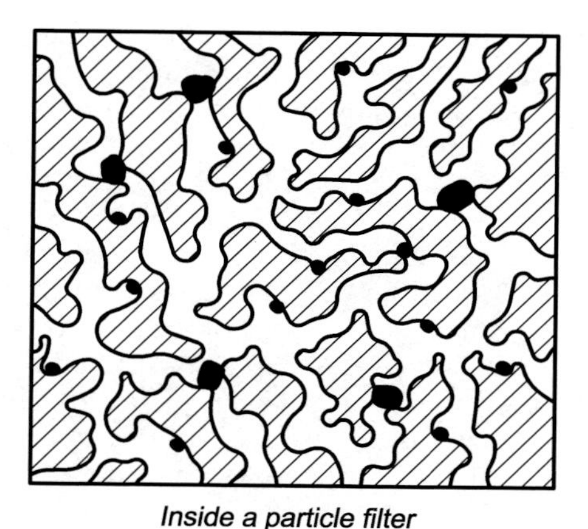

Figure 5.4 Filtration in a depth filter. The black spots are trapped particles.

The first process will occur on the top layer of the filter medium; particles are trapped because they are too large to pass into the channels or pores in the filter media, which will then become blocked. It is, however, of some concern that a few large particles block surface channels of the medium causing a large reduction in filter capacity. The head loss increases and the filter becomes blocked more quickly. Therefore it is more important to utilize settling, adsorption and other effects inside the filter medium; the total depth of the filter medium is then utilized, not just the top layer. However, this requires a low water velocity through the filter medium, which again means that the hydraulic load on the filter surface must be low. When the channels inside the filter are full of particles the head loss will increase.

The maximum particle size that will pass through a depth filter is determined by the grain size of the medium; if the grain size is small, the filter will clog easily. The flow rate through filter medium and the rate of clogging depend not only on the grain size of the filter medium, but also on the characteristics of the particulates in the water to be purified. The performance of a depth filter is dependent both on the type of filter and operating procedures, and also significantly on the characteristics of the filter material.

Depth filters can be classified depending on the direction of the water flow through the filter medium. In an up-flowing filter, the water flows up through the filter medium, while in a down-flowing filter the water flows downwards; the latter type is the most common. Good filtration can be achieved, but the filter medium becomes clogged after a period. In an up-flowing filter the same will occur; here there may also be a break-through of waste-water at one or several places in the filter medium if there is excessive clogging, provided that the water pressure is high enough. In this case almost all the water will go through the filter via these zones and there will be virtually no purification.

A depth filter should be equipped with back-flushing facilities, or regular manual purification of the filter medium will be necessary. In a traditional sand filtration system (on-site system, see later) used for purification of municipal wastewater, this can be the case. Such systems can be used for months or even years without doing anything because the wastewater is only discharged onto the soil surface. Of course, this depends on the hydraulic load on the surface area and the concentration of particles in the water.

If a filter is to maintain capacity, back-flushing is necessary. Even when this is carried out, the hydraulic capacity of this filter type is lower than that of, for instance, rotating microsieves. While doing the back-flushing operation, the filter must be stopped; water is then sent the opposite way through the filter media, so removing particles that have settled. This back-flushing water is sent directly to the farm outlet. Most of the running problems with depth filters are the result of improper back-flushing. To achieve proper purification of the filter medium it is important that the back-flushing water suspends trapped particles. Typical amounts of back-flushing water are in the range 1–5% of the bulk flow.[7]

Depth filters can be separated into those operating at normal atmospheric pressure or those having an over pressure inside the filter, i.e. pressurized filters (Fig. 5.5). In a pressurized filter the medium may be put inside a sealed chamber, with the same overflow arrangements as for an unpressurized filter. When a pressurized filter starts to clog, the pressure will increase and the particles will be pushed further down in the filter medium; ulti-

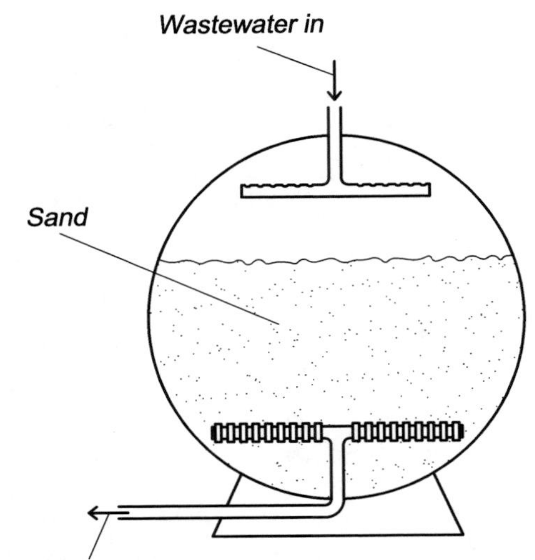

Figure 5.5 A pressurized depth filter with back-flushing.

mately such a filter will become totally blocked, but it takes longer than with an unpressurized filter. The advantages with pressurized filters are that a larger part of the filter medium height is utilized and the back-flushing interval is increased. To use a pressurized filter it is necessary to pressurize the water to be purified, normally up to 6–8 bar. Usually the water passes through a pressure increase pump to achieve this, and consideration must be given to avoiding damage to the particles. If such an arrangement is used on the outlet water, the pump will break the particles up and the proportion of small particles which are more difficult to remove will increase. Therefore pressurized filters are not recommended for use on outlet water; however, if there is no alternative the outlet water must be handled as carefully as possible.

The normal filter medium is sand or gravel. The size of the sand or gravel particles in the filter medium depends on the characteristics of the particles in the water to be purified. By using small particles in the medium the head losses and clogging velocity of the filter are increased whilst the capacity is decreased. In a one size sand filter most of the filtration occurs in the first few centimetres. Instead of using one type of medium over the whole filter depth, i.e. a single size medium, media of several sizes (multimedia) may be used to increase utilization of the filter. In such filters, provided the water enters from the top of the filter and flows down, the largest media are on the top and the smallest on the bottom. The largest particles are then removed in the top layer and increasingly small particles are removed through the lower layers of the filter. This will ensure a more even collection of particles distributed over a larger part of the filter and not just in the top layer. To avoid mixing of the media during back-flushing, media with different relative density can be used. The largest media must have the lowest density and the smallest the highest. When back-flushing from below this will ensure that the largest particles stay on the top. In aquaria, a two-media filter comprising a layer of crushed anthracite or ilmenite above a layer of fine sand has been used.[20] A third bottom layer with even finer particles, such as crushed ilmenite or garnet, could also be included. Regardless of the filter media, it is of great importance that the back-flushing is done correctly so that all the media are back-flushed, otherwise there will be zones that are still clogged and that will not be utilized when the filter is returned to operation.

Each medium can be characterized by its grain size, uniformity of grain size, grain shape and relative density, all factors that are important for filtering performance. Appropriate grain parameters can considerably reduce the resultant head losses.

Several classifications can be used on granular media filters, and these can also be classified with respect to their hydraulic capacity giving (1) slow sand filters, (2) rapid sand filters, and (3) continuously back-flushed filters.[21] Type 1 is unpressurized and type 2 is pressurized; type 3 is constructed to enable continuous back-flushing without interrupting the water flow as is necessary for back-flushing types 1 and 2. Typical reported loading rates are 0.68 l/s/m^2 or less for slow sand filters, up to 1.4 l/s/m^2 for rapid sand filters and up to 5.4 l/s/m^2 for continuously back-flushing sand filters.[21]

In slow sand filters 2–5 mm grains are utilized;[10] pressurized rapid single sand filters have grain sizes ranging from 0.3 to 4 mm.[22] Reported removal rates are for 0.3 mm sand about 95% of particles down to 6 μm, and for 0.5 mm sand about 95% of particles down to 15 μm.[22] A typical pressurized rapid sand filter has a capacity of 136 kg sand, a surface area of 0.29 m^2, a maximum pressure of 3.5 kg/cm^2 and a design flow rate of 238 l/min.

Depth filters can also be of the cartridge type. The cartridge can be made of different materials, such as plastic, ceramic, spun fibre or resinbound fibre,[22] often specific to the supplier. The cartridge has a defined depth, and the waste is collected either inside or outside the cartridge, or has to pass through the cartridge. The typical cartridge filter is used once and then replaced. Cartridge filters can be used for removal of small particles and are available for sizes below 1 μm. They are used on small water flows with low particle loads; otherwise the cost of operation becomes too high.

Diatomite (DE) or pre-coated filters may also be classified as depth filters.[23] Diatomaceous earth filters comprise a filter with a screen coated with DE (fuller's earth), a granular material of fossil diatoms. A cake of DE is formed and put on the filter screen that functions as a skeleton. The water to be purified has to pass through this filter cake and the particles are trapped. When the filter is dirty it can be back-flushed, or a new filter cake can be installed. DE filters can tolerate higher water flows than cartridge filters, but are best suited to small water flows.

Filter bags are another simple type of filter for use with small water flows. The filter bag is woven and so forms a surface, the mesh size of which depends on the closeness of the weave. Particles are trapped on the surface of the bag. When no more water will pass through, the filter bag has to be removed and replaced with a new or clean bag. The bags can be cleaned in a washing machine, for instance.

5.3.3 Settling or gravity filters

Settling basing

Settling is a simple method for removing particles from the water.[24] The principle utilized is that particles have a higher relative density than water (1.005–1.2 compared to 1 for freshwater),[11,12] so they will sink. This phenomenon can easily be observed when water containing suspended particles is allowed to stand for a period. A natural separation process will occur and the particles will sink to the bottom. The difference in relative density between the particles and the water controls the velocity of the separation process.

For small particles (0.1–1 mm) and unobstructed settling, the sinking velocity of the particles is given by Stoke's law:

$$V_s = \frac{g(\rho_p - \rho_w)d_p^2}{18\mu}$$

where:

V_s = sinking velocity
ρ_p = density of particle
ρ_w = density of water
g = acceleration due to gravity
d_p = diameter of particle
μ = dynamic viscosity of the water.

This also demonstrates why the particles with the highest density are the simplest to remove.

The simplest way to utilize gravitational force for separation is to send the effluent water through a basin with a large surface area where the water velocity is reduced: separation will then occur in the basin provided that the sinking velocity of the particle created by the gravitational force does not exceed the horizontal velocity component created by the water flow through the basin (Fig. 5.6), in which case the particle will flow out of the basin in the water. The following equation can be set up to achieve settling:

$$V_s > Q/A$$

Where:

V_s = sinking velocity of the particle (m/h)
A = surface area of the basin (m^2)
Q = water flow through the basin (m^3/h)

The relation Q/A is called the surface load or overflow rate for the settling basin. Because difference in density between water and faeces is small quite a long hydraulic residence time is required. In fish farming the normal surface load is between 1 and

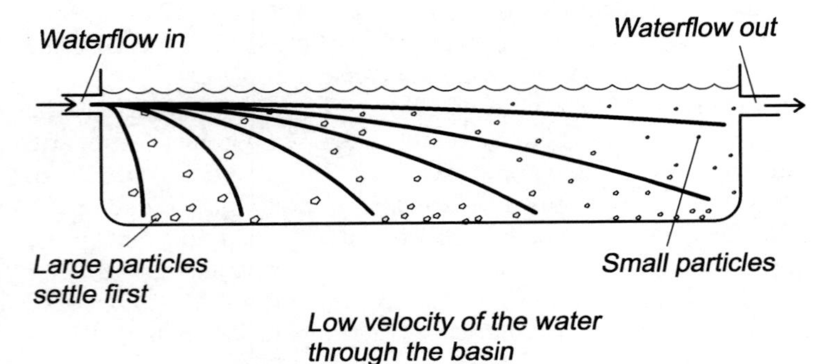

Figure 5.6 A settling basin.

5 m/h (actually $m^3/m^2/h$).[2] Experiences show that a favorable depth for the settling basin is 1 m and the width : length ratio is in the range 1 : 4 to 1 : 8.

Example
A water flow of 1000 l/min (60 m³/h) is to be purified by the use of a settling basin. Determine the size and design of the basin.

Expect a surface load of 3 m/h

$$V_s > Q/A$$

$$A > Q/V_s$$
$$> \frac{60\,m^3/h}{3\,m/h}$$
$$> 20\,m^2$$

This means that the surface area of the pond needs to be more than 20 m² to allow proper settling. If a width : length ratio of 1 : 3 is chosen for the basin, this means that the exterior measurements of the basin will be about 2.6 m × 7.7 m. Two settling basins are needed so that one can be running while purification and sludge removal can be done in the other.

Other designs of settling basin take account of hydraulic retention time (τ) or mean fluid velocity (v_m).

$$\tau = V/Q$$

$$v_m = L/\tau = Q/A_{cs}$$

where:

V = volume of the settling basin
L = length of settling basin
A_{cs} = cross-sectional area of the basin.

Reported values for retention times are 15–40 min, while recommended mean fluid velocities are in the range 1–4 m/min.[2]

The flow pattern in a settling basin used in fish farming is normally horizontal. However, it is also possible to use settling with a vertical flow pattern. Such filters will have a tower like design. Water flows slowly upwards and because of the gravitational forces the particles will sink with a greater velocity. Specially designed filters, such as the lamella separation filter are used to improve settling conditions, and also the investment requirements. Sending the water over biofilm can improve the settling and filtration efficiency. Small particles will be attracted to and settle on the biofilm.[25] Addition of polymers or other chemicals before settling can be used to increase the particle size by flocculation, and this also removes smaller particles, possibly in the settling basin.[16]

The great disadvantage with settling basins is that the settled particles remain lying in the water flow, so there is a possibility that nutrients will leak from the particles, especially phosphorus which is weakly attached to the particles.[2,4] Additionally, resuspension of settled particles into the water may occur, even if particles are regularly removed from the basin; for this reason there must be adequate depth in the settling basin. Continuous removal of settled particles from the basin is impossible from a cost perspective. The removal of settled particles must, however, be carried out regularly to optimize basin function. This can be done by various methods, for instance by using a vacuum pump. It is important that mixing of sludge and water is kept to a minimum to avoid resuspension of particles with the result that the nutrients go directly to the outlet. It is quite common to have two settling basins, so that one can be used while the other is purified.

Compared to using a micro strainer, settling basins require a much greater area and this can be a disadvantage. However, if the surroundings and ground conditions are suitable, settling basins are simple and cheap to construct.

Results from testing shows reduced removal rates with small particles.[26] It is quite difficult to remove particles smaller than 100 μm using a settling basin.[27] Inlet values of less than 10 mg TSS per litre are difficult to treat, and those below 6 mg TSS per litre are also difficult to obtain.[27] To process values in this range special methods must be used, such as addition of polymers or use of biofilm to attract particles.

Swirl separators, hydrocyclones

In a swirl separator or a hydrocyclone the principle that the particles are more dense than water is also used, but here centrifugal forces are used in addition[28–30] to increase this difference. To illustrate this, the water inside a cup can be rotated and the particles will be hurled out towards the edges (Fig. 5.7). This is also the reason why this filter is sometimes called a tea-cup settler. In a swirl separator the water enters along the periphery of a circular tank,

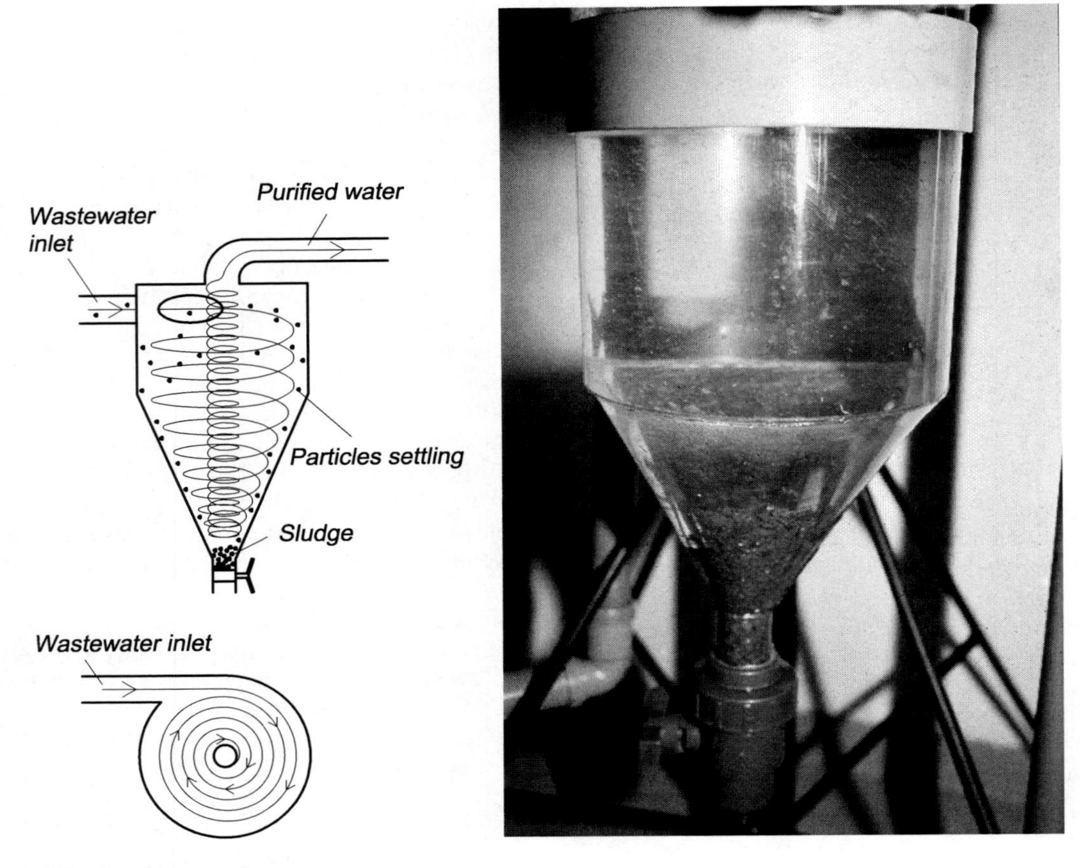

Figure 5.7 A swirl separator.

so the particles follow the periphery and the purified water will be pressed into the centre. The particles will then sink to the bottom, while the purified water is drained out of the centre. Because the centrifugal forces are greater than the gravitational force, a smaller area is needed than for a traditional settling basin to remove the same amount of particles. Re-suspension or leakage is also less than with a settling basin, because the area where this occurs is smaller. However, in comparison to the settling basin, the hydraulic load on a swirl separator may be much higher, 20–25 m/h.[24]

An advantage of this type of filter unit is its quite simple construction with no movable parts; in addition is it cheap to buy and area intensive. The great disadvantage with swirl separators is that they require uniform water flow for optimal efficiency. If the flow is higher than the filter is designed for, the particles will flow out of the unit with the outlet water in the center.

Example
An outlet water flow of 10 l/min (0.6 m³/h) is to be purified using a swirl separator. The acceptable hydraulic load is set to 20 m/h. Find the necessary size of the separator.

$$V_s > Q/A$$

$$A > Q/V_s$$

$$> \frac{0.60\ \text{m}^3/\text{h}}{20\ \text{m/h}}$$

$$> 0.03\,\text{m}^2$$

The radius of the separator must be more than 9.8 cm.

5.3.4 Integrated treatment systems

Nature-based wastewater treatment technology covers several methods that can be used alone or in combination.[31,32] Three mains systems can be used for treatment of effluent water from land-based aquaculture farms: ground filtration, constructed wetland, and pond systems. Ground filtration systems can further be divided into open ponds or subsurface trenches (Fig. 5.8). In a filtration system the soil is used as a filter medium, and is actually the same as the depth filter described earlier. The water is distributed above the filter bed and trickles through the soil in which filtration, adsorption/precipitation and biological degeneration will be major processes. Nutrients, organic matter and micro-organisms will be removed. The local soil, if suitable, is normally used as a filtration medium and therefore it is also called on-site treatment. However, soil (e.g. sand) or other suitable porous media such as Leca may also be trucked to the site. The main problem with ground filtration for treatment of effluent water from aquaculture is the large amount of water to be treated and the low hydraulic capacity of soil-based systems. Even if some improvement can be achieved by using suitable porous media the area needed is still large. Normal values for domestic wastewater when using subsurface trenches are about 10000 l/m^2/day, with some higher values in open ponds. Effluent water from aquaculture facilities has been less well studied and there is a lack of available values. Because the pollutant concentrations are lower, the load may be somewhat higher, but if the system is overloaded it will function sub-optimally and purification will be reduced. Results from treatment of domestic wastewater in a cold climate latitude (69°N) with an annual mean temperature of 1.2°C show the following results: 70% nitrogen removal, 99% phosphorus removal, 70% chemical oxygen demand (COD) removal and close to 100% removal of faecal coliforms. The normal flow was 750 m^3/day in a 2000 m^2 open basin, but during snow melt it can be up to 3500 m^3/day.[33]

If the system becomes totally clogged a tractor with a shovel can remove the upper 5–10 cm of sand, after which the system can be used again. A typical value for ground filtration systems for wastewater is 100 l/m^2/day, but of course this varies with the soil conditions. If using such values for outlet water from fish farming, a tremendous area is necessary for water purification. When using depth filtration in fish farming there is a need for high hydraulic capacity.

Constructed wetlands are commonly used for various types of polluted water such as domestic

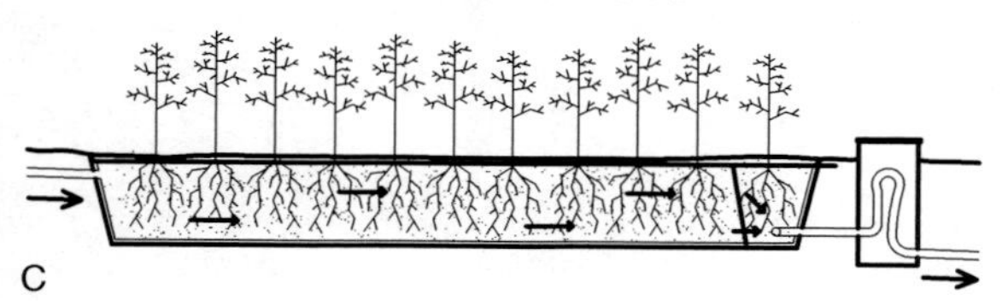

Figure 5.8 Different types of integrated treatment systems. (A, B) Filtration systems where the outlet water from the fish farm is sent to filter through the bottom of a trench: (A) a newly constructed filtration trench; (B) a trench filled with water. (C) Sketch showing a constructed wetland through which the outlet water from the fish farm has to pass before it reaches the recipient water body. The plants will take up nutrients from the outlet water.

and industrial wastewater, and have also been utilized to treat wastewater from aquaculture.[4] The outlet water is sent into the wetland that contains wetland-adapted plants and a porous earth medium. The plants take up and utilize the nutrients in the outlet water; in addition, the roots of the plants create an environment that increases the purification processes. By sending outlet water from a salmon hatchery to an abatement pond coupled to a constructed wetland, the following removal rates were attained: solids above 98%, ammonia above 84%, biodegradable organics and phosphorus above 90%.[34]

Both ground filtration systems and constructed wetlands will, in addition to removal of nutrients and organic matter, inactivate micro-organisms. The great disadvantage with the systems is the low hydraulic capacity in relation to the amounts of water in aquaculture.

5.4 Hydraulic loads on filter units

It is important to be aware that a filter system is designed for a given flow of water with a given characteristic. If either less or more water than the filter unit is designed to treat is used, the filter will not functional optimally. The ability of a filter to tolerate varying water flows, for example when tapping down a fish tank, depends on its design. Equipment using settling as a principle is especially intolerant of variations in the water flow, particularly high flows. For a mechanical filter, variations in load are normally not so critical. However, if the loads are too high the filter cloth may become so clogged that breakdown can occur.

A common fault on aquaculture facilities is that the tanks and outlet pipes are the incorrect design and size, so settling of particles occurs in the system. Shock drainage of the outlet system is used, often once or twice a day, to remove settled particles and avoid total blockage. If shock drainage is necessary to keep the outlet pipes open, something is wrong with the design and construction of the outlet (see Chapter 11). When shock draining the fish production tanks, the water flow in the outlet pipes is increased and so is the particle concentration, because particles that had settled in the outlet system will now go into suspension as a result of the higher velocity. If the filter system does not tolerate variation in water flow, reduced purification results. This is critical, because it is in these situations that the number of particles is highest, and where good purification is necessary. Here the importance of choosing an appropriate filter system, and of doing everything correctly before the filter system, is apparent. It is also necessary to be aware of the interaction between the different parts in the farming system.

5.5 Purification efficiency

As shown, there are several methods for removing particles from a water flow. To evaluate how effectively a filter is functioning and to compare filter systems, the term purification efficiency is commonly used (Fig. 5.9). This can be defined as follows:

$$C_e = ((C_{in} - C_{out})/C_{in}) \times 100$$

where:

C_e = efficiency (%)

C_{in} = concentration of the actual substance entering the filter

C_{out} = concentration of the actual substance exiting the filter.

Example

The concentration of suspended solids entering the filter units is measured and found to be 20 mg/l; on exiting the filter the concentration is measured as 5 mg/l. Find the purification efficiency of the filter.

$$C_e = ((C_{in} - C_{out})/C_{in}) \times 100$$
$$= ((20 - 5)/20) \times 100$$
$$= 75\%$$

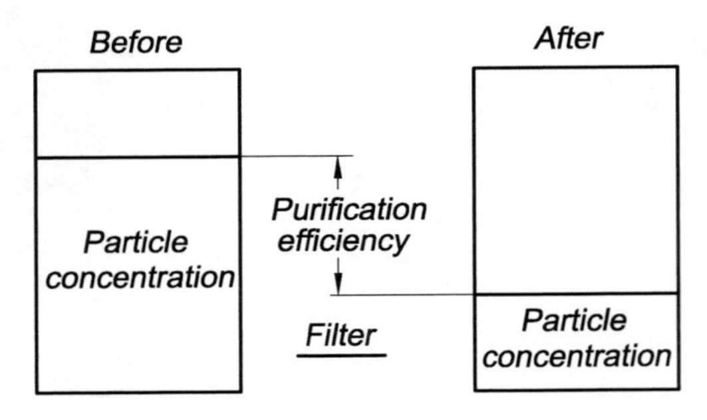

Figure 5.9 The purification efficiency indicates how much of an actual substance is removed by the filter.

In aquaculture the efficiency is normally expressed as the percentage of the incoming TSS removed by the filter. However, it may also be used for other substances, such as the amount of removed nutrients (total phosphorus, TP or total nitrogen, TN) or as reduction in chemical oxygen demand (COD) or biological oxygen demand (BOD) in the water passing through the filter. The last two measurements tell us how the outlet water will affect the oxygen concentration in the recipient water body.

When considering removal of nutrients, it is important to know the proportion attached to particles and not dissolved, since it is not possible to remove dissolved nutrients with a particle filter. These values depend on the species and the feed composition and utilization. Experiments on salmonids have shown that more than 80% of nitrogen compounds are dissolved in the water,[10] whereas the situation is reversed for phosphorus with up to 80% being attached to particles. However, phosphorus leach easily from particles lying in the water; this is a major reason to remove the particles from the water flow as quickly as possible.

The efficiency of the particle filter of course depends on the particle concentration and characteristics of the water to be purified. For a good comparison of filter systems, they must be tested on exactly the same water. Typical values for removal of TSS from wastewater from fish farming range from 30 to 80%[1,7,24,35] while some lower values have been reported for water re-use systems.[7]

5.6 Dual drain tank

Today, circular tanks with a dual drain outlet system are used on some farms (Fig. 5.10).[36–38] Here the tank is used as the first purification step. In a dual drain system gravitational forces, and the fact that the waste particles are denser than water, are utilized to separate the particles and collect them at one point in the tank where a particle outlet is placed and a small amount of water can be withdrawn to flush the particles out, while the main water flow can be withdrawn elsewhere, normally higher up in the water mass. Several methods can be used to do this, and a number of different dual drain systems exist.[39]

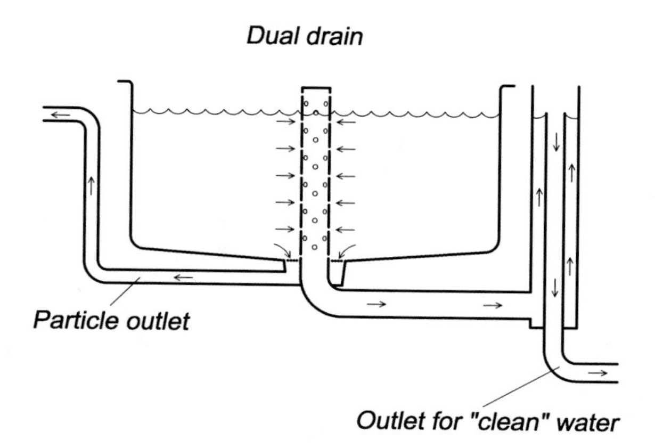

Figure 5.10 A tank with a dual drain outlet is used for the first purification step.

By separating the water flow in this way, purification is achieved inside the tank. The particle concentration in the particle outlet is considerably higher than in the outlet withdrawn higher up in the water mass. This can be the first purification step, and may be the only one. Since the amount of water coming through the particle outlet is much lower than the total flow, only a small part of the water flow has to be treated (Fig. 5.11), so a much smaller filter can be used. As the water flow through the particle outlet can be stable, a swirl separator is very suitable. The amount of water sent through the particle outlet in the dual drain system varies from 0.5 to 20% depending on the system used.

5.7 Sludge production and utilization

Removal of particles from the water creates sludge comprising water and particles. The water content and dry matter (DM) or total solids (TS) in the sludge depend on the particle filter used. To give an idea of the amount of sludge (faeces) created by the fish, the following estimate can be used: for each kilogram of feed eaten by rainbow trout 20% faeces are produced, measured on 100% DM basis.[40] This value, however, depends on a number of factors including feed composition, feed type, fish size and fish species, and will therefore vary. If the feed conversion rate is above unity and traditional dry feed is used, there will be feed loss that goes directly to the outlet in addition to the sludge produced from the faeces. This also shows the importance of correct feeding and avoiding feed losses.

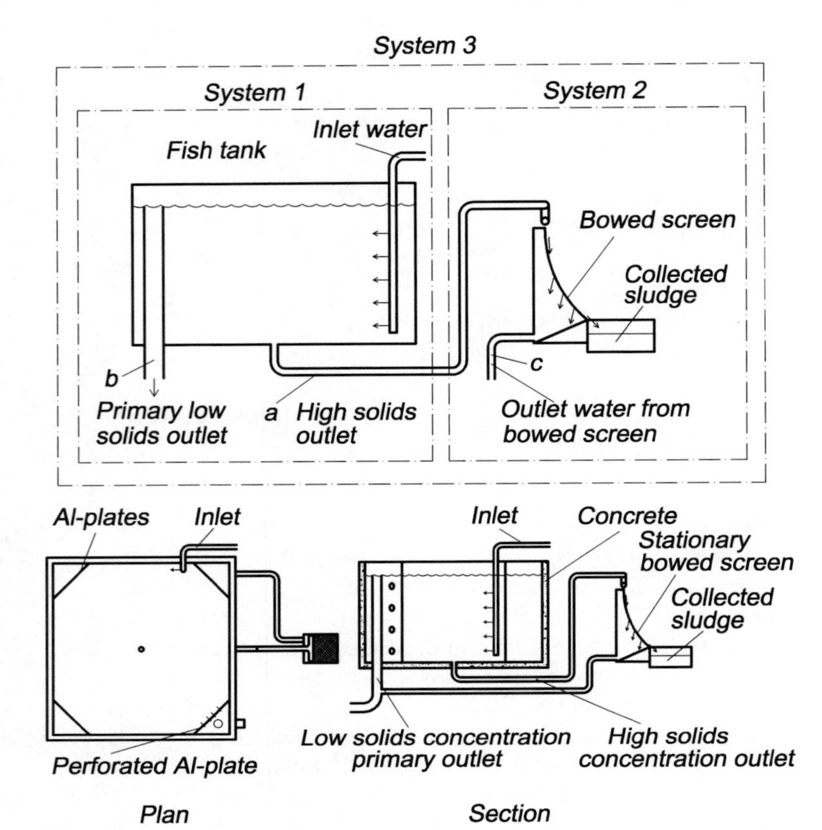

Figure 5.11 A dual drain tank can be coupled with a small and cheap purification system.

Example
How much sludge is produced per kg commercial dry feed supplied to rainbow trout if the purification efficiency of the filter is set to 50% measured as DM? What happens if the feed conversion rate increases to 1.2?

Per kg feed supplied, 200 g of sludge is produced and the filter collects 50% of this. This means that the amount of sludge collected per 1 kg feed supplied is 100 g.

If the feed conversion rate is 1.2 only 0.83 kg of the 1 kg feed supplied will be utilized for growth, while the rest will be feed loss. Calculating that 20% of the feed eaten is converted to faeces, this represents 0.17 kg. If this is added to the feed loss, the amount is 0.17 + 0.17 = 0.34 kg. If collecting 50% of this, the amount collected per kg feed supplied will be 170 g.

The actual amount of sludge can be much higher; the percentage DM in the collected sludge depends on the filter system used. In a mechanical filter it is mainly a consequence of how the straining cloth is back-washed, whether using air or water, and if water, the amount. Normally the percentage DM in the sludge is 0.1–1%, but in special filters it might be up to 5%. It is advantageous to have as much DM in the sludge as possible, which means that back-flushing with water is disadvantageous. The only reason for back-flushing with water is that it is an effective system. A large proportion of DM in the sludge reduces the amount that must be further treated and transported; the sludge is therefore often dewatered to increase the percentage DM. Filter presses and special centrifuges can be employed for dewatering.[16,23] The sludge may also be sent to a settling system for further separation of particles.[26] Vertical sedimentation in a cone has been used and increased the DM in the sludge to 7–10%.

Normally the sludge must be stored for a period to accumulate enough so that it can be collected economically. A particle removal system will therefore include storage tanks for the sludge.

To make it possible to store the sludge, it must be stabilized. If the sludge is immediately placed in an open container with access to air at the surface, an uncontrolled decomposition process (rotting) will take place. This will smell and have negative effects on the development of bacteria and the content of nutrients. Subsequent use of the sludge can be inhibited because of this. Correct decomposition makes the nutrients in the sludge available for plants so that it can be used as fertilizer on agriculture land. Sludge from fish farms is rich in organic nitrogen (3–9% of DM) and phosphorus (1–4% of DM). In addition, the concentration of heavy metals is usually below regulatory limits.[41] This makes the sludge useful as a fertilizer.

Untreated sludge may contain pathogenic (negative) micro-organisms such as viruses, bacteria and parasites.[42] When infected sludge is spread on agricultural land and there is drainage to lakes or rivers, pathogenic micro-organisms could be transferred to the local fish strains. Birds could also transfer pathogenic micro-organisms from the sludge to lakes. Therefore the sludge must be treated to inactivate the negative micro-organisms before it is spread; this is not achieved with uncontrolled decomposition.

There are several ways to inactivate pathogenic micro-organisms in the sludge; wet or dry composting is commonly used. Another method is to add lime to raise the pH in the sludge and hence inactivate the micro-organisms. Both these methods will also stabilize the sludge so that it can be stored, and make it suitable for use as fertilizer on agricultural land, a very important use for sludge, normally a good fertilizer because of the high nutrient content.

When composting sludge, controlled aerobic decomposition occurs.[42] Due to the low content of DM, wet or liquid composting is employed for fish farming sludge (Fig. 5.12). Before composting the

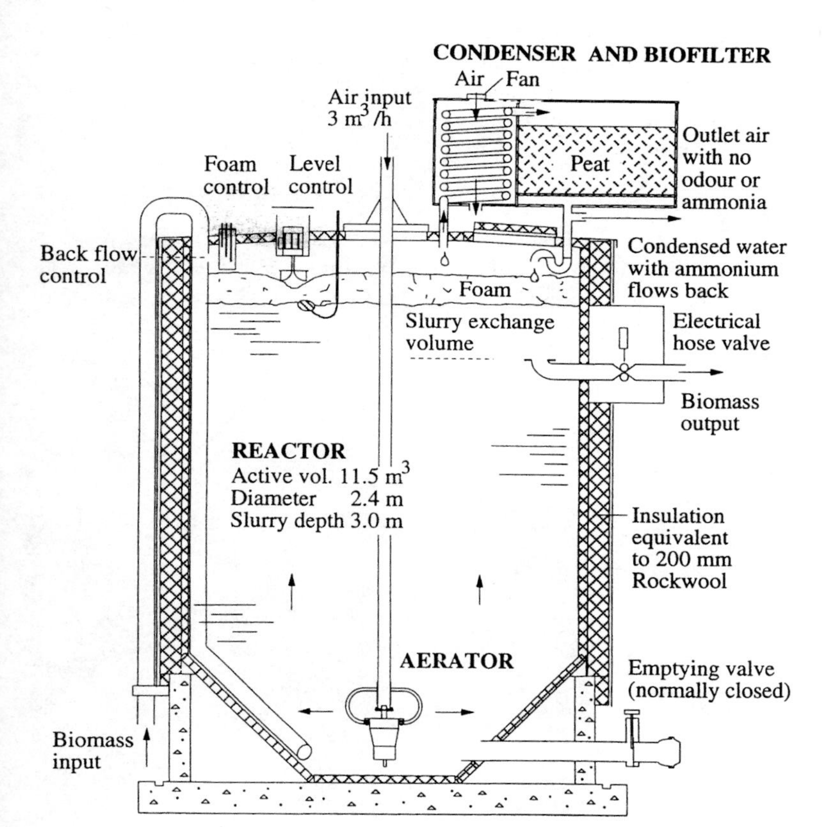

Figure 5.12 A wet composting reactor for treatment of sludge.[44]

sludge may be mixed with manure or municipal sludge. In a community there may be a centrally installed reactor. The sludge is poured into a container where air is added, for instance through an injector pump. In addition the sludge is circulated around in the tank so air comes into contact with all the sludge. This results in controlled bacterial development in the container which decomposes the organic matter. The process is thermophilic, so the temperature increases depending on the sludge. For fish farming the sludge is energy-rich, experimentally producing 3.1 kWh/kg DM,[44] and the temperature rises to 60–70°C (Fig. 5.13). This high temperature is maintained for some time so that the pathogenic micro-organisms are inactivated and the sludge is stabilized for storage. To get rid of the smell developed during the composting process it is an advantage to include a smell filter, for instance made of peat, through which all gases emitted by the composting process have to pass.[43–45] Overall, the composting process will result in biological stabilization of the sludge, removal of the major odour compounds, increased availability of some plant nutrients, nitrification and denitrification, and improved waste consistency.[42]

If the process is run without access to oxygen, anaerobic digestion, also known as fermentation, will occur. As for composting, naturally occurring micro-organisms in the sludge are utilized. It is important that no air is supplied, so anaerobic degeneration of the organic substances will occur. Normally some heating of the sludge will be necessary to allow the micro-organisms to develop. The sludge is stored in a closed digester during this process. For animal manure it takes 10–35 days; a high fat content reduces the time necessary. This fermentation produces methane gas; the process is therefore also called biogas fermentation. In addition to the production of biogas, there will also be a reduction in offensive odours, a breakdown of organic mass, a reduction of pathogens and an improved fertilizing value due to easier availability of the nutrients.[42] The biogas can be used for heating or electricity production, so there is actually a positive output from the process. Sludge that has gone through this process can be stored for later use as fertilizer for agricultural land.

Adding lime (CaO) or slaked lime ($Ca(OH)_2$) to the sludge to increase the pH, will also stabilize and disinfect the sludge. In experiments with sludge from fish farming it was shown that by increasing the pH to 12 and maintaining this value for 7 days, more than 99.9% of the pathogenic viruses and bacteria were killed.[41] The sludge produced is well suited for use as organic fertilizer.

When establishing a system for particle removal it is therefore necessary to think not only about the particle filter itself, but also sludge production and its utilization. This includes tanks for sludge collection (Fig. 5.14).

5.8 Local ecological solutions

It is also possible to devise total ecological production methods for treatment of effluent water from intensive fish farming, not just by using it in more

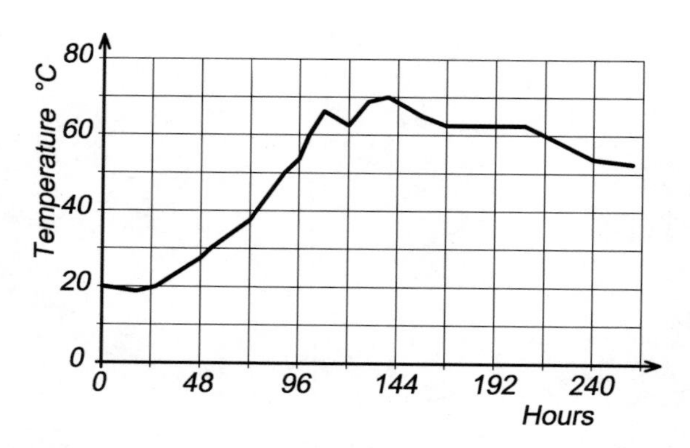

Figure 5.13 Temperature increase in liquid composted sludge from aquaculture.

Figure 5.14 A tank for sludge collection for later treatment is necessary on farms.

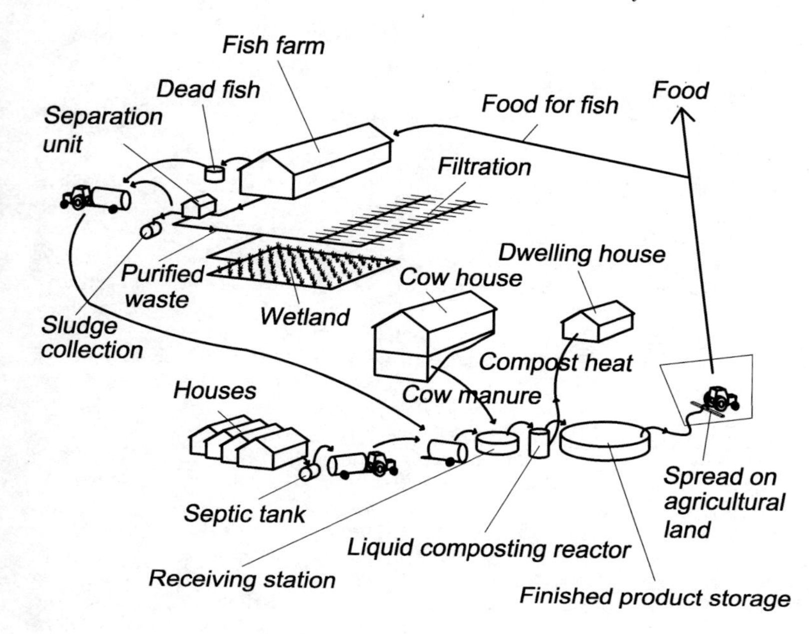

Figure 5.15 Design of a fish farming system with no direct outlet: a local ecological solution.

extensive forms of aquaculture such as polyculture. Such a system is shown above using integrated treatment systems for the water and local use of the sludge (Fig. 5.15).

The fish farm must employ water re-use technology to reduce the amount of outlet water. A microscreen is used as the first step to remove the larger particles. Then the purified outlet water may be sent to a ground filtration unit or constructed wetland for further purification. The sludge produced by the microscreen can be sent to a locally installed liquid composting unit together with animal manure and eventually be used as fertilizer on nearby agricultural land, so avoiding long distance transport of the sludge. A solution with no effluent, either of nutrients, organic matter or micro-organisms is achieved in this way.

References

1. Chen, S., Stechey, D., Malone, R.F. (1994) Suspended solids control in recirculating aquaculture systems. In: *Aquaculture water reuse systems, engineering design and management* (eds Timmons, M.B., Losordo, T.M.). Elsevier Science.
2. Summerfeldt, S.T. (1999) Waste-handling systems. In: *CIGR handbook of agricultural engineering, part II aquaculture engineering* (ed. Wheaton, F.). American Society of Agricultural Engineers.
3. Liltvedt, H., Hansen, B.R. (1990) Screening as a method for removal of parasites from inlet water to fish farms. *Aquacultural Engineering*, 9: 209–215.
4. Midlen, A., Redding, T.A. (1998) *Environmental management for aquaculture.* Chapman & Hall.
5. Bergheim, A., Brinker, A. (2005) Water pollution from fish farms. In: *Water encyclopedia 3: Surface and agricultural water* (eds Lehr, J.H., Keeley, J.). pp. 579–581. Wiley-Interscience.
6. Lekang, O.I., Fjæra, S.O. (2002) *Teknisk utstyr til fiskeoppdrett.* Gan forlag. ISBN 82-492-0353-4.
7. Brinker, A., Bergheim, A. (2005) Waste treatment in fish farms. In: *Water encyclopedia 1: Domestic, municipal and industrial water supply and waste disposal* (eds Lehr, J.H., Keeley, J.). pp. 681–684. Wiley-Interscience.
8. Bergheim, A., Aasgaard, T. (1996) Waste production from aquaculture. In: *Aquaculture and water resource management.* (eds Baird, D.J., Beveridge, M.C.M., Kelly, L.A., Muir, J.F.). Blackwell Science.
9. Chen, S., Coffin, D.E., Malone, R.F. (1993) Production, characteristics, and modeling of aquacultural sludge from a recirculating aquacultural system using a granular media filter. In: *Techniques for modern aquaculture* (ed. Wang, J-K.). pp. 16–25. Proceedings of Aquacultural Engineering Conference, Spokane, Washington. ASAE Publication 02–93.
10. Cripps, S.J. (1995) Serial particle size fractionation and characterisation of an aquacultural effluent. *Aquaculture*, 133: 323–339.
11. Wong, K.B., Piedrahita, R.H. (2000) Settling velocity characterization of aquacultural solids. *Aquacultural Engineering*, 21: 233–246.

12. Warrer-Hansen, I. (1982) The right treatment of wastewater. *Fish Farming International*, Aug., 36–37.
13. Droste, R.L. (1997) *Theory and practice of water and wastewater treatment.* John Wiley & Sons.
14. Davis, M.L., Conwell, D.A. (1998) *Introduction to environmental engineering.* McGraw-Hill.
15. Letterman, R.D. (1999) *Water quality and treatment.* American Water-Works Association McGraw-Hill.
16. Tchobanoglous, G., Burton, F.L., Stensel, D.H. (2002) *Wastewater engineering.* McGraw-Hill.
17. Salvato, J.A., Nemerow, N.L., Agardy, F.J. (2003) *Environmental engineering.* Wiley-Interscience.
18. Trussel, R., Hand, D.W., Howe, K.J., Tchobanoglous, G., Crittenden, J.C. (2005) *Water treatment: principles and design.* Wiley-Interscience.
19. Cripps, S.J., Bergheim, A. (2000) Solids management and removal for intensive land-based aquaculture production systems. *Aquacultural Engineering*, 22: 33–56.
20. Spotte, S. (1979) *Fish and invertebrate culture. Water management in closed systems.* John Wiley & Sons.
21. Huguenin, J.E., Colt, J. (2002) *Design and operating guide for aquaculture seawater systems.* Elsevier Science.
22. Tucker, J.W. (1998) *Marine fish culture.* Kluwer Academic Publishers.
23. Montgomery, J.M. (1985) *Water treatment, principles and design.* John Wiley & Sons.
24. Lekang, O.I., Bomo, A.M. (1999) *Alternative sedimentation methods for effluent treatment in aquaculture.* ITF rapport 101, Norwegian University of Life Science (in Norwegian).
25. Lekang, O.I., Bomo, A.M., Svendsen, I. (2001) Biological lamella sedimentation used for wastewater treatment. *Aquacultural Engineering*, 24: 115–127.
26. Bergheim, A., Liltvedt, H.S., Cripps, G., Indrevik, Nygaard Austerheim, L. (1996) *Avvanning, stabilisering og utnyttelse av våtslam fra fiskeoppdrett.* Rapport Rogalandsforskning 280 (in Norwegian).
27. Henderson, J.P., Bromage, N.R. (1988) Optimising the removal of suspended solids from aquacultural effluent in settlement lakes. *Aquacultural Engineering*, 7: 167–181.
28. Wheaton, F.W. (1977) *Aquacultural engineering.* R. Krieger.
29. Davidson, J., Summerfelt, S.T. (2005) Solids removal from coldwater recirculating system – comparison of a swirl separator and a radial flow settler. *Aquacultural Engineering*, 33: 47–61.
30. Veerapen, J.P., Lowry, B.J., Couturier, M.F. (2005) Design methodology for the swirl separator. *Aquacultural Engineering*, 33: 21–45.
31. Crites, R., Tchobanoglous, G. (1998) *Small and decentralized wastewater management systems.* McGraw-Hill.
32. Mander, U., Jensen, P.D. (2003) *Constructed wetlands for wastewater treatment in cold climate.* WIT Press.
33. Jenssen, P.D. (1992) *Oppfølging av Setermoen renseanlegg.* Jordforsk rapport 7.2400-05 (in Norwegian).
34. Michael, J.H.J. (2004) Nutrients in salmon hatchery wastewater and its removal through the use of a wetland constructed to treat off-line settling pond effluent. *Aquaculture*, 226: 213–225.
35. Cripps, S.J., Kelly, L.A. (1996) Reductions in wastes from aquaculture. In: *Aquaculture and water resource management* (eds D.J. Baird, M.C.M. Beveridge, L.A. Kelly, J.F. Muir), pp. 166–201. *Proceedings of a Conference at the University of Stirling.* Blackwell Science.
36. Ulgenes, Y., Eikebrokk, B. (1994) *Fish farm effluent treatment by microstrainers and a particle trap.* EIFAC Workshop, University of Stirling. June 1994.
37. Timmons, M.B., Summerfelt, S.T., Vinci, B.J. (1998) Review of circular tank technology and management. *Aquacultural Engineering*, 18: 51–69.
38. Lekang, O.I., Bergheim, A., Dalen, H. (2000) An integrated waste treatment system for land-based fish-farming. *Aquacultural Engineering*, 22: 199–211.
39. Timmons, M.B., Riley, J., Brune, D., Lekang, O.I. (1999) Facilities design. In: *CIGR handbook of agricultural engineering, part II aquaculture engineering* (ed. F. Wheaton), pp. 245–280. American Society of Agricultural Engineers.
40. Cho, C.Y., Hynes, J.D., Wood, K.R., Hynes, H.K. (1991) Quantification of fish culture wastes by biological (nutritional) and chemical (limnological) methods; In: *Nutritional strategies for aquaculture waste.* University of Guelph.
41. Bergheim, A., Cripps, S.J., Liltvedt, H. (1998) A system for the treatment of sludge from land based fish-farms. *Aquatic Living Resources*, 11: 279–287.
42. Burton, C.H., Turner, C. (2003) *Manure management: treatment strategies for sustainable aquaculture.* Silsoe Research Institute.
43. Donantoni, L., Skjelhaugen O.J., Sæther, T. (1994) Combined aerobic and electrolytic treatment of cattle slurry. Paper given at *CIGR and AgEng Congress*, Milan, 1994.
44. Skjelhaugen, O.J., Sæther, T. (1994) Local ecological waste water solution for rural areas, based on aerobic treatment and recycling of nutrients into agricultural land. Paper given at *CIGR and AgEng Congress*, Milan, 1994. ITF trykk nr 59/1994. Norwegian University of Life Science.
45. Lekang, O.I., Jenssen, P.D., Skjellhaugen, O.J. (1995) Local ecological solution for treatment from aquaculture. In: *Technical solutions in the management of environmental effects of aquaculture* (ed. J. Makkonen), Seminar 258. Scandinavian Association of Agricultural Scientists.

6
Disinfection

6.1 Introduction

Disinfection can be described as the reduction of micro-organisms such as bacteria, viruses, fungi and parasites to a desired concentration. This is not the same as sterilization where all micro-organisms are eliminated. The aim of disinfection of water in fish farming is to reduce to an acceptable level the risk of transfer of infectious disease from the water to the fish. When disinfecting water for fish farming, selective inactivation of fish pathogenic micro-organisms is required in addition to overall reduction in the total number of micro-organisms. Pathogenic micro-organisms infect the fish and cause disease. Transmissions of infectious diseases is possible in two ways, horizontal and vertical. Horizontal transmission includes direct or indirect contact between individuals or populations. Direct contact occurs between individuals or urine or faeces, while indirect contact occurs through contact with water, equipment and personnel with pathogens. Vertical transport includes transmission from one generation to the next through roe or milt, for example.

Disinfection can be performed in different situations in aquaculture. Water, equipment, buildings and effluent can all be disinfected. Equipment includes tanks, nets, pipes and shoes. Disinfection of buildings includes, for instance, disinfection of the hatchery after seasonal use. Effluent may include sludge and by-products. Disinfection of water actually occurs at several places in an aquaculture plant. Usually the inlet water is disinfected, whether it is seawater or freshwater. At the larval stage it is particularly important to reduce the number of micro-organisms because larvaes are more vulnerable to infections. In a water re-use plant, the water may also be disinfected before being used again to avoid increasing the micro-organism burden. The outlet water may also be disinfected to avoid transfer of micro-organisms to fish species in the recipient water body.

There are several methods for disinfecting water and a number of general textbooks are available (for example, refs 1–4). Disinfectants can be separated into chemical agents and non-chemical agents.[1] Alternatively, a four-group classification can be used: 1, chemical agents; 2, physical agents; 3, mechanical agents; 4, radiation.[4] The first group includes chlorine and its compounds, bromine, iodine, ozone, phenol and phenolic compounds, alcohols, heavy metals and related compounds, soaps and synthetic detergents, quaternary ammonium compounds, hydrogen peroxide and various alkalis and acids. The second group includes heating and the use of sunlight, especially the ultraviolet (UV) end of the spectrum. The third group includes particle separation; although particle separation is the main objective, there will also be a reduction in the number of micro-organisms because many are attached to particles. Larger parasites such as *Costia* and *Gyrodactylus* will also be removed with a particle filter with small (20 μm) mesh size. To the fourth group belong different types of radiation including electromagnetic, acoustic and particle. For example, gamma rays are used to disinfect and also sterilize water and food, although this method is expensive.

Many of the chemical agents employed oxidize the organic materials, including the micro-organisms. The oxidizing potential indicates how effective the agent is likely to be: ozone has the highest potential, while bromine and iodine have the lowest potentials.

Regardless of the method chosen for disinfection the quality of the water to be disinfected is of major importance. Pure inlet water is much simpler to disinfect than outlet water because the latter contains more particles. Turbid water and water with a high content of organic substances, such as re-use water, are also more difficult to disinfect and therefore not so commonly disinfected. Before disinfecting contaminated water, it is essential to carry out some kind of pre-treatment, normally comprising removal of particles.

For disinfection of water supplies to aquaculture facilities, UV light and ozone are most often used. Later in this chapter there is a survey of methods employed, with emphasis on UV light and ozone.

6.2 Basis of disinfection

6.2.1 Degree of removal

The term percentage removal of actual micro-organisms is used in environmental engineering. In microbiological terms $\log_{10}$ removal or inactivation (decimal removal) is used to define the disinfection yield; normally a reduction of between 99 and 99.99% of the total number of bacteria is wanted, which corresponds to a log disinfection of 2–4. However, these terms do not give exact values of the number of micro-organisms left; they only indicate by how much numbers are reduced from the starting concentration.

Example
The normal concentration of bacteria is 10^7/ml and a reduction of 99.9% is required. Find the concentration of bacteria present after disinfection.

Solution

$$\begin{aligned}\text{concentration of bacteria after infection} &= 10^7(1-0.999)\\ &= 10\,000\,000 - 9\,990\,000\\ &= 10\,000\\ &= 10^4/\text{ml}\end{aligned}$$

Example
The starting concentration of bacteria is 10^7/ml. A log disinfection of 3 is wanted. Calculate the new concentration of bacteria.

Solution

Let the starting concentration be N_1 and the end concentration N_2; log(disinfection) = 3.

$$\begin{aligned}\log(\text{disinfection}) &= \log(N_1/N_2)\\ &= \log N_1 - \log N_2\end{aligned}$$

$$\begin{aligned}\log N_2 &= \log N_1 - \log(\text{disinfection})\\ &= 7-3\\ &= 4\end{aligned}$$

$$N_2 = 10^4/\text{ml}$$

6.2.2 Chick's law

Inactivation of micro-organisms in a disinfection plant depends on the time that the micro-organism are exposed to the disinfectant. This is described by Chick's law:

$$\frac{dn}{dt} = kN$$

where:

dn/dt = necessary time to inactivate n micro-organisms
k = time constant depending on disinfectant, type of micro-organism and water quality
N = number of live micro-organisms
t = time.

This differential equation can be integrated within limits to give the following equation:

$$N_1 = N_0 e^{-kt}$$

where:

N_0 = number of micro-organisms at the start
N_1 = number of micro-organisms after time t.

6.2.3 Watson's law

Based on the results of Chick's law, Watson's law can be developed:

$$\ln\frac{N_1}{N_0} = -\Lambda C^n t$$

where:

Λ = coefficient of specific toxicity
C = concentration of disinfectant
n = exponent (normally around 1)
t = time after start-up.

This means that the relation between the number of active and inactive micro-organisms is a product of the concentration of the disinfectant and the exposure time.

6.2.4 Dose-response curve

Based on Watsons's law a dose-response relation may be established for specific types of micro-organism. This gives the proportion of micro-organisms inactivated by fixed doses of disinfectant over various time periods. Exact dose-response relationships are difficult to determine in practice for several reasons. It is often difficult to isolate new pathogens and the response to a certain dose depends, amongst other factors, on the immune status of the organism, environmental conditions and population density.

6.3 Ultraviolet light

6.3.1 Function

Ultraviolet (UV) light is electromagnetic radiation with a wavelength of 1–400 nanometer (1 nm = 10^{-9} m) located at the lower end of the visible spectrum and beyond (Fig. 6.1) (the spectrum of visible light extends down to 380 nm). At the opposite end of the visible spectrum is the infrared (IR) region, heat radiation with longer wavelength which cannot be detected by the human eye.

The ability of the UV light to inactivate and destroy micro-organisms varies with both wavelength and the micro-organisms to be inactivated.

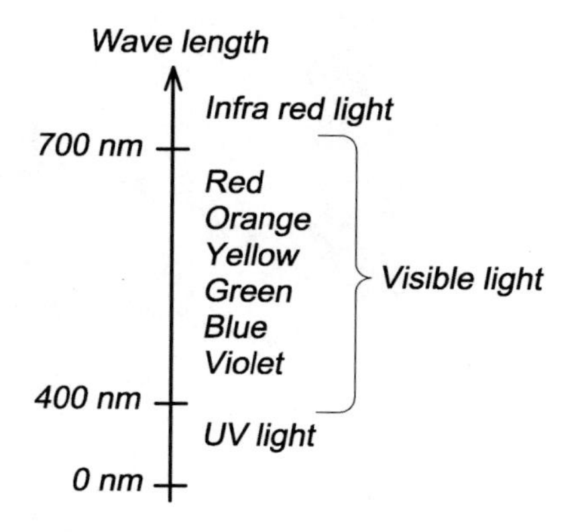

Figure 6.1 Different wavelengths of light.

The most effective wavelength for general disinfection is 250–270 nm.[4] UV light created by mercury vapour lamps will have a wavelength of 253.7 nm, which is effective for disinfection.

6.3.2 Mode of action

UV light will damage the genetic material (DNA and/or RNA) in the micro-organism by disruption of the chains which results in inactivation and death. Inactivation (D) is proportional to the dose of radiation per unit area (intensity) of the UV light (I) and the exposure time (t):

$$D = It$$

The radiation dose is normally given in units of μWs/cm^2 (microwatt second per centimetre squared), i.e. radiation intensity (energy) per unit area.

The effectiveness of the UV light depends on a number of factors including lamp intensity, age of the lamp, cleanliness of the lamp surface, distance between the lamp and the organism to be inactivated, type of organism to be inactivated, duration of UV exposure and purity of the water.

UV lamps become less efficient with use, and need to be replaced regularly, normally at least once a year. Normally the lamp is changed when its output has diminished to 60% of the original.[5] The intensity of the lamp should be measured to ensure sufficient exposure to UV radiation.

How well UV light passes through water depends on the characteristics of the water. UV transmission depends on the particle content (turbidity) of the water, for example. Transmission will be lower for re-used water than for new good quality inlet water. When dimensioning UV systems, it is therefore important to be aware of this.

6.3.3 Design

UV lamps can be placed either in the water (Fig. 6.2a,b) or above the water surface[6] (Fig. 6.2c,d). Usually the lamps are placed in a chamber through which the water flows. UV chambers may be equipped with reflectors or turbulence discs to irradiate the total water flow more effectively. The UV lamp is normally placed inside a quartz glass pipe to protect it from direct cooling by the water and fouling of the lamp surface.

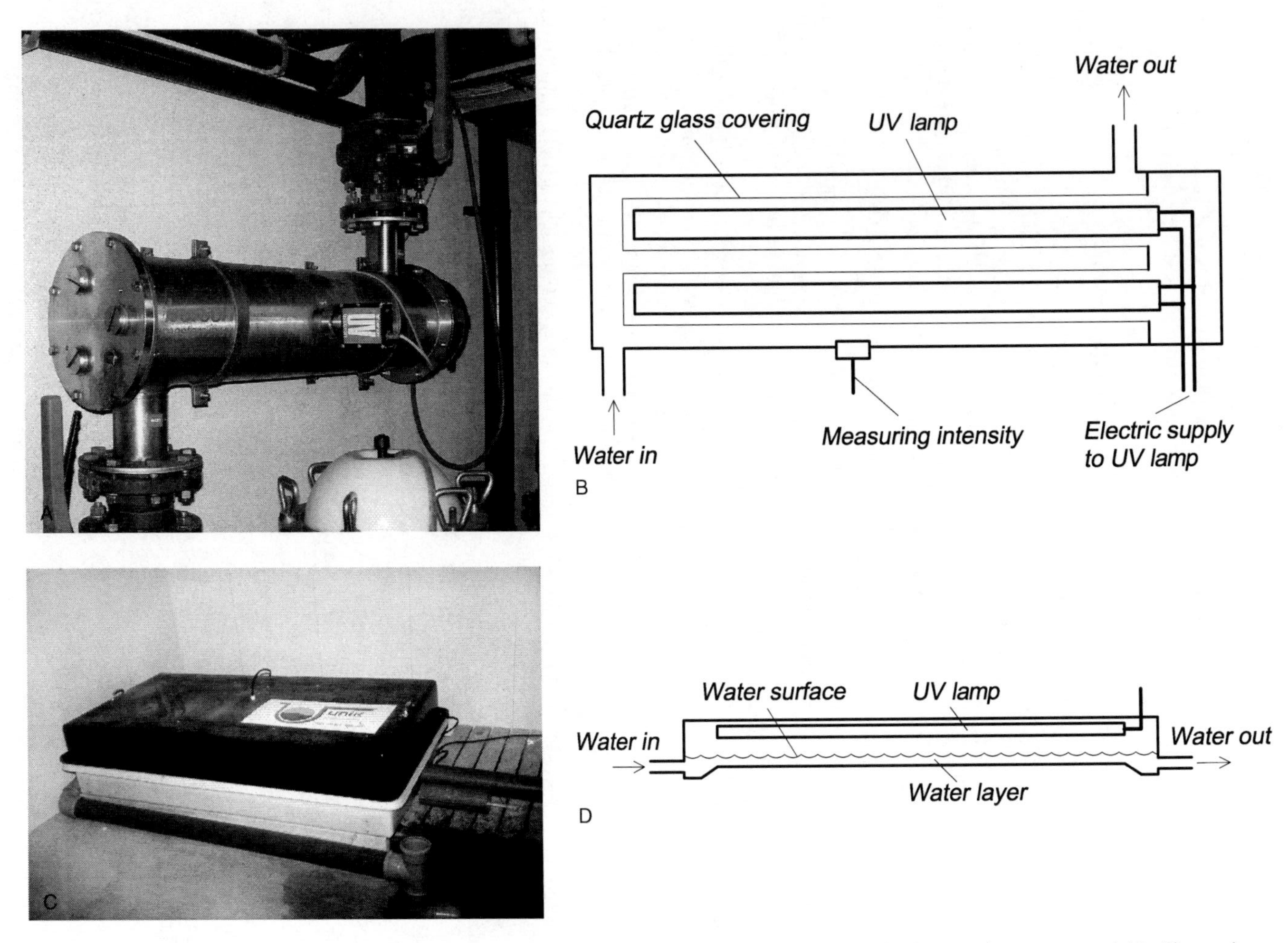

Figure 6.2 A UV plant can be constructed with UV lamps placed in the water flow, which is the usual arrangement (A, B), or above the water flow (C, D).

Fouling may occur on the quartz glass pipes which therefore must be cleaned regularly, either manually or automatically by washing with brushes, to maintain optimal UV intensity. Fouling will decrease UV transmission.

The intensity of the UV radiation can be measured continuously and readings linked to a regulator, so that sufficient radiation intensity can be maintained in water of varying turbidity and with various extents of fouling on the pipes. If the radiation intensity decreases to a limiting value, an alarm can be activated so that remedial measures can be taken.

If the UV lamps are placed above the water surface, the water flows in a thin layer directly below them. In this system the UV rays must pass through the water surface and therefore must be more intense than in a submerged system. For the same reason, the layer of water is thin. As the lamps are above the water surface, fouling will not be so critical for this kind of system, but the problem of varying water quality will be the same as in plants where the UV lamp is placed inside the water flow.

6.3.4 Design specification

When designing the UV plant, the radiation dose, water retention time in the chamber, and the UV transmission must be specified. The transmission is given as the percentage of the known transmission for distilled water. If the water is contaminated with particles, humus, etc. and is coloured, the transmission will decrease and must therefore be measured. This is normally performed for one distance, for instance 5 cm. The transmission for other distances may then be found from the following equation:[7]

$$T_L = T_0^{L/L_0}$$

where:

T_L = transmission through L cm of water
T_0 = transmission through L_0 cm of water
L = thickness of the water layer (cm)
L_0 = thickness of the water layer set at 1 cm.

Example
The UV transmission for re-used water of a given quality is found to be 90% through a 1 cm thick layer (i.e. 90% of the UV radiation passes through a water layer 1 cm thick). Find the UV transmission for a 5 cm thick water layer.

Using the equation

$$T_L = T_0^{L/L_0}$$

where:

$$L = 5\,\text{cm}$$
$$L_0 = 1\,\text{cm}$$
$$T_0 = 90\% = 0.90$$

$$T_L = 0.90^{5/1} = 0.59$$

Thus 59% of the initial UV radiation passes through a 5 cm thick water layer, i.e. the UV transmission is 59%.

The UV intensity can be defined as the amount of radiation per unit surface area and the following equation can be used to find the radiation dose at distance L from the UV radiation source[7]:

$$D = \frac{P}{S} T_0^L \text{P}t$$

where:

D = radiation dose (mWs/cm^2)
P = radiation effect (W)
S = area of radiated surface (cm^2)
T_0 = the transmission of the water through 1 cm (%)
L = thickness of the water layer that is radiated (cm)
t = necessary time for radiation (s).

Example
A UV tube is mounted in the middle of a cylindrical chamber of radius 5 cm. The length of the UV lamp is 1 m and therefore the largest radiated area will be 0.31 m^2. The UV transmission through a 1 cm layer of water is 95%. The time from when the water enters the chamber to when it goes out is 10 s, so it is exposed to UV radition for 10 s. The UV radiation effect of the tube is 16 W.

Find the lowest UV radiation dose to which the water is exposed (this is close to the interior walls of the chamber, almost 5 cm from the UV tube).

Using the equation

$$D = \frac{P}{S} T_0^L t$$

where:

$P = 16\,W$
$S = 0.31\,m^2 = 3100\,cm^2$
$T_0 = 95\%$
$L = 5\,cm$
$t = 10\,s$

$$D = \frac{16}{3100} \times 0.95^{5/1} \times 10$$
$$= 0.0400\,\text{Ws/cm}^2$$
$$= 40\,\text{mWs/cm}^2$$

To obtain good disinfection it is necessary to have a low content of particles and humus in the water, so these must be removed before the water enters the UV chamber, otherwise they will shield the micro-organisms from the UV radiation.

6.3.5 Dose

The dose required to kill pathogenic micro-organisms depends on the organism. In commercial plants a normal UV dose is in the range 30–35 mWs/cm^2, and this is adequate for a log disinfection of 3 of most of the common aquaculture bacteria.[5,6,8,9] However, some viruses, such as IPN, are much more difficult to inactivate and doses of 100–200 mWs/cm^2 have been suggested.[10] Care should be taken when reading papers where the effective dose rates are given, as the methods used in the laboratory are often different from those used in commercial aquaculture where, for instance, UV transmission through the water is variable.

6.3.6 Special problems

As previously described, particles can be a problem because they shade the micro-organisms from the UV light, so it is absolutely necessary to remove them before UV irradiation. If the water is very turbid, as may occur in systems with a high degree of water re-use, UV transmission may be so reduced that it is impossible to use a UV lamp for disinfection.

6.4 Ozone

6.4.1 Function

Ozone (O_3) is a colourless gas with a boiling point of –112°C, that is sometimes called trioxygen. Ozone gas is unstable and will quickly be broken down to O_2; the half-life of O_3 is around 15 minutes. It is therefore necessary to produce the ozone on site.

Ozone is produced by the corona method; air or pure oxygen gas is passed through a high voltage electric field[11] (Fig. 6.3). This is actually the same process that happens with lightning in a thunderstorm, where ozone gas can also be smelled. Energy is added to the oxygen molecule and ozone is created: $3O_2 + \text{energy} = 2O_3$. An ozone generator is shown in Fig. 6.4, which can use either pure oxygen or air to produce ozone. If using air, for the highest cost effectiveness, it must be as dry as possible before entering the ozone generator. An air drier must therefore be used. If using air, the water may become super-saturated with nitrogen. Using pure oxygen to generate ozone is more expensive. The best source of oxygen to use must be decided on a case by case basis. Use of air result in 0.5–3% ozone in the gas stream, whereas the corresponding values for oxygen are 1–6%.[4] Energy requirements for producing ozone are typically in the range 3–30 kWh/kg.[7,11] Only a small amount of (5–10%) the supplied energy is used for ozone production.[7] Small quantities of ozone may also be produced by radiation of air or oxygen gas with UV light of specific wavelength; this is named photozone (see section 6.5.1).

6.4.2 Mode of action

Ozone is a very strong oxidizing agent, highly toxic to all forms of life. When dissolved in water it starts two reactions, slow direct oxidation of organic

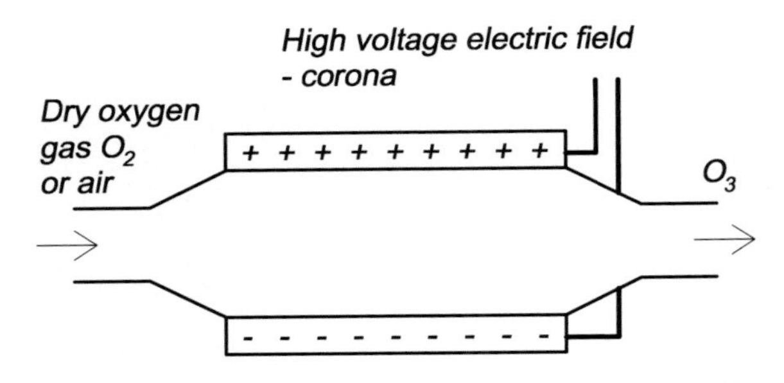

Figure 6.3 In the corona method, ozone gas is created by allowing air or oxygen to flow through a high voltage electric field.

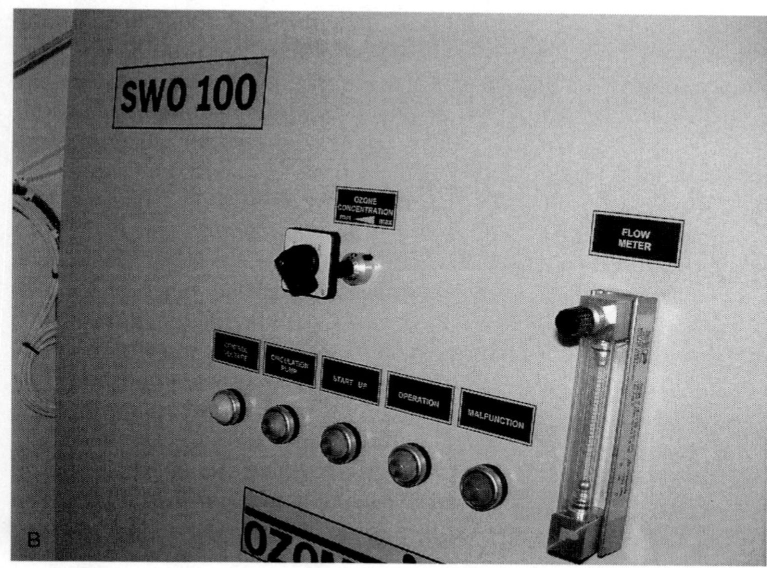

Figure 6.4 (A) An ozone generator for production of ozone gas. (B) Ozone generator control panel.

substances by O_3 and a chain reaction with formation of different free radicals based on hydroxyl radical (OH^-). Ozone acts by damaging cell membranes and nucleic acids, breaking long chain molecules down into simpler forms which may be further degraded in the biological filter. This inactivates the micro-organisms. Ozone has another effect that could be an advantage in aquaculture systems: by oxidation it reduces the amount of NH_3, NO_2 and BOD biofilm on surfaces.[11,12] This can be seen in water re-use systems where disinfection with ozone improves water quality and may therefore be more beneficial than use of other disinfectants. Oxidation by ozone eliminates the yellow/brown coloration of the water that is built up in a water re-use system with very high degrees of re-use. When using ozone as a disinfectant it is recommended that particles be removed from the water before the ozone is added, otherwise much of the ozone will be used to oxidize the particles.

6.4.3 Design specification

When adding ozone gas to water a special injection system has to be used to ensure good gas water mixing. The method is similar to those for mixing oxygen gas into water and use of a venturi is quite common (Fig. 6.5), 90% transfer of ozone has been achieved with a properly designed diffuser system.[4] An ozone disinfection system therefore consists of a production system, the generator and an injection system.

Since there is a dose–time relation for ozone disinfection, the ozone needs to have a certain working time in the water to function and oxidize the micro-organisms. A water retention tank is quite commonly used. Ozone is added to the water which then enters the retention tank. This must be large enough to ensure a satisfactory contact time for the ozone to achieve disinfection. Alternatively, the ozone can be added at the beginning of the inlet pipe to the aquaculture plant. Because the water takes some time to reach the fish tanks, this might be enough to achieve sufficient contact between ozone and the micro-organisms to disinfect the water.

When designing an ozone disinfection plant, it is necessary to include an injection system to get the ozone gas into the water and a system that gives sufficient retention time between the ozone and the water. It is important that the residual concentration of ozone is above the value that is needed for disinfection, this is, of course, less than the inlet concentration.

6.4.4 Ozone dose

To inactivate pathogens with ozone a dose–time relation applies, as for many other disinfectants.

Figure 6.5 A venturi for adding ozone gas to water.

Either a high dose can be used for a short time, or vice versa. Over-dosing must be avoided because this may kill the fish.

Example 3
Ozone is being used to inactivate the bacterium Vibrio anguillarum. *The dose is set to 0.1 mg/(l min) for a log disinfection of 3. This can be achieved by having a residual concentration of 0.1 mg/l after 1 minute working time or 0.5 mg/l after 2 minutes.*

Most pathogens are killed by an ozone dose of 0.1–1.0 mg/l and contact time of 1–10 minutes, but this varies with the organism (for more information see refs 7, 9, 13).

The water quality will have large impact on the residual ozone concentration after a given time; factors such as concentration of dissolved organics, particular organics, inorganic ions, pH and temperature will affect the concentration (for example, see refs 1, 13). An increase in temperature results in a reduced lethal dose. Because of these variations it is important to add enough ozone to obtain a satisfactory residual concentration to achieve disinfection.

6.4.5 Special problems

The great problem with the use of ozone is that it is highly toxic for fish and humans. For fish, ozone is toxic even at relatively small concentrations, because it oxidizes the gill tissue. Recommended safe values for fish are generally below 0.002 mg/l, but there are large variations in tolerance.[14] Therefore after adding ozone to the water and leaving it to react for the necessary time, any residual ozone must be removed or destroyed. Having an adequate retention time ensures that most of the ozone has reacted and the product is mainly oxygen gas (O_2): this time is normally much longer than that necessary to achieve satisfactory disinfection, but of course depends on the decomposition rate. The retention time must therefore be long enough for two processes: (1) to get the ozone to destroy the micro-organisms and (2) to remove residual ozone toxic to the fish. Non-toxic ozone concentrations are normally achieved after 10–20 minutes.[15] This can be done either by increasing the size of the retention tank or by increasing the rate of ozone decomposition to oxygen; methods here include aeration of the retention tank, ozone removal by sending the water through a carbon filter, stripping of the water in a packed column and addtion of chemicals.[11]

Ozone is also toxic to humans, even a very low concentration of ozone in the air being harmful. In the USA, the maximum during an 8-hour work shift in an enclosed area is set to 0.1 mg/l O_3; for 10 minutes exposure time the maximum level is 0.2 mg/l.[2] Humans can detect levels of 0.02–0.05 mg/l by the smell, so when ozone is smelt the building must be evacuated immediately. The possibilities of ozone gas entering the air which humans breathe must therefore be minimized. Proper ventilation in the room where the ozone is produced and added is absolutely essential. In addition there must be safety equipment to monitor the concentration of ozone in the air continuously, and equipment that automatically turns off the ozone supply if the concentration rises above the recommended values; in addition, warning signals must be given (Fig. 6.6).

Another problem with the use of ozone results from it being a very strong oxidizing agent. It is so effective that it will oxidize all materials with which it comes into contact, whether as a gas or dissolved in water. Plastic and metals are examples of materials that the ozone gas will try to oxidize. Ozone will destroy most plastics to various extents. Polypropylene pipes are recommended for transporting water with a high content of ozone. Ozone will oxidize metals, causing significant corrosion problems. Special additives must be used in fibreglass if it is to be used in retention tanks for ozonated water, for instance.

6.4.6 Measuring ozone content

Control of the content of ozone is essential: enough must be injected but overdosing avoided to prevent harm to the fish and to humans as a result of degassing of ozone to the air in an enclosed space. The ozone content of water can be measured either by a chemical method employing indigo triosulfonate or by using a probe for online measurement employing a potentiometric principle.[11] However, online equipment is quite expensive. Measurement of the redox potential which changes with the amount of ozone in the water is therefore quite commonly used instead.

Figure 6.6 Warning equipment must be used to avoid toxic ozone concentrations in the production rooms.

6.5 Other disinfection methods

6.5.1 Photozone

Using photozone gas for water treatment is similar to the use of ozone gas. If air is blown through a UV light chamber photozone will be produced. The UV wavelength should be less than 200 nm. Photozone includes the following substances: ozone, atomic oxygen, hydrogen peroxide and hydrogen dioxide. All of these substances are oxidants and in water are strong disinfectants. Concentrations of these oxidizing substances must be below toxic levels when the water reaches the fish. Use of photozone is not common, but it could be employed where the water requirements are small, such as for disinfecting the water supply to hatcheries.

6.5.2 Heat treatment

All micro-organisms can be destroyed if the water is heated and the high temperature maintained for a certain period of time. The necessary water temperature and contact time with hot water depend on the organism and must be determined by experiment. Heating of the water is, however, an expensive disinfection method. The high costs of heating water (see Chapter 7) mean that the water or the heat must be re-used, for instance by installing heat exchangers.

It is not necessary to boil the water (100°C) to inactivate micro-organisms. The temperature and time needed vary depending on the specific micro-organisms. A large number of micro-organisms will be inactivated at temperatures of 60–80°C if the retention time is correctly chosen. The water must of course be chilled again before it reaches the fish.

Heat or steam is commonly used for disinfecting boots, nets, tanks, pumps and other equipment.

6.5.3 Chlorine

Chlorine is a very effective disinfectant for water and the most common method used for disinfection of municipal drinking water worldwide. It is normally obtained by adding liquid sodium hypochlorite (NaOCl) to the water, but solid calcium hypochlorite ($Ca(OCl)_2$) mixed into the water or pure chlorine gas (Cl_2) may also be used.[1,4] All these compounds are strong oxidizing agents and have the ability to break up organic molecules.

As for ozone, there is a need for a certain contact time to achieve the necessary effect. This includes time for dissociation in water, time for diffusion through cell walls and time to inactivate selected enzymes. Chlorine concentrations of 0.2–0.5 mg/l with 20–30 min contact time or 3–5 mg/l for 1–5 minutes have been reported.[16] To kill parasites in wastewater, typical values of free residual chlorine concentration of 1–3 mg/l with a contact time of 10–15 minutes have been reported.[17] As for ozone, the residual concentration after a certain retention time is also of interest, and overdosing must be avoided. As an example the minimum residual concentration of chlorine in the drinking water supply in Norway is 0.02 mg/l after 30 minutes contact time.

Methods for achieving this contact time are as for ozone: use of a retention tank, or to add it at the start of the transfer pipe, so a natural retention time is achieved when the water is flowing through the pipeline. However, a higher retention time is needed for chlorine than for ozone.

Water containing free chlorine is very toxic for fish. Concentrations of chlorine should not exceed 3–5 µg/l, although for shorter periods of up to 30 minutes concentrations up to 0.05 mg/l can be tolerated by most species.[18] When disinfecting a tank or other equipment with chlorine, it is important that enough clean water is used to wash away the chlorine residues produced. Therefore chlorine is not normally used for disinfection of inlet water for aquaculture facilities. If chlorine is to be used, a method for dechlorination must also be included which can, for instance, be use of aeration, UV light, activated carbon, or reducing agents such as $Na_2S_2O_3$ or Na_2SO_3.[17] However, this will be a sub-optimal method for fish farming purposes, because of the costs.

Effluent water could be disinfected with chlorine, but here also it might be necessary to dechlorinate before discharge to the recipient water body because it might be toxic for the fish it contains. If effluent with a high content of chlorine were to be discharged into recipients with high content of organic substances, chlorinated organic substances might be created: trihalomethanes (THMs) are found in drinking water which has been disinfected by chlorination; chloroform is the most common THM and its presence is well correlated with the dosage of chlorine.

6.5.4 Changing the pH

Increasing or decreasing the pH may also be used for disinfections of water. The pH can be increased by adding lye, or decreased by adding some kind of acid. This is unsatisfactory, however, because the pH has to be normalized again whether the water is going to be used on the fish farm or whether it is the outlet water sent to the recipient. This method is sometimes used for treatment of purified process water and bleeding water from slaughter houses.

6.5.5 Natural methods: ground filtration or constructed wetland

On-site methods such as ground filtration and constructed wetlands may also be used for inactivation of micro-organisms. Put simply, naturally occurring micro-organisms in the on-site systems destroy the pathogenic micro-organisms in the water.[19,20] It is important to have adequate residence time in the systems to ensure that pathogenic micro-organisms are destroyed.

References

1. Montgomery, J.M. (1985) *Water treatment: principles and design.* John Wiley & Sons.
2. Langlais, B., Reckhow, D.A., Brink, D.R. (eds) (1991) *Ozone in water treatment: application and engineering.* Lewis.
3. Spellman, F.R. (1999) *Choosing disinfection alternatives for water/wastewater treatment plants.* CRC Press.
4. Tchobanoglous, G., Burton, F.L., Stensel, D.H. (2002) *Wastewater engineering.* McGraw-Hill.
5. Rodrigues, J., Gregg, T.R. (1993) Consideration for the use of ultraviolet in fish culture. In: *Techniques for modern aquaculture* (ed. Wang, J.K.), Proceedings Aquacultural Engineering Conference, Spokane, Washington. ASAE Publication 02-93, American Society of Agricultural Engineers.
6. Weaton, F.W. (1977) *Aquacultural enginering.* R. Krieger.
7. Gebauer, R., Eggen, G., Hansen, E., og Eikebrokk, B. (1992) *Oppdrettsteknologi – vannkvalitet og vannbehandling i lukkede oppdrettsanlegg.* Tapir Forlag (in Norwegian).
8. Brown, C., Russo, D.J. (1979) Ultraviolet light disinfection of shellfish hatchery sea water. *Aquaculture,* 17: 17–23.
9. Liltved, H., Hektoen, H., Efrainsen, H. (1995) Inactivation of bacterial and viral fish pathogens by ozonation or UV radiation in water of different salinity. *Aquacultural Engineering*, 14: 107–122.
10. Yoshimiza, M., Takizawa, H., Kimura, T. (1986) UV susceptibility of some fish pathogenic viruses. *Fish Pathology,* 21: 47–52.
11. Colt, J., Cryer, E. (2000) Ozone. In: *Encyclopedia of aquaculture* (ed. Sickney, R.R.). John Wiley & Sons.
12. Tango, M.S., Gagnon, G.A. (2003) Impact of ozonation on water quality in marine recirculation systems. *Aquacultural Engineering*, 29: 125–137.
13. Lawsons,T.B. (2002) *Fundamentals of aquacultural engineering.* Kluwer Academic Publishers.
14. Wedemeyer, G.A., Nelson, N.C., Yasutaka, W.T. (1979) Potentials and limits for the use of ozone as a fish disease control agent. *Ozone: Science and Engineering*, 1: 295–318.
15. Huguenin, J.E., Colt, J. (2002) *Design and operating guide for aquaculture seawater systems.* Elsevier Science.
16. Johson, S.K. (2000) Disinfection and sterilization. In: *Encyclopedia of aquaculture* (ed. Sickney, R.R.). John Wiley & Sons.
17. Wedemeyer, G. (2000) Chlorination/dechlorination. In: *Encyclopedia of aquaculture* (ed. Sickney, R.R.). John Wiley & Sons.
18. Wedemeyer, G. (1996) *Physiology of fish in intensive culture systems.* Chapman and Hall.
19. Stevik, T.K. (1998) *Retention and elimination of pathogenic bacteria percolating through biological filters.* PhD thesis, Norwegian University of Life Science.
20. Bomo, A.M. (2004) *Application of natural treatment systems on fish farm wastewater.* PhD thesis, Norwegian University of Life Science.

7
Heating and Cooling

7.1 Introduction

In aquaculture heating of the water may be necessary for several reasons, for example to increase the growth rate, to get the fish reach a specific size at a certain time, to get them to mature or to spawn. Different species have different optimal temperatures; if the ambient water temperature is cooler than the optimal temperature, it can be useful to heat the water.

The principles used for heating in aquaculture are normally the same as those used in houses or industrial facilities; however, systems used in aquaculture facilities must heat large amounts of water and therefore be efficient. Important factors when choosing a system are the total heating requirements and the necessary temperature increase. In this chapter a survey of methods and equipment is given. It starts with some basic physical laws and ends with some simple specifications and calculated examples.

Instead of purchasing all the heat necessary, it could be taken from other available sources, such as geothermic water, or the water could be re-used. For species needing much warmer water than is available from source, both these methods could be used.

In some cases it is necessary to chill the water, for instance in connection with storing the brood stock, to get the fish to mature, and for storing eggs and fry. Heating and chilling both involve energy transfer. When heating water, energy is added to the system, while chilling removes energy from the system. In this chapter the focus is on heating systems used for aquaculture; much of the basic information applies to both heating and cooling. A great deal of general engineering literature is available on this subject and also on basic thermodynamics; see, for example, refs 1–6.

7.2 Heating requires energy

A supply of energy is needed to heat water. Energy can be transferred in three different ways: by radiation, by conduction or by convection. Electromagnetic radiation from the sun or an electric heater, for example, will be partly absorbed when it falls on a substance and become internal energy. With conduction, the energy is transferred between solids or liquids as vibrational energy of the atoms or molecules. Additionally, in materials with a supply of 'free' electrons, e.g. metals, these electrons share in any energy gain resulting from a temperature rise and their velocities increase more than those of atoms or molecules, so energy is quickly transferred to other parts. For example, when the end of an iron bar is heated, the energy will be transferred to the whole bar and after a short while it will be hot along its entire length. Convection is heat transfer resulting from mixing of substances with different temperatures. Convection can be natural or forced. Natural convection occurs in water as a result of density variation caused by temperature. Water is of maximum density at 4°C. If heated water is sent in below colder water it will move upwards because it is less dense and natural convection will take place. Forced convection occurs when a medium such as water is exposed to an exterior force, for instance a pump or mixer.

The power (P) required for heating water is proportional to the flow and the temperature, and is given by the equation:

$$P = mc_p dt$$

where:

m = water flow (kg/s)
c_p = specific heat capacity (kJ/(kg °C))
dt = temperature increase for the water (°C).

The specific heat capacity is the amount of energy required to heat 1 kg water by 1°C: c_p = 4.18 kJ/(kg °C) for freshwater and 4.0 kJ/(kg °C) for seawater. The temperature increase for the water is the difference between inlet and outlet temperatures.

A continuous power supply is required to heat the flowing water; the unit of power is normally the kilowatt (kW) and 1 kW = 1 kilojoule (kJ) per second. Therefore, it can be seen that power is rate of energy transfer.

Example
A freshwater flow of 10 l/min (0.17 l/s) is heated from its original temperature of 2°C to 10°C. What is the rate of energy transfer to the water, i.e. the power supplied?

$$P = mc_p dt$$

The mass of 1 l of water is 1 kg.

$$\begin{aligned} P &= 0.17\,\text{kg/s} \times 4.18\,\text{kJ/(kg °C)} \times (10°\text{C} - 2°\text{C}) \\ &= 5.7\,\text{kJ/s} \\ &= 5.7\,\text{kW} \end{aligned}$$

The total amount of energy that has to supplied during a given period or that has been used during a given period can be calculated from the power supplied multiplied by the time for which this power is used.

$$Q = Pt$$

where:
Q = total amount of energy (kilowatt-hour, kWh)
P = power (kW)
t = time over which heating takes place (h).

If this is compared to flowing water, the power (kW or kJ/s) corresponds to water flow rate (l/s), while the total energy consumed corresponds to the total amount of water which has flowed past a certain point during a given period of time.

It is the energy consumed (Q) that is paid for, be it electricity, oil or another energy source. Electricity is charged per kilowatt-hour, consumed and oil is sold by volume. The power (P) gives the energy rating of the heating equipment.

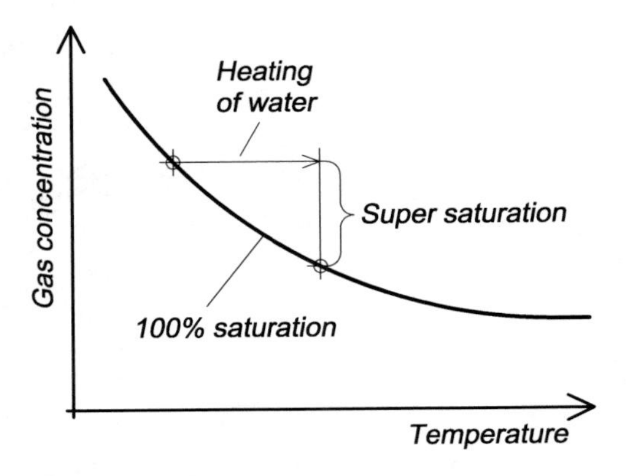

Figure 7.1 Heating of water that is in gas equilibrium with the surrounding air (100% saturation) will result in supersaturation with gases that are harmful to the fish.

Example
Calculate the daily cost of heating the water in the previous example. The price of electricity is 0.1 €/kWh.

$$\begin{aligned} Q &= Pt \\ &= 5.7\,\text{kW} \times 24\,\text{h} \\ &= 136.8\,\text{kWh} \end{aligned}$$

Therefore the cost is

136.8 kWh × 0.1 €/kWh = 13.68 €.

When the water is heated it must always be aerated before it is used on fish because it may become supersaturated with gas. When the water temperature increases the amount of dissolved gas is reduced. If there was equilibrium between the gases in the water and the surrounding air before the water was heated, the water will be supersaturated with these gases afterwards (Fig. 7.1). Nitrogen levels will be harmful (see Chapter 8). It is very important to be aware of this, because it means that water containing fish must not be heated directly.

7.3 Methods for heating water

Several methods are used for heating water, either directly or indirectly. The latter case requires available sources of hot water that can be used to meet all or a part of the heating requirements. When using direct methods, all necessary energy must be added to the water. Electricity, oil or gas are the usual sources of energy for direct heating of water.

It is better, however, if other available energy sources can be used to meet part or all of the energy requirement. These can be separated into low and high temperature sources. High temperature sources can be used to meet the entire need for heating, if the amounts are large enough as they are much hotter than the required final temperature. Hot industrial effluent water or geothermal water are examples of such sources.

Low temperature sources are hotter than the raw water to be heated but their temperature is insufficient to meet total energy needs. The source can therefore be used as a part supply to the total water heating requirements. Seawater or groundwater are examples of sources that can be used to meet part of the energy needs; industrial effluent water can also be included in this category.

Whether industrial effluent water or geothermal water can be used directly for the fish, or whether heat exchange is necessary depends on its quality. Normally it is necessary to use a heat exchanger, because the quality of industrial effluent water is not continuously good enough for direct use as growing water for the fish. Low temperature sources are also suitable for use with heat pumps (see Section 7.6).

When supplying energy to heat water, it must be utilized as efficiently as possible. Recovery of energy from the outlet water is therefore common, either by direct or indirect systems. Re-use systems are direct, with the water being used again directly, while heat exchangers where only the energy is re-used, represent indirect systems. It is not possible to recover all the energy that is added by using heat exchangers. Even if it were theoretically possible, it is not cost effective because the heat exchanger would need to be very large and expensive.

7.4 Heaters

7.4.1 Immersion heaters

An immersion heater is an electrical element (heat element) placed in the water (Fig. 7.2). Electricity is supplied to the element which heats up. The basic principle is the same as that is used in a heater in a house: the electricity passes through a thin resistant filament where heat develops. It is important that the element is sealed to avoid a short circuit; therefore it is typically placed in a tube. When water passes the heated element, energy is transferred from the element to the water, the temperature of which increases. Normally a thermostat is attached to the heater so that heat is transferred until the water reaches a set temperature when the electricity supply is automatically/switched off; when the temperature drops below this value, the electricity to the element is automatically switched on. When using an electric heater almost all of the electric energy supplied is transferred to the water. For greatest efficiency when heating requirements is vary, it is an advantage to have several elements in the heater. In addition to a thermostat, the heater is normally equipped with a system to avoid overheating, for instance when the water flow is reduced or stops. An immersion heater must be correctly installed. There must be no possibility for air pockets inside the heater, because the element may overheat in such places. It is important to choose correct materials for the heating element to avoid corrosion resulting in an electrical short circuit. Materials used are acid-proof steel or stainless steel, depending on the quality and composition of the water to be heated.

The size of heater depends on the energy requirements. For large water flows or temperature gradients, large heaters are necessary; these consume much electricity as can be inferred from the high cross-sectional area of the wires connected to the heater.

Example
The ambient water temperature is 3°C while that in the hatchery is 8°C. The hatchery water flow is 60 l/min (1 l/s). Find the size of the heater needed to achieve the required temperature increase.

$$
\begin{aligned}
P &= mc_p\mathrm{d}t \\
&= 1\,\mathrm{l/s} \times 4.18\,\mathrm{kJ/(kg\ ^\circ C)} \times (8^\circ\mathrm{C} - 3^\circ\mathrm{C}) \\
&= 20.9\,\mathrm{kW}
\end{aligned}
$$

The necessary size of the heater is 20.9 kW. The heater is ordered from a supplier and adjusted as required.

Example
On a land-based freshwater production farm the supply of energy necessary, to increase the water temperature to at least 12°C throughout the year must be calculated. Water consumption is 2 m³/min.

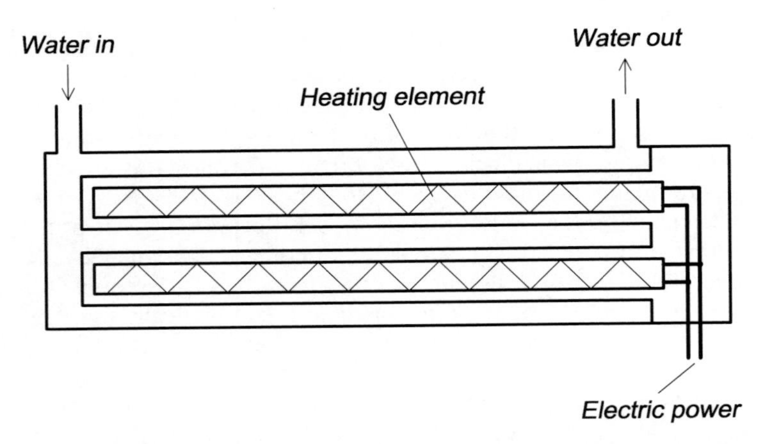

Figure 7.2 An electric immersion heater.

The incoming freshwater has the following ambient average monthly temperatures:

Month	*Ambient water temp. (°C)*	*Temp. increase (°C)*
Jan	*2*	*10*
Feb	*2*	*10*
Mar	*3*	*9*
Apr	*4*	*8*
May	*8*	*4*
Jun	*12*	*0*
Jul	*14*	*0*
Aug	*14*	*0*
Sep	*12*	*0*
Oct	*10*	*2*
Nov	*6*	*6*
Dec	*3*	*9*

In the months January, February, March, April, May, October, November and December the temperature is so low that it is necessary to heat the water.
Amount of energy required for January is:

$$\begin{aligned} P &= mc_p dt \\ &= 33.3\,\text{l/s} \times 4.18\,\text{kJ/(kg °C)} \times (12\text{°C} - 2\text{°C}) \\ &= 1392\,\text{kW} \end{aligned}$$

The energy needed for the whole month is

$$\begin{aligned} Q &= P \times \text{number of hours} \\ &= 1392\,\text{kW} \times 24\,\text{h} \times 31\ \text{days} \\ &= 1\,035\,648\,\text{kWh} \end{aligned}$$

This calculation is repeated for the other months giving the following values:

Month	*Power (kW)*	*Total electricity consumption per month (kWh)*
January	*1392*	*1 035 648*
February	*1392*	*935 424*
March	*1253*	*932 232*
April	*1114*	*802 080*
May	*557*	*414 241*
June	*0*	*0*
July	*0*	*0*
August	*0*	*0*
September	*0*	*0*
October	*278*	*206 832*
November	*835*	*601 200*
December	*1253*	*932 232*
Total		*5 859 889*

If the price of the electricity is 0.1 €/kWh, the yearly cost for heating the water with an immersion heater is

$$5\,859\,889\,\text{kWh} \times 0.1\,€/\text{kWh} = 585\,989\,€$$

As this example shows, the cost of heating water for fish farming is very high because of the large amount of water. This is also the reason for the great interest in alternative hot water sources and recovery of energy from the outlet water.

7.4.2 Oil and gas burners

When oil burns a great deal of energy is released, and in an oil burner this energy is used to heat water. The amount of energy released when burning 1 l oil depends on the characteristics and quality of the oil; a typical value is 41 800 kJ/kg. Letting the oil burn in a combustion chamber around which water flows ensures transfer of energy from the burning oil to the water (Fig. 7.3). To keep the oil burning, air must be supplied to the combustion chamber and a chimney is necessary to get rid of the flue gas. There will always be energy losses in an oil-fired boiler from the flue gas and due to incomplete burning and incomplete transfer of heat to the water. Depending on the system, oil-fired boilers are usually between 60 and 90% efficient. A shunt valve may be used to regulate the temperature of the water flowing out from an oil-fired boiler.

Instead of oil, gas can be used as an energy source. The construction of a gas-fired system is similar to an oil-fired system, with a combustion chamber around which the water circulates. However, the gas-fired boiler is slightly simpler to construct, because gas is more flammable than oil. Normally gas-fired appliances are slightly more efficient than oil-fired boilers.

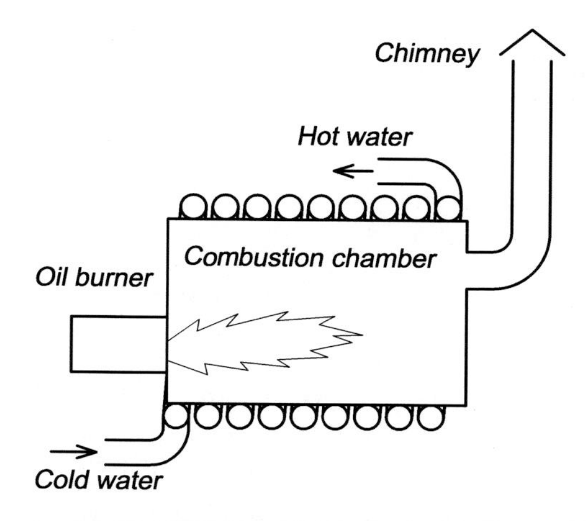

Figure 7.3 An oil-fired boiler.

7.5 Heat exchangers

7.5.1 Why use heat exchangers?

A heat exchanger is used to transfer energy from one medium to another and to achieve this a temperature gradient is necessary. Energy transfer can take place via direct contact between the two media, or indirectly where the media are separated by a barrier. In indirect exchangers the heat must be transferred through this barrier.

An illustration of the direct method is as follows. Water is fed in at the top of a tower where is it spread on a perforated plate; it falls through as drops which pass through the air into a basin placed below. The water drops have a large surface area exposed to the surrounding air; if this is warm, energy will be transferred from the air into the drops and the water will be heated. The indirect method is illustrated by having water of different temperatures separated by a thin metal plate. Energy will be transferred through the plate from the warm side to the cold side because of the temperature gradient until the temperature on both sides is the same. This method is the main one used in fish farming.

Heat exchangers can be used both for heating and chilling. In heat pumps and cold-storage plants, a refrigerant may be used as one of the media (see Sections 7.6 and 7.8). Heat exchangers may also be air to air as in ventilation systems in houses, or air to water as in air coolers.

Heat exchangers can be used to recover energy from the outlet water; for example, to recover the energy from hot industrial effluent water and use it to heat the water supply to a fish farm. Heat exchangers are also commonly used to recover the energy from the outlet water from the fish farm and transfer it to the inlet water. Seawater may also be used as a heat source on one side of the heat exchanger, to supply part of the heat energy to the inlet water supply to a freshwater fish farm.

7.5.2 How is the heat transferred?

The energy transfer in a heat exchanger having liquid on both sides of a metal plate occurs in three stages (Fig. 7.4):

(1) Transfer of heat from the liquid with the highest temperature to the fixed material, the plate
(2) Transfer of heat through the fixed material
(3) Transfer of heat from the fixed material to the liquid of lower temperature

Transfer of heat from the liquid to the fixed material is normally via convection and conduction.

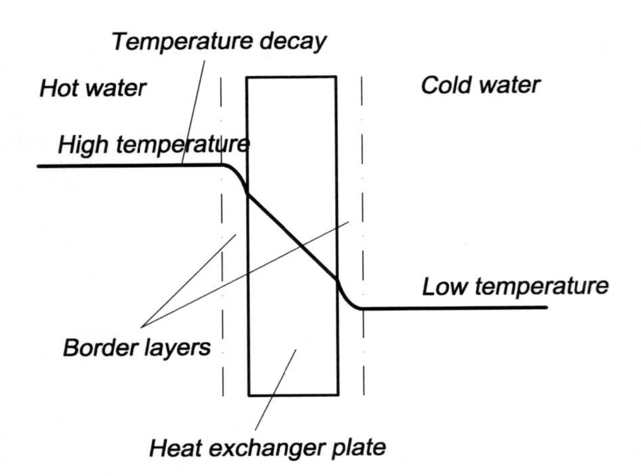

Figure 7.4 In a heat exchanger the heat is transferred from the hot to the cold side.

Close to the fixed material there will be a layer with laminar flow through which heat is transferred by conduction more slowly than for convection. It is advantageous to reduce the thickness of the laminar layer and, to improve convection, have turbulent conditions in the flowing liquids. Heat transfer through the fixed material is by conduction; a material with good conductivity (e.g. metal) improves this. Conduction and convection cause heat transfer into the liquid on the other side of the fixed material.

7.5.3 Factors affecting heat transfer

The rate of energy transfer (P) of the heat exchanger depends on the temperature difference between the media, the thermal conductivity of the material in the heat exchanger and the area over which the energy is transferred; it can be calculated from the equation

$$P = kA\,\text{LMTD}$$

where:

k = heat transfer coefficient (W/(m^2 °C))
A = heat transfer area (m^2)
LMTD = log mean temperature difference (°C).

The value of k gives the quantity of energy transferred per square metre surface area and degree temperature difference. Various factors affect k; in practice, values of up to 8 kW/(m^2 °C) are achieved. k can be calculated from the following equation:

$$\frac{1}{k} = \frac{1}{\alpha_1} + \frac{1}{\alpha_2} + \frac{t_p}{\lambda} + R_f$$

Here, α_1 and α_2 are the heat transfer coefficients on each side of the material in the heat exchanger. They give the quantity of energy transferred from a liquid or gas to or from unit area of a fixed material and per degree temperature difference. The values depend on the condition for convection and conduction and can be improved by optimizing operational conditions.

t_p is the thickness of the material that separates the two flowing media; increasing t_p will decrease k because the heat must be transported a greater distance through the fixed material. Use of a thin fixed material results in a low k value.

λ is the heat convection factor for the material; steel and other metals have a high factor, while glass and plastics have a lower factor. This why metal with good conductivity is used for the heat transfer plates.

R_f is the fouling factor, which gives the amount of fouling on the material of the heat transfer plates. Fouling reduces k, and therefore the rate of energy transfer will be reduced. High turbulence close to the surface of the heat transfer plates will reduce the amount of fouling, in addition to improving α_1 and α_2. Cleaning of the exchange surfaces will also reduce the fouling, R_f. The reduction in k is because the conductivity of the layer of fouling is low and the thickness of the transfer material is increased.

Manufacturers of heat exchangers normally give kA as a single value. This is because each manufacturer will have their own design for the heat transfer area to create optimum flow with turbulence; to improve conditions for turbulence and increase the heat transfer area, various patterns are used on the exchange surface, including grooves and corrugations.

The temperature gradient between the warm and the cold side in the exchanger (Fig. 7.5) ensures energy transfer. It is expressed as the LMTD. A logarithmic expression is used because temperature equalization between the media through the exchanger may not be linear. LMTD can be calculated from the following equation:

$$\text{LMTD} = \frac{\Delta T_1 - \Delta T_2}{\ln(\Delta T_1/\Delta T)}$$

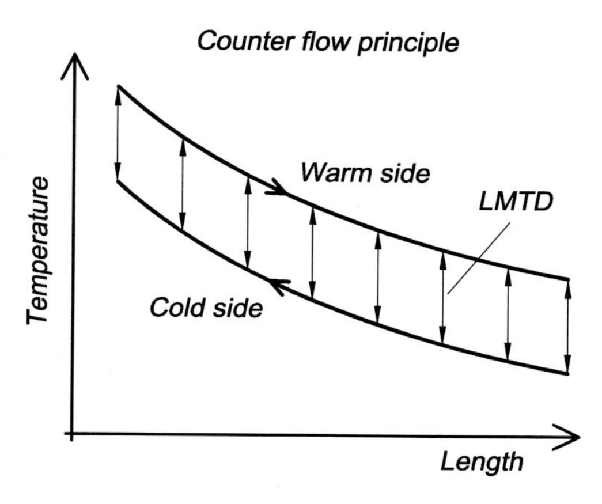

Figure 7.5 Log mean temperature difference (LMTD) represents the mean temperature difference between the warm and cold side in a heat exchanger. It is this gradient that ensures heat transfer.

where:

$T_1 = t_{in}$ (hot water) – t_{out} (heated water)
$T_2 = t_{out}$ (cooled water) – t_{in} (water to be heated).

If the amount of water and the heat exchange area are the same on both sides of the heat exchanger, $\Delta T_1 = \Delta T_2 =$ LMTD. LMTD will vary depending on whether it is a parallel-flow or counterflow exchanger (see sections 7.5.4 and 7.5.6).

7.5.4 Important parameters when calculating the size of heat exchangers

Number of transfer units

Number of transfer units (NTU) indicates how much energy can be transferred per unit LMTD. Indirectly, it gives an idea of what the exchanger will look like, its size, etc. The following calculation can be used to find the NTU of the exchanger:

$$\text{NTU} = (t_{in} - t_{out})/\text{LMTD}$$

where:
NTU = number of transfer units
t_{in} = temperature of water flowing into the exchanger
t_{out} = temperature of water flowing out of the exchanger
LMTD = log mean temperature difference (°C).

Example
Calculate the NTU for the warm side in a heat exchanger, i.e. the side where the hot water is chilled. Water flows into the exchanger at 1000 l/min at a temperature of 10°C and out at 4°C; the LMTD is 1.5°C.

$$\text{NTU} = \frac{10 - 4}{1.5} = 4$$

The NTU can be calculated both for the warm and cold side of the exchanger. Depending on the flow conditions and the transfer area, it could be the same (if the flow and heat transfer area are the same on both sides of the exchanger).

In a plate exchanger described in section 7.5.5 the NTU is seldom above 5, or the plates will be unreasonably large. If the LMTD is estimated and the NTU known, the highest possible out temperature that can be reached can be calculated. If NTU is above 5, heat exchangers can be connected in series, or several-stroke exchangers can be used.

The design of the heat exchange surface area varies. A closed design results in a large NTU with good heat transfer but a high head loss, whereas a more open design results in a lower NTU.

Specific pressure drop

The pressure drop in the liquid flowing through the exchanger is necessary to achieve heat transfer. If a high head loss through the exchanger is accepted, the size of the exchanger can be reduced. Higher pressure creates more turbulence and improved contact between the exchange plates, but a higher input pressure to the exchanger is necessary.

The specific pressure drop (J) for a heat exchanger gives the pressure loss for every transfer unit, and can be represented by the following equation:

$$J = \frac{\Delta P}{\text{NTU}}$$

where:

J = specific pressure drop
ΔP = total head loss through the exchanger
NTU = number of transfer units.

The economic optimum head loss through the heat exchanger varies depending on the situation and has to be calculated in every case. Normally it lies in the range 2–10 mH_2O per NTU for exchangers used in fish farming. When looking at the different areas where exchangers are used, the following approximate specific pressure drop per transfer unit can be used as an approximate base to start a simulation process to optimize the size of a plate heat exchanger:

- Seawater exchanger 5–10 mH_20
- Outlet water exchanger 10–12 mH_20
- Evaporator in a heat pump 3 mH_20
- Condenser in a heat pump 4 mH_20

Other important measurements

The maximum pressure (design pressure) in heat exchangers used in fish farming is normally in the range 1–2 MPa or 10–20 mH_2O. Usually the plate exchangers have up to 2000 m^2 of exchange surface area and the water flow can be 1000 l/s or more.

Exchanger price of course depends on size: a large exchanger will cost more than a small one. If higher head loss can be accepted the size of the exchanger can be reduced. Also the size of the LMTD is of great importance. If a small LMTD is needed the size of the exchanger will increase (Fig. 7.6).

When installing a heat exchanger, the energy profit must therefore always be evaluated against the price of the exchanger. To have a low LMTD is normally very costly. The necessary transfer area, or

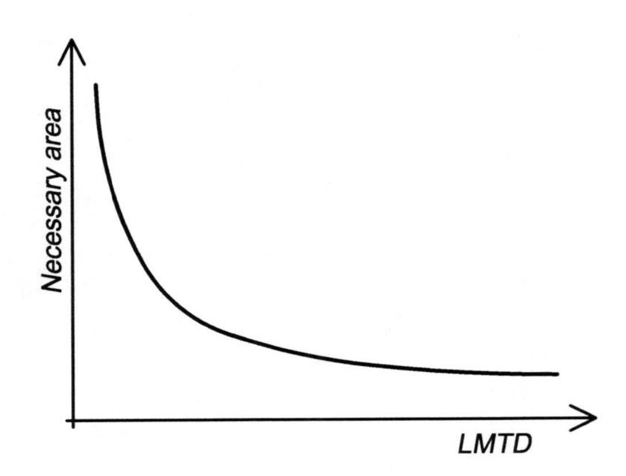

Figure 7.6 If the log mean temperature difference (LMTD) is small, the size of the heat transfer area must be increased.

size of the exchanger increases when the LMTD is reduced. Suppliers of equipment normally have their own programs to simulate the temperature of water discharged (their own kA values) to optimize the size of their exchangers.

7.5.5 Types of heat exchanger

When designing a heat exchanger the aim is to create a large area where exchange of energy between the two media can take place. Two types of heat exchangers are common in fish farming: plate, and shell and tube exchangers.

Plate exchangers

A common plate exchanger consists of the following components: a rack, two end-plates with pipe connection in one or both sides, the heat transfer plates and gaskets that are used between the plates to avoid leakage (Fig. 7.7). In plate exchangers that cannot be dismantled, it is possible to omit the gaskets and braze or solder the plates together instead. The two media flow into the exchanger at one end; the hot and cold water flow in two separate circuits divided by the heat transfer plates. Hot and cold water will flow in parallel through the whole length of the exchanger, assuming the countercurrent principle is used as is usual. The size of the energy transfer area can be changed by adding and removing heat transfer plates as long as there is sufficient space in the rack. The exchanger can be opened to add or remove plates and for cleaning unless the parts are brazed together. The former type is normally used in fish farming because they are easy to open for cleaning and removal of fouling. The latter type can be used as evaporators and condensers in a heat pump, or in places where both the flowing liquids are pure and fouling of the heat transfer surfaces does not occur.

Some kind of corrugation on the heat transfer plates is normal (Fig. 7.7). This increases the area were heat transfer occurs and hence the NTU. In addition the corrugation will increase the turbulence and because of this improve the heat transfer, so increasing the value of k. The gasket between the plates which inhibits leakage, is either glued or clipped to the plates and made of a resistant rubber material.

To increase the heat transfer area, several heat exchangers can be connected in series, one after another. The alternative is to use the so called 'several-stroke' exchanger. Normally an exchanger has one stroke, but by adding dense plates in the middle where the direction of the water flow is changed, a several-stroke exchanger is achieved (Fig. 7.8). The result is the same as adding two exchangers in series, but only one rack is used. Whether a plate exchanger is a one-stroke or several-stroke type can easily be seen from the connection points. A one-stroke exchanger has all pipe connections on one side to, one of the end-plates, while several-stroke exchangers have pipe connections on both sides to, both end-plates. To be able to dismantle a two-stroke exchanger, e.g. for cleaning, it must be possible to disconnect the pipe connections in a simple way.

The value of k for plate exchangers varies significantly depending on the corrugation pattern on the surface; normal values are from 3.5–8 kW/(m^2 °C). The plates are very thin, in the region of 0.5 mm.

Shell and tube exchangers

Another common type of heat exchanger is the shell and tube. This is widely used in condensers and evaporators in heat pumps and cold storage plants in fish farming. It is constructed with a shell covering a number of small tubes (Fig. 7.9). The shell is normally a large tube. One medium flows in the small tubes while the other flows around the tubes, in the shell. The small pipes and the large pipe constitute the two different circuits where the two different media flow and heat can be transferred. Shell and tube exchangers are seldom used for traditional water-to-water heat exchange in fish farming because they are quite difficult to clean manually. They can be opened by removing the cap at one end of the shell, but it is impossible to reach all heat transfer surfaces for simple manual brushing. They must be adapted for automatic chemical cleaning.

Special types, pipes in seawater

A pipe that is laid down in water will have a higher or lower surrounding temperature than the water flowing through the pipe and will function as a heat exchanger. This principle has been used in fish farming. If the temperature in the surrounding

Figure 7.7 A typical plate heat exchanger consists of a rack, two end-plates with pipe connections, heat transfer plates and gaskets between the plates to avoid leakage. (A) An exchanger *in situ*. (B) An open heat exchanger showing the corrugated surfaces of the transfer plates.

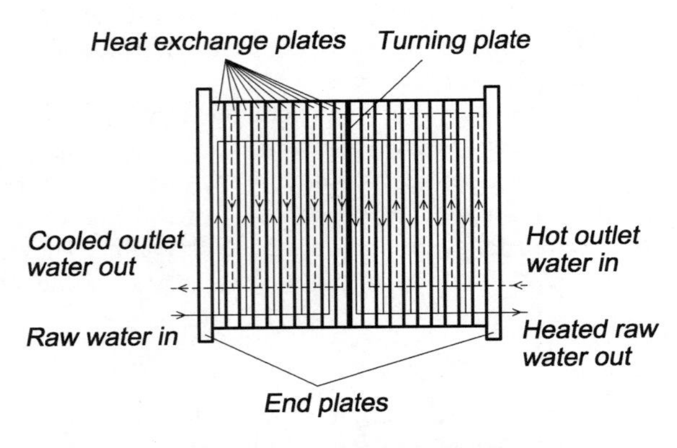

Figure 7.8 Multi-pass exchangers include one or more dense plates to change the direction of the water flow inside. In a multi-pass exchanger there are pipe connections at both ends of the rack.

water is higher, heat will be transferred into the water flowing inside the pipe. To obtain a noticeable temperature increase in the water flowing inside the pipe, the following factors are of importance: length of the pipe, temperature difference between the water inside the pipe and around the pipe, and the heat transfer coefficient (k) of the pipe material. This again depends on the thickness of the pipe, material of which the pipe is constructed, contact area between water and the pipe material on both sides, and the convection conditions inside and around the pipe.

In practice, plastic seems to be the material most often used. Polyethylene does not, however, transfer energy particularly effectively as it has, quite a low k value, but PE pipes are cheap and easy to lay.

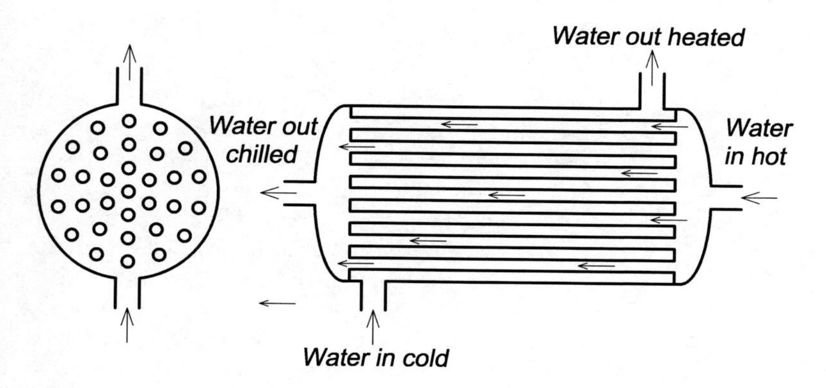

Figure 7.9 The shell and tube exchanger is constructed with a shell covering a number of small pipes.

The pipes must be moored; normally concrete blocks are used. If in seawater, it is normally to lay pipes at depths below 20–30 m, too avoid the most critical depth that results in much fouling of the exterior surface. In fish farming the system may be used for heating freshwater in the winter. In this case the inlet pipeline must be laid into seawater where the temperature is normally higher than in freshwater. A substitute for a traditional on-shore heat exchanger where seawater is pumped through, is then achieved. What is found in practice is that the pipe has to be several hundred metres long to achieve a noticeable temperature increase. Depending on the site conditions and the geographical distance between the site and the water source, this might be a good solution if the water transferred from the source to the fish farm is near the sea.

7.5.6 Flow pattern in heat exchangers

Two flow principles are used in heat exchangers: with-current or countercurrent, the latter only in plate exchangers of particular construction (Fig. 7.10). In the with-current system, the liquids on both sides of the exchange material flow in the same direction, or approximately in the same direction as in the shell and tube exchanger. In this case the temperature gradient between the media is high at the start but gradually decreases. The highest possible temperature that can be achieved in the cold liquid being heated is the mean temperature between the two flowing liquids. This requires equal flow of both liquids; otherwise the temperature depends on the flow ratio of the two liquids. In a countercurrent exchanger the cold media flows in the opposite direction to the hot media. The temperature of the cold media thus gradually increases and the hot water is correspondingly chilled. With a countercurrent exchanger, the temperature of the cold media can be raised to almost the temperature

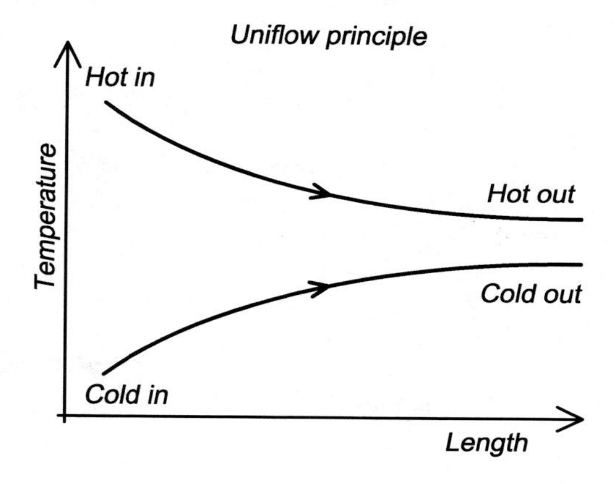

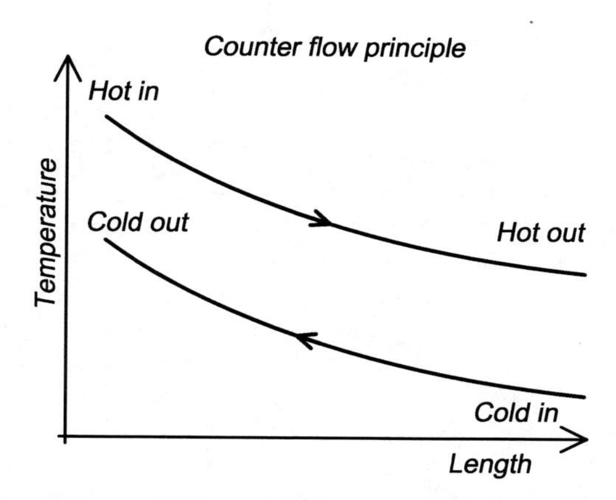

Figure 7.10 The flow pattern in a plate heat exchanger can be either uniflow or countercurrent.

of the hot media, depending of the size of the exchanger.

Example
Find the LMTD for a countercurrent heat exchanger having the following temperatures: hot water into the exchanger, $t_1 = 11°C$; hot water out of the exchanger, $t_2 = 5°C$; cold water into the exchanger, $t_3 = 3°C$; cold water out of the exchanger, $t_4 = 7°C$.

$$\Delta T_1 = t_1 - t_4 = 11 - 7 = 4$$

$$\Delta T_2 = t_2 - t_3 = 5 - 3 = 2$$

$$\text{LMTD} = \frac{4-2}{\ln(4/2)} = 2.89$$

With parallel/linear temperature equalizing the answer would be slightly higher at 3.0. The way to achieve the conditions shown in the example is to have different media flow rates on the two sides of the transfer plates. If there are no energy losses to the surroundings and the water flow in both circuits is equal T_1 will equal T_2 and this value can then be used for further calculations.

7.5.7 Materials in heat exchangers

Heat exchangers are built of different materials for different purposes. It is important that the material has good conductivity, such as a metal, and that the heat transfer surfaces are corrosion-resistant. In fish farming, it is also important that the materials do not release substances that are toxic for the fish, e.g. copper. The more contaminated the flowing media are, either by acids or bases, the more resistant to corrosion the heat transfer surface areas need to be; however, this will increase the price of the material.

If the media are water and it is uncontaminated, stainless steel could be used. If the water is contaminated to some extent, for instance with humus substances, acid-proof steel at least should be used. If the water is salt or brackish, titanium covered plates ought to be used. For ground water titan-covered plates are also recommended; the price of these is, however, much higher than for stainless steel plates.

7.5.8 Fouling

Fouling is a problem with the use of heat exchangers in fish farming. This occurs particularly when using outlet water on one side in the exchanger, but there are also problems associated with using seawater on one side. Fouling of the transfer surfaces will reduce the value of k and the heat transfer. Fouling will cover the transfer surfaces, and the conduction through the layers of fouling is dramatically reduced compared to surfaces with no fouling. Since fouling occurs normally and it is impossible to remove it continuously, this must be taken into consideration when designing heat exchangers for use in fish farming. This is done by including the fouling factor (R_f) in the calculations which again decreases the value of k (see section 7.5.3).

It is important to reduce the amount of fouling as much as possible, but of course within economic limits. Washing of the surfaces will reduce the fouling, whether by the use of chemicals, manual brushing, or both. To make chemical washing possible in a simple way, the exchangers should be equipped with a backwashing circuit (Fig. 7.11). For the washing procedure the heating system is stopped and the exchanger backwashed several times with water containing a detergent: caustic

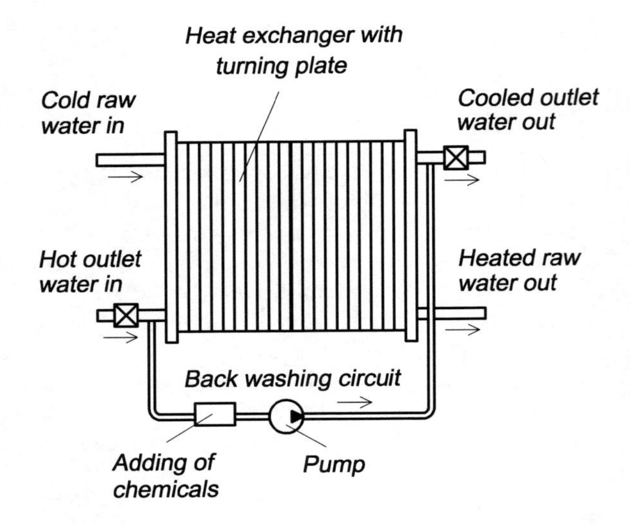

Figure 7.11 To remove fouling, exchangers should be equipped with a washing circuit which is used to backwash the system using detergents to dissolve fat.

soda or lye may be used as a detergent in fish farming. Chemical washing, however, is normally not enough and the exchangers have to be opened and the heat transfer surfaces brushed manually. It may not be necessary to do this every time, but only for some of the washings, for example once a week. This of course depends on the degree of fouling which varies with the characteristics of the water flowing through the exchanger, i.e. whether it is new water or outlet water.

When cleaning the exchangers the heating system must be switched off. Because of this, it can be advantageous to have at least two exchangers, so at least half the heating capacity is functioning during the cleaning procedure. It is important to have enough valves in the pipelines to be able to change the flow direction to permit this (see Chapter 2).

The degree of fouling decreases with increased water velocity through the exchanger, because more turbulence is created. However, this will also increases the head loss, so there is a balance to finding the optimal water velocity. What is certain is that a low velocity through the exchanger increases the degree of fouling; this will occur if the water flow through the exchanger is reduced compared to what it is designed for; in the worst cases total blockage of the exchanger can result. The normal water velocity through exchangers is around 2 m/s.

Fungus clots and larger particles may block the exchanger totally. To reduce the risk of blockage, for instance when using outlet water, the water must always be filtered before entering a heat exchanger. This will to some extent, also help to reduce the amount of fouling. A particle removal filter, for instance a rotating screen filter, is commonly used for this purpose.

7.6 Heat pumps

7.6.1 Why use heat pumps?

The use of heat pumps is beneficial in many aquaculture facilities because the temperature increase is relatively low and the amount of energy that can be transferred is quite large. Today heat pumps are finding increasing use in all type of industry, in greenhouses and housing, as a result of the high cost of electricity and oil.

The great advantage with a heat pump is that the large amount of low temperature energy available is transferred to smaller amounts of media with higher temperature. The energy transfer between the two conditions and the equipment itself must be paid for.

A heat pump is basically the same as a refrigerator or a cold-storage plant. The difference is the method of utilization. A refrigerator is used to remove energy while the heat pump is used to add energy. Theoretically a heat pump and a refrigerator can be the same unit that is used both for cooling and heating, for instance cooling of the water supplied to spawning fish and heating of water supplied to fry production. It is however difficult to optimize the heat pump for both purposes.

7.6.2 Construction and function of a heat pump

A heat pump consists of four main components (Fig. 7.12):

- Evaporator
- Compressor
- Condenser
- Expansion valve.

Between the four components there is a closed pipeline, the transport circuit, in which the working medium or refrigerant circulates. The medium is adapted so that it performs a phase transfer: it will change phase between liquid and gas when circulating between the components in the circuit. The system utilizes the energy needed for evaporation of the working medium, which is released when the medium is condensed.

To explain the construction and function of the heat pump or refrigerator, the working medium can be followed for one lap round the closed pipeline circuit. When the medium enters the evaporator, it is a liquid of low temperature and with relatively low pressure. The boiling point of the medium is quite low. In the evaporator, which is actually a heat exchanger, the temperature is higher than in the working medium. Energy is therefore transferred from the surroundings and into the working medium, i.e. heat exchange occurs. The temperature increases up to the evaporation point for the working medium which starts to change phase from liquid to gas. When a medium changes from liquid to gas much energy is needed which is stored in

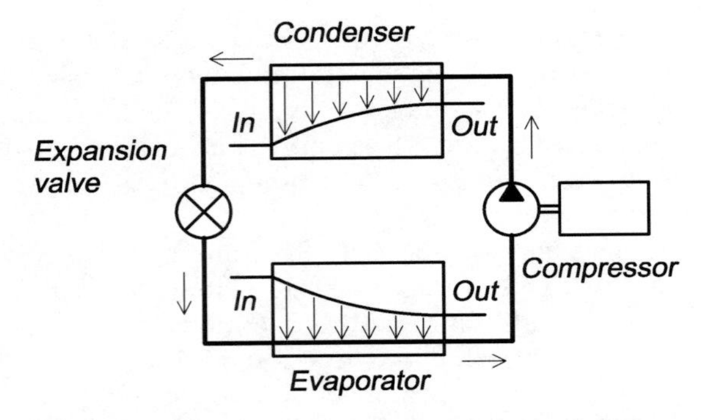

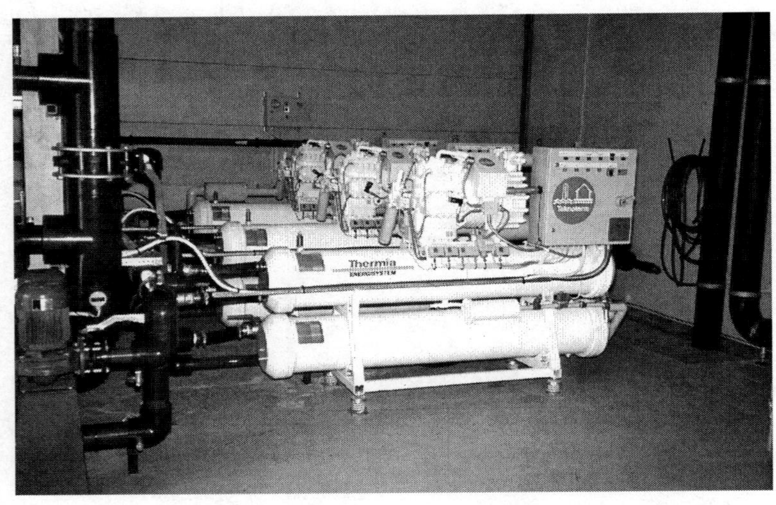

Figure 7.12 The main components and function of a heat pump.

the gas. The liquid–gas mix is then sucked into the compressor where the pressure increases together with the temperature so that all the medium is transformed into gas. The pressure in the working medium is now much higher than before.

The compressor is supplied with additional energy (normally electricity) to function. Available compressors either use a piston principle or the screw principle, the latter having continuous delivery of gas. Not all the energy supplied to the compressor is transferred to the gas, because there are some energy losses. The efficiency of a compressor is normally about 70–80%, meaning that 70–80% of the added electricity is transferred to the gas.

From the compressor the gas is pushed into the condenser with the help of the created pressure. The condenser is another heat exchanger where energy is transferred from the gas, which has the highest temperature, into the surrounding water. The gas is now chilled and it reaches a temperature where phase transfer occurs again, the dew point, and changes into liquid. The evaporation energy that has been stored in the gas is now released and transferred to the cold medium. The gas–liquid mix (or just the liquid, depending on the construction and conditions) now flows to the expansion valve where the pressure is reduced, for instance by increasing the cross-sectional area of the pipe where the medium is flowing. The gas–liquid mix now expands. (A simplification of this is that the liquid is pressed through a small hole and out into a pipe with larger diameter.) All the medium now changes into liquid. The pressure decrease of the medium is accompanied by a drop in temperature. The working medium now enters the evaporator again and can then start a new lap round the circuit; thus it circulates continuously.

The working medium or refrigerant must be adapted to the temperature and pressure conditions in the evaporator and condenser. It must evaporate and condense at temperatures that fit the temperature to which the inlet water is to be heated, and of the water from which the heat is collected. Neither must the pressure and pressure difference be too large. There has been much discussion about refrigerants, because the most suitable have negative effects on the environment by contributing to the greenhouse effect. New more environmentally friendly refrigerants have therefore been developed during the past few years. In fish farming ammonia is much used, which is a relatively environmentally friendly material regarding the greenhouse effect but highly toxic for the fish, so it is important to avoid leakage of refrigerant into the inlet water.

7.6.3 Log pressure–enthalpy (p–H)

The heat pump process is often illustrated in a log pressure (p)–enthalpy (H) diagram (log p–H diagram) (Fig. 7.13). Enthalpy is a parameter that is a measure of the energy content of refrigerant; units are kJ/kg. The log p–H diagram illustrates clearly the changes of phase in the refrigerant. The pressure of the refrigerant is constant through the evaporator, but its energy content (H) increases because it gradually changes phase from liquid to gas and in doing so takes up energy from its surroundings. When the refrigerant enters the compressor it is in the gas phase. Electric energy is supplied to the compressor, most of which is transferred to the refrigerant, so further increasing its enthalpy. In the compressor the pressure of the gas increases and it is then fed into the condenser. Here energy is released because the gas changes phase to liquid and the enthalpy drops, but the pressure remains stable: all the energy that was stored when the liquid changed phase into gas is now released. The refrigerant exits the condenser as liquid and then enters the expansion valve where its pressure drops, but no energy is removed or added (assuming ideal conditions). Therefore the enthalpy is the same, as shown by a vertical line.

The p–H diagram is specific for each medium, and shows the phase of the medium in relation to its pressure and enthalpy content. It is also known as a nose diagram, depending on the presentation. On one side of the 'nose' the phase of the working medium is liquid; on the other side it is gas, and in between both gas and liquid. The heat pump processes therefore happen inside the 'nose', because it is here that the phase transfers occur.

7.6.4 Coefficient of performance

The coefficient of performance (COP) is an important parameter when talking about heat pumps and heating systems; it indicates how much energy is transferred to the water in relation to the supply of energy (normally electricity). For heat pumps the only electricity supplied is to the compressor. The COP (ε) will therefore be the relation between

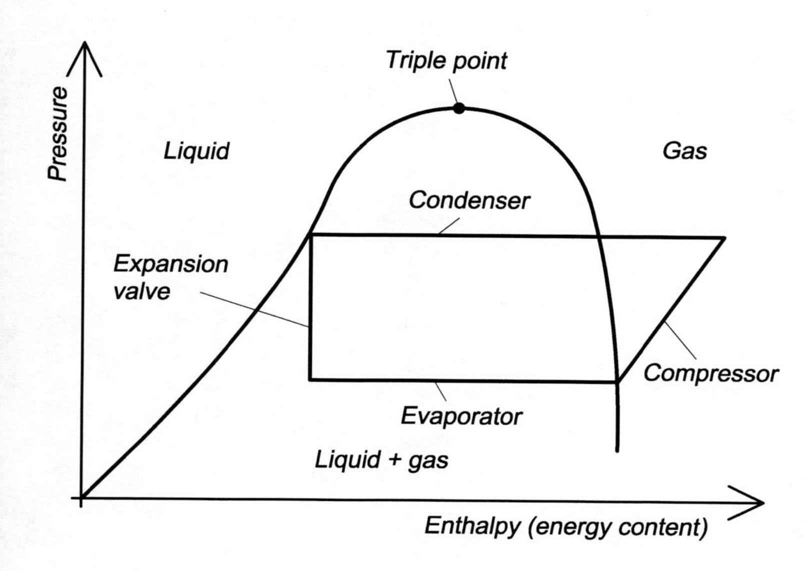

Figure 7.13 The heat pump process illustrated in a log p–H diagram.

the energy released in the condenser and the energy added to the compressor.

$$\varepsilon = Q_{\text{released in condenser}}/Q_{\text{added to compressor}}$$

Normally for commercial heat pumps used in fish farming, COP can be up to 5.

7.6.5 Installations of heat pumps

When installing a heat pump location of the condenser is very important, because it is here that the energy is released. For a refrigeration plant location of the evaporator will be important because here removal of energy occurs. There are a number of places for installing the condenser in the heat pump in a fish farm; all have advantages and disadvantages (Fig. 7.14). One solution is to place it directly into the inlet water, which gives very good heat transfer. However, leakage from the condenser could lead to contamination of the inlet water by the refrigerant, which is dangerous because it could be toxic to the fish. To avoid this, an extra closed circuit is more commonly used between the condenser and the inlet water consisting of a pump and a heat exchanger through which a non-toxic liquid (water or glycol, for instance) circulates. Glycol has a high thermal capacity so is a good choice of medium. Between 10 and 15% reduction in the COP is normal when using the extra circuit, because the heat is transferred twice.

Another method is to install the condenser in the outlet water and afterwards have a heat exchanger between the outlet water and the inlet water, also a two-step heat transfer process. The disadvantage with this method is that increased fouling will present problems. Cleaning of the condenser, which is normally of the shell and tube type, is difficult.

The evaporator is normally placed in the outlet water. Here there will also be problems with fouling, and a closed circuit can also be used to advantage in this situation. A plate exchanger is used in this circuit to transfer the heat, because it is easier to open for cleaning.

When using the evaporator in the outlet water and the temperatures are low freezing of the outlet water may be a problem because the temperature is reduced as much as possible through the evaporator to recover the stored energy in the outlet water, meaning that the temperature in the working medium is 0°C or less to ensure effective heat transfer. It is therefore possible that the water in the evaporator will freeze, for instance if there is a reduction in the water flow or drop in the inlet water temperature. To avoid breakage of a shell and tube evaporator due to ice, it is normal to use glycol which functions in the same way here as in the condenser circuit; glycol is a liquid with a very low freezing point.

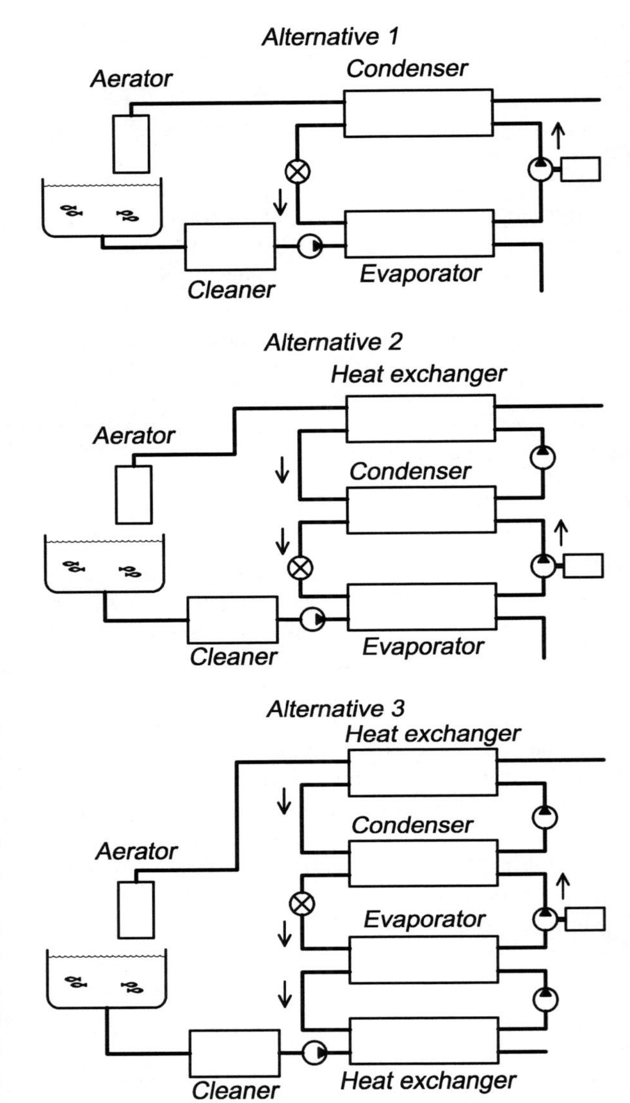

Figure 7.14 There are several alternatives for installing the condenser when using a heat pump. Either it can be placed directly in the inlet water, or a circulation circuit including a heat exchanger can be used.

In practice, there is not usually enough energy in the outlet water to get a heat pump to function with a good COP. Therefore the heat pump will usually be an integral part of a total energy system where additional energy from seawater or another low temperature source may be used. Energy may also be added directly, for instance by using an immersion heater; this will of course require a much smaller heater compared to using an immersion heater alone and not in combination with a heat pump.

Example
Inlet water to a fish farm is heated from 2 to 8°C by transfer of energy from the condenser. The water flow is 500 l/min (8.33 kg/s). Fifty kW of electric energy is supplied to the compressor. Find the COP for the heat pump.

Start by calculating the total energy input to the inlet water:

$$\begin{aligned} P &= mc_p \, dt \\ &= 8.33\,\text{kg/s} \times 4.2\,\text{kJ/(kg °C)} \times 6\,°\text{C} \\ &= 210\,\text{kJ/s} \\ &= 210\,\text{kW} \end{aligned}$$

The COP is therefore

$$\begin{aligned} \varepsilon &= 210/50 \\ &= 4.2 \end{aligned}$$

This means that for every kW of electric energy that is supplied to the compressor, the water is heated by 4.2 kW.

7.6.6 Management and maintenance of heat pumps

Fouling is always a problem when using heat pumps in fish farming. Plate exchangers with glycol circuits, as mentioned earlier, can therefore be advantageous because they are easier to clean and dismantle. Chemical washing circuits should also be installed.

Normally, heat pumps are thermostatically controlled to prevent the water freezing and damaging the evaporator. If the temperature falls too low, circulation of refrigerant is stopped allowing any ice that has formed to melt. A back-up heating system might be installed to safeguard against freezing but is expensive. Alternatively the water can be pre-warmed by using another heat source, such as groundwater, before it enters the heat pump.

7.7 Composite heating systems

A composite heating system is normally used to heat water for use in fish farming. The system comprises several components that all have some heating effect on the inlet water. Usually there are one or several heat exchangers in addition to either a heat pump on large farms, or a heater or an oil burner on smaller farms. The COP is calculated for the entire heating system and are usually in the range 15–25, which means that for each kW of electric energy supplied, the water is heated by 15–25 kW. Examples given below include heaters, heat pumps and heat exchangers to illustrate the profitability of using a composite heating system.

Example
Heater and heat exchanger (Fig. 7.15)

Calculate the profitability of adding a heat exchanger compared with using only an electric immersion heater in a small heating system. A water flow of 180 l/min (3 l/s) is to be heated from 4 to 8°C. The first calculation is for an electric heater alone. Size of the heater:

$$\begin{aligned} P &= mc_p \, dt \\ &= 3\,\text{l/s} \times 4.18\,\text{kJ/(kg °C)} \times (8 - 4\,°\text{C}) \\ &= 50.2\,\text{kJ/s} \\ &= 50.2\,\text{kW} \end{aligned}$$

The daily cost of using this system with an electricity price of 0.1 €/kWh is:

$$50.2\,\text{kW} \times 24\,\text{h} \times 0.1\,\text{€/kWh} = 120.5\,\text{€}$$

Now a heat exchanger is included in the circuit to recover the energy in the outlet water; 75% recovery is quite normal. This value of course depends on the cost of the heater, heat exchanger and electricity; simulations should be done to find the most economical combination. If 75% of the total heat increase above 4°C is provided by the heat exchanger, 3°C of the temperature rise results from its use. This gives the following temperatures in the heat exchanger as the same amount of water flows on both sides:

Water entering heat exchanger	*4°C*
Water leaving heat exchanger	*7°C*

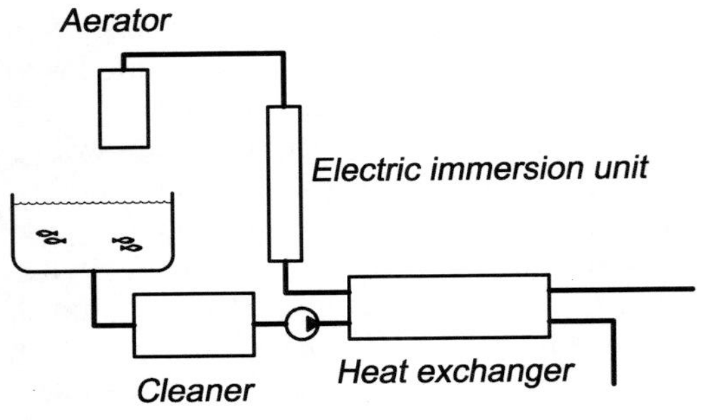

Figure 7.15 A heating system with a heater and a heat exchanger.

Water entering heater	*7°C*
Water leaving heater	*8°C*

A smaller heater is therefore needed as the water is only going to be heated from 7 to 8°C:

$$P = 3 \times 4.18 \times (8 - 7)$$
$$= 12.5\,\text{kW}$$

This gives the following new daily running costs:

$$12.5 \times 24 \times 0.1 = 30.0\ €$$

As can be seen, the daily cost of heating is reduced from 120.5 to 30.0 €, a saving of 90.5 € per day, by adding a heat exchanger. This clearly illustrates the advantage of using a heat exchanger.

The necessary size of the exchanger will now be found.

Since the same amount of water is circulating on both sides of the heat exchanger, it has the following temperature programme:

Water entering heat exchanger	*4°C*
Water leaving heat exchanger	*7°C*
Water entering heater	*7°C*
Water leaving heater	*8°C*

The energy to be transferred from the warm to the cold side in the heat exchanger given by

$$P = mc_p\,\text{d}t$$
$$= 3 \times 4.18 \times 3$$
$$= 37.6\,\text{kJ/s}$$

For a heat exchanger the following equation applies:

$$P = kA\,\text{LMTD}$$

The k value for the plates in the exchanger is set to 6 kW/(m² °C) and the LMTD (temperature difference that drives the heat transfer) is 1.0°C, which gives the following area:

$$\begin{aligned} A &= P/(\text{k LMTD}) \\ &= 37.6/(6 \times 1.0) \\ &= 6.3\,\text{m}^2 \end{aligned}$$

Assuming a plate size of width 0.4 m and height 1 m, the area is 0.4 m² (this depends on the size of plate supplied). The number of plates required = 6.3/0.4 = 15.8; this is an exchanger with 16 plates.

Example
Heater with heat exchangers in outlet water and in seawater (Fig. 7.16).

The inlet water has a temperature of 2°C and this is increased to 4°C when the water passes the seawater exchanger. The water then enters an outlet exchanger where the temperature is further increased to 9°C. The last increment up to 10°C, which is the temperature in the fish tank, is supplied through an electric heater. Calculate the COP of the system.

$$\begin{aligned} \varepsilon &= Q_{\text{deliverd}}/Q_{\text{supplied}} \\ &= (t_{\text{tank}} - t_{\text{raw water}})/(t_{\text{tank}} - t_{\text{before heater}}) \\ &= (10 - 2)/(10 - 9) \\ &= 8 \end{aligned}$$

The COP is 8 for this system, so that for every kW of electric energy supplied 8 kW is supplied to the inlet water.

Example
Heat pump and heat exchanger (Fig. 7.17)

Find the profitability of installing a heat pump, compared with a total energy system including a heat pump and heat exchangers. A water flow of 300 l/min (50 l/s) is to be heated from 2 to 8°C. The heat pump has a COP of 5. How much electric energy must be supplied?

First, the total amount of energy that has to be transferred to the water is calculated.

$$\begin{aligned} P &= mc_{\text{p}}\,\text{d}t \\ &= 50\,\text{l/s} \times 4.18\,\text{kJ/(kg °C)} \times (8 - 2\text{°C}) \\ &= 1254\,\text{kJ/s} \\ &= 1254\,\text{kW} \end{aligned}$$

The COP is 5, meaning that the amount of energy transferred to the compressor is

$$1254/5 = 250.8\,\text{kW}$$

The daily cost of using a heat pump with an electricity price of 0.1 €/kWh is therefore

$$250.8\,\text{kW} \times 24\,\text{h} \times 0.1\,\text{€/kWh} = 601.9\,\text{€}$$

Now a heat exchanger that recovers the energy in the outlet water and transfers it to the inlet water is added to the circuit. The heat exchanger is assumed to meet 75% of the heating requirement. Of the total temperature increase of 6°C, 75% is provided by the heat exchanger, i.e. 4.5°C, which gives the following temperatures in the heat exchanger:

Water entering heat exchanger	*2.0°C*
Water leaving heat exchanger	*6.5°C*
Water entering heat pump	*6.5°C*
Water leaving heat pump	*8.0°C*

The new size of the heat pump can now be calculated:

$$\begin{aligned} P &= 50 \times 4.18 \times (8 - 6.5) \\ &= 313.5\,\text{kW} \end{aligned}$$

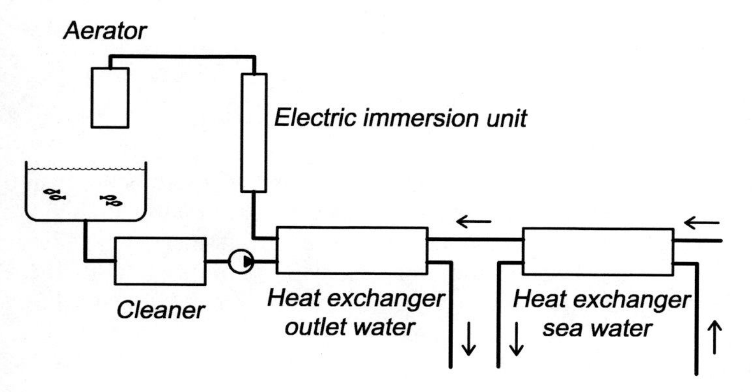

Figure 7.16 A heating system with a heater and heat exchangers both in the outlet water and to seawater.

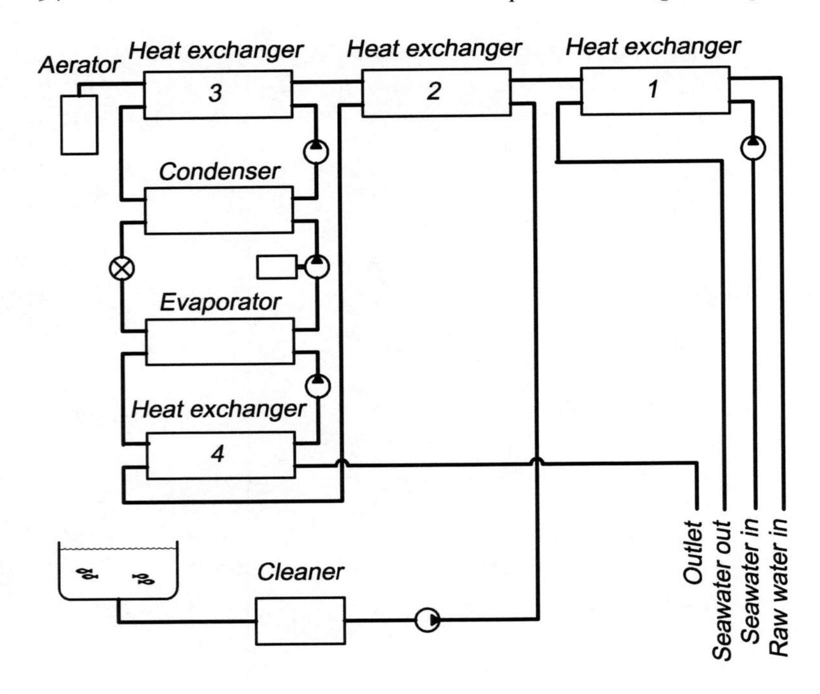

Figure 7.17 A heating system with a heat pump and a heat exchanger.

With a COP of 5, the amount of energy that must be supplied to the compressor is 313.5/5 = 62.7 kW. Therefore new daily costs are:

$$62.7 \times 24 \times 0.1 = 150.7\text{€}$$

As can be seen the daily cost of heating is reduced from 601.9 € to 150.7 €, a saving of 451.2 € per day, by using a heat exchanger in addition to the heat pump. This illustrates how useful it is to utilize a heat exchanger together with a heat pump. Heat pumps are nearly always used together with one or several heat exchangers because of the large reduction in energy costs compared to the investment costs of heat exchangers.

The overall COP can now be calculated:

$$\begin{aligned} \varepsilon &= \text{total energy transferred to the water/electric energy supplied} \\ &= 1254\,\text{kW}/62.7\,\text{kW} \\ &= 20.0 \end{aligned}$$

This means that of the total energy transferred to the water, only 1/20 (5%) is supplied as electric energy.

7.8 Chilling of water

When water is to be chilled, energy has to be removed; the amount is the same that has to be added when heating for a given temperature difference, so of course the same equation can be used.

$$P = mc_p\text{d}t$$

where:

m = water flow (kg/s)
c_p = specific heat capacity (kJ/(kg°C))
$\text{d}t$ = temperature decrease for the water (°C), i.e. the difference between inlet and outlet temperatures.

Example
A water flow of 60 l/min is to be chilled from 4°C to 2°C. Find the rate of energy removal.

$$\begin{aligned} P &= 1\,\text{kg/s} \times 4.18\,\text{kJ/(kg °C)} \times (4\text{°C} - 2\text{°C}) \\ &= 8.36\,\text{kJ/s} \\ &= 8.4\,\text{kW} \end{aligned}$$

Heat exchange can be used if there is an available water source that could be used as a chilling medium. For instance, freshwater can be chilled using bottom water from the sea during the summer when freshwater is warmer. Direct mixing of colder water may also be used to avoid the use of a heat exchanger, but requires satisfactory water quality.

Figure 7.18 A water cooling plant.

Alternatively, ice may be added directly to the water, or one circuit of a heat exchanger can pass through a basin to which ice is added. If this method is used, it is necessary to have an ice machine on the farm; however, this is not very satisfactory because it is a costly method involving two stages: first production of ice and then chilling of the water. Only when small amounts of water are to be chilled at specific times of the year, is it a viable solution, for example in small slaughter houses, and where ice is available for chilling the fish.

A cooling plant is therefore used to chill the water. This is basically the same as a heat pump, but is optimized in another way (Fig. 7.18). The evaporator is placed in the water or liquid to be chilled. When the working medium or refrigerant evaporates, energy is taken from the water or liquid, the temperature of which falls. This is made possible by choosing a suitable refrigerant (working medium). After this the working medium goes into the compressor and then on to the condenser where it condenses, releasing heat to the surroundings. A cooling plant often contains an air–liquid heat exchanger (the condenser) so the heat is released directly to the air. This is the same principle that is used in a refrigerator where the energy is also released to the air. As seen from the set-up, the main component in a cooling plant is the evaporator that 'removes' energy. When talking about the efficiency or COP when a heat pump is used as a cooler, it is the relation between the energy (Q) that is added to the compressor and the energy removed from the water in the evaporator that is critical. The equation to use is therefore:

$$\varepsilon = Q_{\text{removed in the evaporator}}/Q_{\text{added to compressor}}$$

The same circulation circuits that are used on heat pump installations to prevent refrigerant from contaminating the inlet water and avoid freezing are also used in cooler units (see Section 7.6.5).

If both chilling and heating need to be performed on a fish farm, it might be possible to utilize the same heat pump for both purposes. Energy can be taken from the inlet water to be chilled and added to the inlet water to be heated. This means placing the evaporator in the water to be chilled and the condenser in the water to be heated. It will, however, normally be a rather unsatisfactory solution because it is difficult to optimize the heat pump with pressure conditions throughout the evaporator and condenser, and to find good refrigerant.

References

1. Killinger, J., Killinger, L. (2002) *Heating and cooling: essentials.* Goodheart-Wilcox.

2. Cengel, Y.A. (1998) *Heat transfer: a practical approach.* McGraw-Hill.
3. Silberstein, E. (2002) *Heat pumps.* Thomson Delmar Learning.
4. Moran, M.J., Shapiro, H.N. (2003) *Fundamentals of engineering thermodynamics.* John Wiley & Sons.
5. Ibarz, A., Barbosa-Cánovas, G.V. (2003) *Units operations in food engineering.* CRC Press.
6. Incropera, F.P., DeWitt, D.P. (2001) *Introduction to heat transfer.* John Wiley & Sons.

8
Aeration and Oxygenation

8.1 Introduction

The purpose of aeration or oxygenation is either to remove gases such as nitrogen (N_2) and carbon dioxide (CO_2) from the water, or to increase the concentration of gases such as oxygen (O_2) in the water. There are several reasons for aerating and oxygenating the inlet water to a fish farm. The water may contain insufficient oxygen or too much nitrogen or carbon dioxide. If the content of oxygen in the water is increased, the less water need be added. Too much nitrogen (super-saturation) will create gas bubble disease (diving disease) in the fish with high possibilities of mortality.[1] Too much carbon dioxide is also toxic for the fish.[2,3] Adding too much oxygen to the water will also be toxic for the fish.[1]

Aeration or oxygenation is carried out in almost all production systems. On land-based farms and in ponds it is common to use either aeration or a combination of aeration and oxygenation. Recent research also shows improved production resulting from adding oxygen to cages.[4] During transport of fish, the addition of air or oxygen is also nessesary.

A great deal of literature is available concerning aquaculture, but the literature regarding wastewater treatment, environmental engineering and water chemistry is also a source of useful information.

8.2 Gases in water

Water contains a certain amount of dissolved gases. When the water has taken up the amount possible under normal atmospheric pressure it is fully saturated (100%) or in equilibrium (see Section 8.3.1). If the water contains less gas than can be taken up at a given temperature, it is less than saturated (<100%); if the water contains more gas than when fully saturated, it is super-saturated (>100%).

The percentage saturation can be calculated by dividing the measured concentration (C_m) by the concentration at saturation (C_s):

$$\text{Percentage saturation} = (C_m/C_s) \times 100$$

All the gases in the atmosphere can be dissolved in water; the sum of the partial pressures of all the dissolved gases is known as the total gas pressure (TGP). The pressure difference ΔP is the difference between TGP in the water and the barometric pressure (BP) of the air above the water:

$$\Delta P = \text{TGP} - \text{BP}$$

If TGP measured in the water is higher than the BP of the air (positive ΔP), the water is super-saturated and gas will be forced out of the water. If TGP is less than BP, the water is under saturated and gases will be forced into the water. TGP may be expressed as a percentage of the BP:

$$\text{TGP}(\%) = \frac{(\text{BP} + \Delta P)}{\text{BP}} \times 100$$

The water vapour pressure, ΔP_{H_2O}, may also be included in the equation:

$$\text{TGP}(\%) = \left(\frac{\text{BP} + \Delta P - \Delta P_{H_2O}}{\text{BP}}\right) \times 100$$

Air and water contains many gases, but in fish farming oxygen, nitrogen, and in some cases also carbon dioxide, are of greatest importance. Since the major reason for adding water is to provide oxygen, the added water should have as high an oxygen concentration as possible, close to 100% saturation. If under saturated water is used, more water must be supplied. Water may also be super-saturated with oxygen, i.e. its concentration in the water is raised above 100% by the addition of pure oxygen, to reduce the amount of water that must be supplied to the fish. If the other gases in the water are in equilibrium, water that is super-saturated with oxygen will have a positive ΔP, i.e. TGP is higher than BP. The concentration of nitrogen in the water must not be above 100%, because this can cause bubble (diving) disease in the fish.

Water treatment and biological processes in the water source can result in both super-saturation and under saturation of the different gases. Super-saturation can result from naturally occurring or man-made processes.[5] Super-saturation can be the result of rapid heating of water, mixing of water of different temperatures, freezing of water when some of the gases remain in the water, mixing air into the water, for instance under waterfalls or by waves, photosynthesis that creates oxygen gas, and changes in the BP.

If the saturation of the gases is 100% or less, they are completely dissolved in the water. However, if the concentration of gas is above 100% saturation, there will be free gas bubbles in the water; these are small and difficult to observe. Only the oxygen actually dissolved in the water is available to the fish: when the water is super-saturated with oxygen, the bubbles will gradually be dissolved in the water and the oxygen made available for the fish as the fish consume the oxygen that is already dissolved in the water.

Air and water have different gas contents; in both nitrogen and oxygen are the major gases. In air the relation between nitrogen and oxygen is approximately 78%–21% by volume (Table 8.1), while in water the relation is 60% nitrogen and 40% oxygen when the gases are in equilibrium. It is important to be aware of this relationship because if air is pressed into the water, super-saturation of nitrogen may occur which is harmful to the fish.

Table 8.1 Characteristics of dry air.[1]

Gas	Weight %	Volume %
Nitrogen (N_2)	75.54	78.084
Oxygen (O_2)	23.10	20.946
Argon (Ar)	1.29	0.934
Carbon dioxide (CO_2)	0.05	0.033
Other	0.02	0.003

The air will exert pressure on the water surface. At sea level this pressure is 1 atmosphere, equal to 760 ± 25 mmHg; it varies depending on atmospheric conditions. The oxygen partial pressure is related to the percentage of the total volume that oxygen constitutes, which in air is 21%; multiplying this value by the BP (1 atm or 760 mmHg) gives the partial pressure of oxygen.

$$\text{Partial pressure } O_2 = 0.21 \times 760\,\text{mmHg} = 159.6\,\text{mmHg}$$

This means that the pressure of oxygen on the water surface is 159.6 mmHg.

The total pressure represented by the BP is the sum of the partial pressures of the gases in the atmosphere. This can be described as follows, neglecting the gases that constitute a very small proportion of the air:

$$BP = P(N_2) + P(O_2) + P(Ar) + P(CO_2)$$

The amount of a gas dissolved in water is referred to as the solubility of the gas, and depends on several factors, such as water temperature, pressure, salinity and substances in the water.[6] At higher temperature gases are less soluble in water because the molecules then need more space. The solubility of oxygen and nitrogen decreases linearly with increasing temperature (see later) and for this reason hot water contains less oxygen than cold water. The solubility for oxygen and nitrogen also decreases with increased salt content of the water. Tables for the content of oxygen in water with different temperatures and salinity is given in Appendix 8.1.

The solubility of gases in water is normally expressed as mg/l of the actual gas, but it may also be expressed as the partial pressure. Equations for conversion are available;[1] for oxygen and carbon dioxide the following may be used:

$$\text{Partial pressure}_i = \left[\frac{C_i}{\beta_i}\right] A_i$$

where:

partial pressure for the actual gas (i) is in mmHg

C_i = concentration of the actual gas in mg/l
β_i = Bunsen coefficient for the actual gas, depending on temperature and salinity (Appendix 8.2)
A_i = constant depending on the actual gas (0.5318 for oxygen and 0.3845 for carbon dioxide).[7]

Solubility may also be expressed on the basis of tension. The oxygen tension may, for instance, be the necessary partial pressure in the atmosphere to keep a certain concentration in the water. If the atmosphere is air at normal pressure, the oxygen tension is 159 mmHg. This creates 100% oxygen saturation in the water. If the pressure is less than this, the concentration in the water will also be reduced.

8.3 Gas theory – aeration

How much and how fast gases are transferred in and out of water depends on two factors:[8]

(1) Equilibrium conditions (also known as the saturation concentration)
(2) Mass transfer.

Equilibrium is when there is no net transfer of gas in or out of the water. Mass transfer occurs before equilibrium is reached when there is transport of gas into or out of the water.

8.3.1 Equilibrium

When equilibrium has been reached there is no net transport of gas into the water or from the water to the air. There is still some transport of gas molecules through the water surface, but what goes in equals what comes out; there are no free gas bubbles in the water. In the same way, when salt is added to freshwater only a certain amount can be dissolved; after reaching this level no more will dissolve, even if it is added. The excess salt will only remain on the bottom as salt crystals. The same happens with the gas as it stays in bubble form when the water is super-saturated.

To find equilibrium conditions, Henry's law can be used. Henry's law can be expressed in several ways;[1,9,10] it illustrates that the amount of gas that can be dissolved is proportional to the partial pressure:[11]

$$P_g = HX_g$$

where:

P_g = partial pressure of gas (atm)
H = Henry's constant (atm/mol fraction)
X_g = concentration of gas in water (mol gas/(mol gas + mol water)).

If SI units are used, P_g is given in pascal (Pa) H in Pa m^3/mol and X_g in mol/m^3.

Henry's constant depends on temperature and gas type; its value increases with temperature. The gas obeys Henry's law if the gas that can be dissolved in water decreases with temperature.

As explained in section 8.2, partial pressure, P_g, depends on how much of the actual gas is in the atmosphere above the water surface. The pressure it exerts on the surface will press the gas into the water. In air there is about 20% oxygen, so P_g is in this case 0.2; if the atmosphere is pure oxygen, i.e. there is 100% oxygen above the water, $P_g = 1.0$.

If the gas above the water surface is above normal atmospheric pressure, there will be new equilibrium conditions given by Dalton's law:[8]

$$p_i = p_t y$$

where:

p_i = partial pressure of gas i
p_t = total pressure of a mixture of gases
y = mol fraction of gas i (mol gas i/mol total gases).

Thus if the total pressure of the gas is increased, the partial pressure will also increase and more gas can be pressed into the water.

Henry's law can be combined with Dalton's law to give the Henry–Dalton law:

$$p_t y = H x_i$$

where:

y = mol gas i
p_t = total gas pressure
H = Henry's constant
x_i = amount of gas i in the liquid.

x_i is the key value because, as the equation shows, by increasing the total pressure or by increasing the amount of gas i in the atmosphere, the gas concentrations in the liquid will be increased.

For instance, when air is under pressure above the water surface, more gas will be pressed into the water which will be super-saturated.

Super-saturation with nitrogen from the air is harmful to fish, so overpressure of air is unwanted and must be avoided. The same may occur if adding air to deep water where the water pressure is higher and the air pressure must be higher than at the surface, i.e. more than 1 atm. To illustrate this, air can be added to the water 2 m deep, but when doing this air is compressed by the water before it dissolves.

Example

A plastic bag containing both air and water is lowered to the bottom of a 2 m deep basin full of water. The air will be compressed by the water pressure and the air inside the plastic bag will be pressed into the water because the pressure on the water surface increases. We can now see what will happen to the nitrogen saturation. The normal pressure on the surface is 1 atm (10 mH_2O); when lowering the bag to 2 m deep the total pressure will increase to 12 mH_2O. The Henry–Dalton law can be used to find the concentration of nitrogen:

$$p_t y = Hx$$

y is constant, but the pressure increases from 10 to 12 mH_2O, a 20% increase. Since H is a constant, x will increase by 20%. Therefore the concentration of nitrogen increases by 20% which is harmful to fish. Under practical conditions, however, some nitrogen will go directly to the surface and escape without having time for gas transfer, but an increase in concentration of 0.5% per 10 cm or 5% per metre depth occurs.

Adding air directly into the water at depths of more than 1 m is not recommended because super-saturation of nitrogen should be avoided.

8.3.2 Gas transfer

To describe the mechanism of gas transfer into water, the two-film theory proposed by Lewis and Whitman in 1924 is the simplest and most commonly used (Fig. 8.1; refs 11–13). The interface between the water and the gas can be divided into two films with laminar flow, one gas film and one liquid film. These films will inhibit the transport of gas molecules either into the water or out of the water. The thickness of the two films with laminar flow depends on the amount of turbulence in the water and in the air; much turbulence results in reduced thickness. To achieve a good transport of gas molecules it is important to have a thin film to increase the velocity of the gas transfer through the interface.

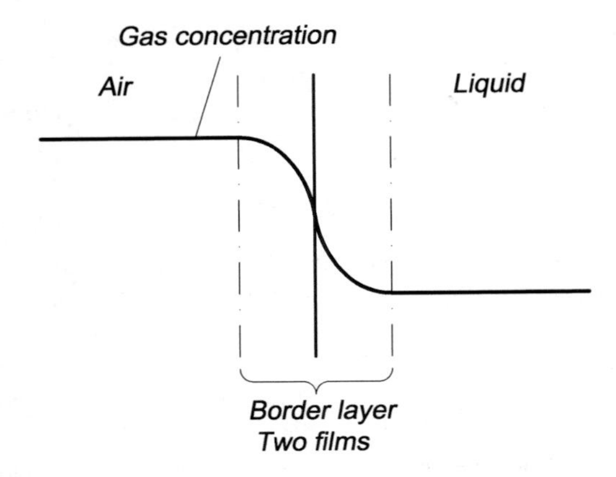

Figure 8.1 The two film theory is commonly used to describe the transfer of gas into water.

When gas molecules from the air are pressed into the water, they at first go from the gas phase into the gas film, also referred to as the surface film. This process is a combination of diffusion and convection, and is quite fast. The next step is through the gas–liquid film interface where the force is diffusion. Diffusion, which results from random molecular movements, is a slow process and this step is rate limiting. The last step is transfer of gas into the liquid; the main force here is convection.

The gas transfer per unit time through the surface can be described by the following differential equation:

$$dc/dt = K_L(A/V)(C^* - C_0)$$

where:

dc/dt = change in concentration per unit time (mg/(l h))
K_L = coefficient for gas transfer (cm/h)
A/V = contact area of the gas–liquid interface (cm^2) in relation to the total liquid volume (cm^3)
C^* = saturation concentration for the gas in the liquid (mg/l)
C_0 = starting concentration of the gas in the liquid (mg/l).

The diffusion constant K_L is proportional to the diffusion rate through the surface film and inversely proportional to the surface film thickness. This shows the importance of a thin surface film. K_L is also dependent on temperature; it will increase with temperature, because the diffusion velocity will increase. The following relationship shows this:[14]

$$K_L = K_{20}\theta^{T-20}$$

where:

K_L = overall mass transfer coefficient
K_{20} = transfer coefficient at 20°C
θ = correction factor (1.024 for freshwater)
T = temperature.

It is normal to express the constant K_L and the contact area (A/V) together as the K_{LA} value, which is known as the overall mass transfer coefficient for an aeration system. It is difficult to calculate K_{LA} for an aeration system; however, if this value could be found it would be easy to calculate the efficiency of the aeration system. Mathematical models have been developed, but they are difficult to use. Therefore the K_{LA} value for an aeration system is usually determined by experiment.

The gas concentration in a system after time t with mass transfer for a given K_{LA} value can be found by integration of the basic equation to give the following result:

$$\frac{C - C^*}{C_0 - C^*} = e^{K_{LA}t}$$

This may also be expressed as:

$$\ln\left(\frac{C - C^*}{C_0 - C^*}\right) = K_{LA}t$$

This shows that the concentration (C) after using an aeration system for time t depends on the starting concentration (C_0), the saturation concentration (C^*) and the K_{LA} value:

$$C = C^* + (C_0 - C^*)e^{K_{LA}t}$$

When all the oxygen has been removed from the water and aeration is started, the content of oxygen will increase rapidly and later level off at the limitation value (100% or full saturation); see Fig. 8.2. The value $(C - C^*)/(C_0 - C^*)$ is known as the oxygen deficit. If the natural logarithm (ln) of the oxygen deficit is plotted on the y-axis in a diagram with time on the x-axis, the K_{LA} value will be given by the slope of the line.

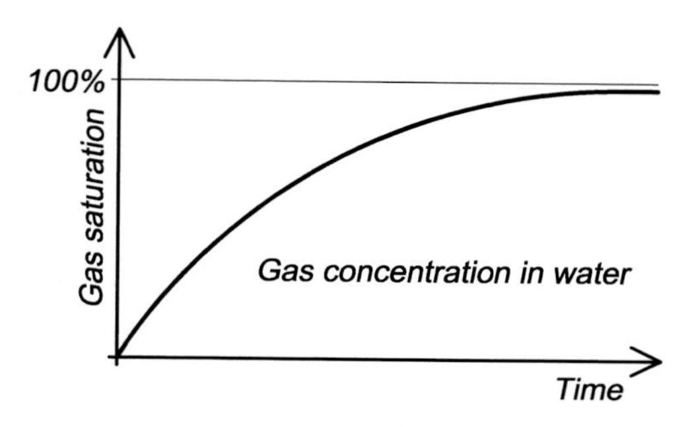

Figure 8.2 When all the oxygen has been removed from the water and aeration is started, the content of oxygen will increase rapidly and later level off at the limitation value (100% or full saturation).

8.4 Design and construction of aerators

8.4.1 Basic principles

With aerator the aim is to create conditions as near equilibrium as possible between the gas in the air and the gas in the water. Eventually super-saturated gas, especially nitrogen, will be aerated out and oxygen will be supplied if the concentration is below saturation. The aim in constructing an aerator is to achieve optimal conditions for exchange of gas between air and water, so that equilibrium can be reached. By creating a large contact area between the water and the air a good gas–water exchange will occur; see equations for gas transfer and the A/V relationship. The layers of air or water ought to be as thin as possible. To create thin layers it is important that both the current of air and the current of water are turbulent so that effective gas transfer can occur at the water–air interface. The aeration process needs time, and effective aerators will need less time to achieve the same degree of saturation than ineffective ones. Mass transfer from the air will occur into a tank of non-saturated water, almost 100% saturation being achieved, but this will take a very long time.

Two main methods are utilized for aeration (Fig. 8.3). Either air can be supplied into a flow of water, or water may be supplied into a flow of air. An example of the first method is bubbling of air through a water column; droplets of water passing through a layer of air illustrate the second method. Since the droplets are small, a large surface area

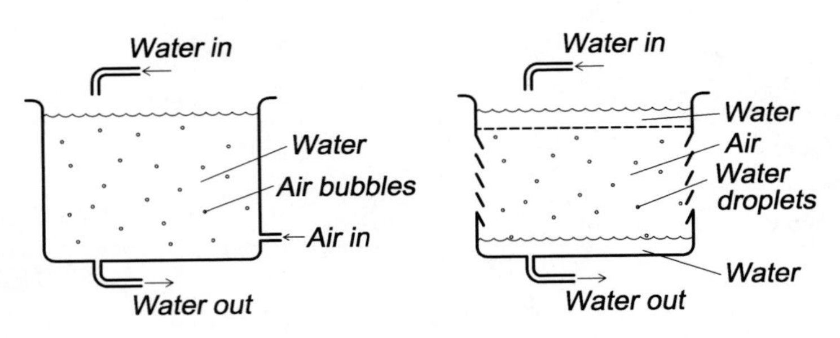

Figure 8.3 Two methods are used for aeration: either air can be supplied into a flow of water or water may be supplied into a flow of air.

between the water and the air is created; the smaller the bubbles or droplets, the larger the surface area.

Aerators can be constructed either for supplying oxygen to the water (gassing) or removing nitrogen or carbon dioxide from the water (degassing). An aerator built for gassing is not necessarily good for degassing; aerators for both O_2 and N_2 are usual.

8.4.2 Evaluation criteria

Many methods are used for aeration, and several designs of aerators are available either for adding oxygen, or removing nitrogen or carbon dioxide. Objective criteria are needed by which to evaluate aerators; these have been developed for both clean water conditions and field conditions.[1,11] Most have been developed for wastewater applications, however, and may not be the most accurate or valid for use on aerators in aquaculture.[13]

Performance testing methods depend on the aerator type. For some, such as gravity aerators, the difference in the gas concentration entering and leaving the aerator can be used. The gas exchange in a tank of a given size is used for testing the performance of surface aerators. Standardized set-ups and equations are available for testing the performance of aerators under laboratory conditions and with clean water. Normally, the aerator is tested for oxygen, but other tests can be carried out, especially for nitrogen or carbon dioxide. A typical way to perform a standardized aeration test for oxygen is to remove all oxygen from the water with, for example, cobalt chloride and sodium sulphite,[15] after which the aerator is started and the increase in oxygen concentration measured in relation to time. By doing this the K_{LA} value for the aerator may be found; this value also shows the performance of the aerator, a high K_{LA} value representing an effective aerator. Performance tests can also be carried out under field conditions, but the results are difficult to compare, because the water quality can vary from site to site, as can the starting saturation.

The oxygen transfer rate gives the amount of oxygen transferred into the water through the aerator. If it is possible to measure the gas concentration entering and leaving the aerator, as for gravity aerators, the oxygen transfer rate can found directly by measuring the water flow (Q) and the difference in the oxygen concentration entering and leaving the aerator ($C_{out} - C_{in}$). The oxygen transfer rate (OTR) can then be calculated from the following equation:

$$OTR = Q_w(C_{out} - C_{in})$$

If Q_w is given in l/min, and C in mg O_2/l, the following equation can be used to find the OTR value in kg oxygen transferred per hour:

$$OTR = 3.6Q_w(C_{out} - C_{in})$$

As said, this can be measured directly for gravity aerators, and also under field conditions (OTR_f).

For basin aerators a standardized test procedure may be used on clean water; the standardized oxygen transfer rate (SOTR) in kg/h is given by the following expression:[15]

$$SOTR = K_{LA20}C_{20}V \times 10^{-3}$$

where:

K_{LA20} = gas transfer coefficient determined for the aerating system at 20°C (kg O_2/h)
C_{20} = equilibrium concentration at 20°C (g/m^3)
V = tank volume (m^3)
10^{-3} = factor converting grams to kilograms.

This test is performed with clean water, but compensating factors for water quality can be used to adapt it to field conditions.[14,16] The following equation[14] can be used to describe this:

$$OTR_f = SOTR\left(\frac{\beta C_s - C_w}{C_{s20}}\right)\theta^{T-20}(\alpha)$$

where:

OTR_f = actual oxygen-transfer rate under field-operating conditions in a respiring system
SOTR = standardized oxygen transfer rate (kg/h)
α = field correction factor (varies with type of aeration device, basin geometry, degree of mixing and wastewater characteristics)
β = field correction factor for difference in oxygen solubility due to constituents of salt, particles and surface active substances
θ = temperature correction factor (between 1.015 and 1.040; typical value 1.024)
C_s = concentration of oxygen at full saturation and measured temperature
C_{s20} = concentration of oxygen at full saturation at 20°C
C_w = concentration of oxygen in the water that is aerated.

All aerators require a supply of energy; this can be added directly (electricity), or it may be energy that is stored in the water (potential energy) as in gravity aerators. The potential energy can be utilized by sending the water from a high level to a lower level through an aerator. When choosing an aerator it is important that it uses the energy supplied as efficiently as possible to transfer the gas. The standardized aeration efficiency (SAE) or field aeration efficiency (FAE) in units of kg O_2/(kWh) is used for this purpose:

$$SAE = \frac{SOTR}{\text{Power input}}$$

$$FAE = \frac{OTR_f}{\text{Power input}}$$

Another indicator variable is the oxygen transfer efficiency (OTE) in units of kg O_2 transferred per hour in relation to the added oxygen gas and the mass flow (m), either measured under standardized (OTE) or in field conditions (OTE_f). This indicates how effectively the added gas is transferred to the water and can be calculated from the following equations:

$$OTE = \frac{SOTR}{m}$$

$$OTE_f = \frac{OTR_f}{m}$$

For some aerators, such as some gravity aerators, it is impossible to measure the oxygen flow rate because they are open to the atmosphere; the OTE value can then be measured instead.

The effectiveness (E) of the aerator is another useful indicator that shows how effectively the gas is transferred into the water (in gassing) or supersaturated gas is removed from the water. The values will vary, depending on the gas. The effectiveness can also be used to compare aerators, but they must be tested in the same system and under indentical conditions:

$$E = (C_{out} - C_{in})/(C_{sat} - C_{in}) \times 100$$

where:

E = effectiveness
C_{in} = concentration of gas in the liquid entering the aerator or before starting the aerator
C_{out} = concentration of gas in the liquid leaving the aerator or after using the aerator for a given period
C_{sat} = concentration of gas at full saturation.

The price of getting gas in or out of the water is the most important factor, and this also makes comparisons of the methods possible. Included are both the aerator purchase price and the costs of running it in relation to the amount of oxygen transferred:

Price for gas transferred in or out of the water
= amount of transferred gas/equipment cost and running cost for the aerator

8.4.3 Example of designs for different types of aerator

Many types of aerators are available, and different classifications are possible depending on the design principles employed. One classification is surface, subsurface and gravity aerators,[1] another is mechanical, gravity and air diffusion systems,[17] while a third is gravity, surface, diffusers and turbine aerators.[18]

In gravity aerators the water falls under gravity and air is mixed into it from the surrounding atmosphere. The simplest type is an artificial waterfall. Gravity aerators require water with natural pressure (head); otherwise the water must be pumped up and into the aerator. In a surface aerator the water is sprayed or splashed into the air with a mechanical device; it functions on the same principle as a water fountain and creates a large surface area where gas exchange can occur. In subsurface aerators air is directed under the water surface, which creates air bubbles that go to the surface; the air bubbles in the water create a large gas transfer surface.

The design of the aerator will, as mentioned before, vary depending on the main purpose of the aerator, whether it is designed for adding oxygen, or removing nitrogen or carbon dioxide. Even if all aerators have some effect on all the gases, the effect is not necessarily optimal. Below is a brief review of some major types of aerator.

Figure 8.4 Packed column aerator.

Gravity aerators

The packed column aerator is one of the most commonly used in intensive fish farming (Fig. 8.4). A column is filled with a medium with a large specific surface area. A dispersal plate (perforated plate) installed at the top of the column, over the aeration medium, ensures proper distribution through the total cross-sectional area of the water that enters the top of the column. There is also a perforated plate in the bottom of the column to keep the medium stable inside. The water trickles down the column on the surface of the aeration medium in a thin film. This arrangement creates a large area between the flowing water and the air around it and ensures effective gas exchange. It is, however, important to get sufficient air into the aerator. An open structured column can therefore be advantageous, or a fan that blows air through the column can be used (see later). The aeration medium is commonly plastic, and it has a design that creates a large surface area per unit volume. The relation between the surface area (A) and the volume (V) is called the A/V condition for the medium. For effective aeration media an A/V value of between 100 and 200 m^2/m^3 is normal; if the value is too high, the possibilities for blockage of the column increase. To give satisfactory aeration under practical conditions the column ought to be up to 2 m in height. This, however, depends on the medium, the flow rate and the water temperature.[18] The diameter of the column cylinder depends on the quantity of water to be aerated, but is normally between 30 and 100 cm. If the diameter is too small, the proportion of the water flow against the wall in the column, where reduced aeration is achieved, will increase; a design rule of thumb value is 0.5–11 water/min/cm^2 cross-sectional area. The advantage with a column aerator compared to other aerators is the low surface requirement. Column aerators are effective with reported SAE values of 1.5–2.0 kg O_2/kWh.[12] This type of aerator also shows quite good results for removal of nitrogen.[20]

To remove carbon dioxide from the water is quite difficult because the amounts are so small (less than

1% of the total gas volume) and it is rather more soluble than oxygen.[7] Therefore, an aerator that is suitable for removing nitrogen and adding oxygen may not be so suitable for carbon dioxide removal. What is normally needed is a large flow of air through the water: 3–10 volumes of air for every 1 volume of water flow treated has been suggested.[21] Specially designed packed column aerators have proved effective for this purpose.[22] Either a long (several metres) packed column aerator or a short packed column with a fan blowing air through to increase the air flow can be used (Fig. 8.5).

A packed column aerator may also be set under low pressure. The principle here is that the water is aerated in a low pressure atmosphere. Thus the amounts of gases that can be dissolved in the water are reduced (Henry–Dalton law states that by reducing the pressure of the gas the amount of dissolved gases will be reduced) and it is possible to reduce the saturation in the water to below 100% by natural pressure. This method is especially useful when growing species that have a low tolerance against super-saturation, such as marine fry.[19] Ejector aerators will also have this advantage.

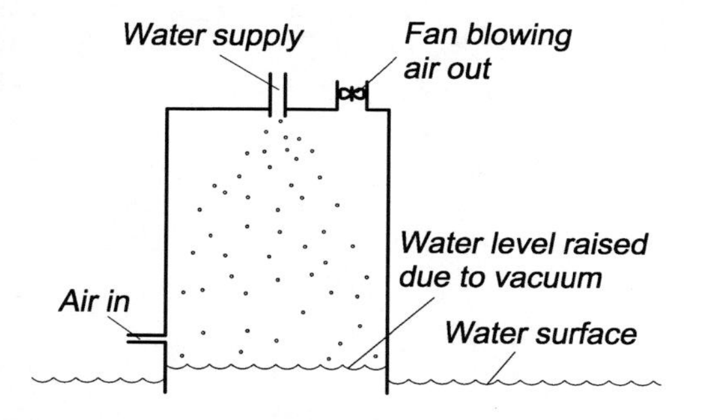

Figure 8.5 Packed column suitable for removal of carbon dioxide.

Packed column aerators are normally placed a certain distance above a level basin to achieve the best possible air transport through the aerator. This is important for the aeration results. Normally the distance is set to 10 cm. Too great a distance between column and basin must, however, be avoided because if the water drops from the aerator fall into the level basin with too high a velocity, air will be dragged into the water together with the water drop and some super-saturation of nitrogen might result. This can be harmful to critical life stages.

Another much used aerator type is the cascade aerator (Fig. 8.6). Water is supplied at the top and flows over a series of horizontal perforated plates or trays placed on top of each other. Vertical distances between the trays are typically 10–25 cm, and the number of trays between 4 and 10.[18] The effectiveness of the aerator increasing with number of

Figure 8.6 Cascade aerator.

trays and distance between them. At least 10 trays are recommended if using 10 cm between them. The water is distributed throughout the whole perforated tray and drops down from tray to tray until it reaches the level basin located underneath the last one. When the water flows down through the holes in the perforated trays, drops will be created and this ensures a large contact area between the air and the water. The space requirement for a cascade aerator is much higher than for the column aerator; the recommended hydraulic rate is down to one-twelfth of what is recommended for a packed column aerator.[19]

Typical SAE values for gravity aerators are in the range 0.6 to 2.4 kg O_2/(kW h).

Subsurface aerators

Diffusers are commonly used as subsurface aerators. A diffuser is a construction where small bubbles of gas are created: the simplest form is a tube with holes. If air is pumped in and the tube lowered under the water surface, the air will come out in the water as bubbles. Tubes with many small holes may also be specially produced for use as diffusers; porous ceramic stones may also be used. A large transfer surface is created by air bubbles going through the water. Bubble size depends on the difference in the pressure inside the bubble and around the bubble. A high pressure difference gives smaller bubbles. Smaller bubbles will give a greater total surface area and by this gas transfer area. The rise velocity is also slower for smaller bubbles which increases the gas transfer rate.[23] SAE values vary from 0.6–2.0 depending on the bubble size,[1] higher values being achieved with smaller bubbles.[11] Bubbles that are too small will, however, be pushed together to create large bubbles, and are therefore not effective. As mentioned previously, when mixing air under pressure, care should be taken to avoid super-saturation with nitrogen which may occur if air is to be added below 1.5–2 m depth, or the pressure is over 0.15–0.2 bar.

In the Inka aerator the incoming water flows over a perforated plate in a thin layer (Fig. 8.7). An air blower supplies air from the underside. The air will create bubbles that flow through the thin water layer and aeration is achieved. An Inka aerator requires a large surface area but the head loss is very low. Therefore it may, for example, be installed under the roof on a fish farm. The aerator type is less used than those described previously, but is useful for special conditions.

Venturi and air-lift pumps are other principles that can be used for aeration where air bubbles are supplied into a water flow; SAE values are in the range 2–3.3 kg O_2/(kW h).[1]

Surface aerators

Surface aerators are commonly use in ponds,[24,25] but may also be used in large tanks, in distribution basins and in sea cages under special conditions (Fig. 8.7). A great number of different designs of surface aerators are available. Normal designs use rotating wheels with a type of paddle, or horizontally placed propellers. They function by throwing the water into the air and creating thin films or bubbles. This establishes a large exchange surface area between water and air where gas exchange can take place. Surface aerators are driven by electricity, solar power or from tractors through a shaft; SAE values are between 1.2 and 2.9 kg O_2/(kW h).[1]

8.5 Oxygenation of water

To increase the amount of oxygen in the water above equilibrium and levels possible with traditional aerators, pure oxygen gas can be added. The addition of pure oxygen gas to the water is used in several cases. One is to increase fish production when there is not enough water. If the water has to be pumped to the farm, regardless of whether it is salt water or freshwater, it may be viable to add pure oxygen to reduce the necessary water flow and hence reduce the pumping costs. Normally it is more economical to add pure oxygen instead of pumping, but a calculation must be made in every single case. In systems with re-use of water it may also be worth adding pure oxygen to reduce the amount of new water and amount of water pumped through the system. When transporting fish pure oxygen is usually supplied.

It is important that the water is fully saturated with oxygen before starting to add pure oxygen gas. An aerator should therefore be installed before the point where pure oxygen gas is added; if there is no aerator, pure oxygen is used to saturate the water up to equilibrium before starting to super-

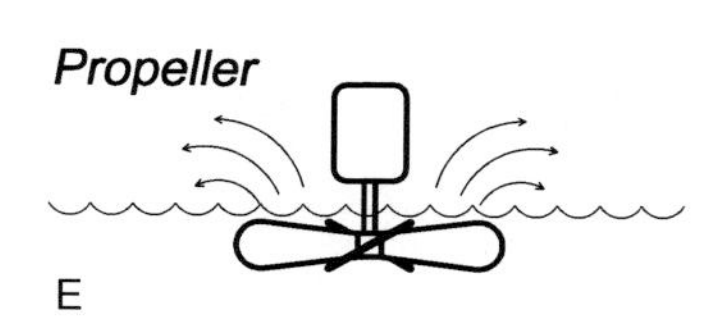

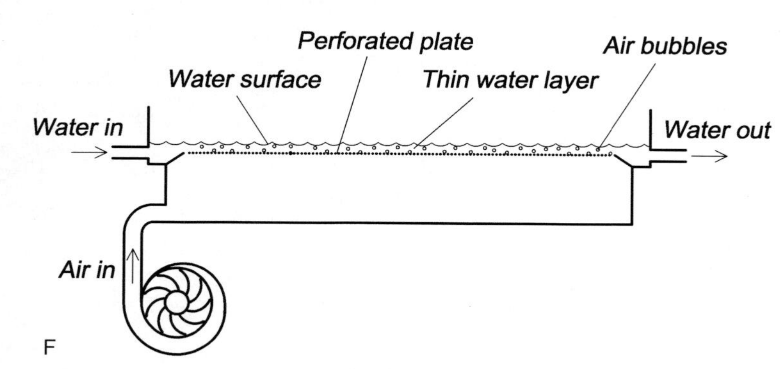

Figure 8.7 Various types of surface aerator are available. (A) and (B) show a paddle wheel aerator, whilst (C), (D) and (E) show a propeller aerator, (F) shows an Inka aerator. (D) A propellor aerator being used in a closed cage.

saturate. This is, in most cases, unnecessarily expensive, but calculations can be made to evaluate cost-effectiveness.

When the water is supersaturated with oxygen free gas bubbles are present; if this water were to have a free surface to the atmosphere there would be gas exchange and the oxygen content would fall to 100% saturation. Therefore free surfaces should be avoided after supersaturating the water with oxygen; the best method is to have a short pipe run from where the oxygen is added to where it reaches the fish.

An oxygenation plant may be designed either to cover the whole oxygen requirement for the fish or as a top-up oxygenation facility. The design criteria for the oxygenation plant are determined by the situation where the biomass and water temperature are highest. A complete oxygenation system on a fish farm includes two parts: (1) the injection system that brings the gas into the water and (2) the source of oxygen gas.

8.6 Theory of oxygenation

When water is oxygenated pure oxygen gas is added; by this the saturation of oxygen in the water can be raised above, the equilibrium level of 100%. The following prossesses are included in oxygenation (cf. aeration): (1) increase of the equilibrium concentration, (2) increase of the gas transfer velocity, (3) addition under higher pressure.

8.6.1 Increasing the equilibrium concentration

Henry's law can be used to describe what is happening:

$$p_i = Hx_i$$

where:

p_i = partial pressure of gas i
H = Henry's constant
x_i = amount of gas i in the liquid (mg/l).

x_i is the value of interest here. The oxygen content of air is 20% compared to a theoretical value of 100% for pure oxygen. The partial pressure (p_i) when having pure oxygen atmosphere is therefore almost five times as high as when using an atmosphere of air. Since Henry's constant does not vary almost five times as much oxygen can be dissolved in the water by oxygenating in a pure atmosphere than when doing it in air. An example of this is the packed column aerator where the atmosphere inside is pure oxygen.

8.6.2 Gas transfer velocity

The same equation as for aeration can be used to describe the gas transfer:

$$dc/dt = K_{LA}(C^* - C_0)$$

where:

dc/dt = change in concentration per unit time, i.e. velocity of gas transfer
K_{LA} = diffusion coefficient
C^* = equilibruim concentration of the gas in the liquid
C_0 = concentration of gas in the liquid at the start point.

The equilibrium concentration C^* for oxygen dissolved in water standing in a pure oxygen atmosphere will be higher than for water in normal atmosphere (see Section 8.3.1, Henry's law). Therefore the velocity of gas transfer into the water is greater, because the difference $C^* - C$ is larger. Less time is then needed to increase the concentration of oxygen.

8.6.3 Addition under pressure

When using pure oxygen gas it is possible to increase the pressure and thereby increase the amount dissolved in the water. As distinct from adding air under pressure, there is no possibility for supersaturation of nitrogen which is toxic. Here the Dalton and Henry–Dalton laws can be used to describe what is happening:

$$p_i = p_t y$$

where:

p_i = partial pressure of gas i
p_t = total pressure of a mixture of gases
y = mol fraction of gas i (mol gas i/mol total gases).

By increasing the total pressure in the oxygen gas above the water surface, the partial pressure also increases.

When combining the Henry and Dalton laws the following is obtained:

$$p_t y = Hx_i$$

where:

y = mol gas i
p_t = total gas pressure
H = Henry's constant
x_i = amount of gas i in the liquid.

The oxygen concentrations in the liquid (x_i) is of interest and will be increased by increasing the total pressure (p_t) of the oxygen gas.

8.7 Design and construction of oxygen injection systems

8.7.1 Basic principles

In the injection system the oxygen gas comes from the source and is injected or mixed into the water. To get as much gas as possible into the water, a special injection system is necessary. Oxygen gas can be mixed into water under normal or high pressure. With high-pressure oxygenation, the water is pressurized up to 4 bar (see conversion factor box below) before the oxygen is added, and more than 500% supersaturation can be achieved without problems. When using normal pressure or low-pressure oxygenation (up to 1 bar), 100–300% saturation can be achieved. Normal or low-pressure oxygenation is used on the main inlet pipe to the farm, while high-pressure oxygenation will normally be carried out in a part flow divided from the main flow because it is too costly to pressurize the main water inlet. Furthermore, it is easy to get a concentration of oxygen higher than either needed or recommended. Too high an oxygen concentration may damage the gills of the fish. When the water enters the fish tanks, it is not recommended to be above 150–200% oxygen saturation, and even these concentrations have been called into question.

The same design principles as for aerators can be used for the injection system. Either oxygen gas bubbles can be supplied into water or water droplets can be supplied into an atmosphere of pure oxygen.

Conversion factors for pressure units are as follows:

- 1 bar = 1×10^5 Pa = 10.19 mH_2O = 0.9869 atm.
- 1 Pa = 1×10^{-5} bar = 1.02×10^{-4} mH_2O = 9.861×10^{-6} atm
- 1 mH_2O = 9806.65 Pa = 0.098 bar = 0.097 atm
- 1 atm = 101 325 Pa = 1.01325 bar = 10.33 mH_2O

8.7.2 Where to install the injection system

On a land-based fish farm oxygen gas can be added to the water in several places (Fig. 8.8):

(1) To the main water inlet pipe to the farm
(2) To a part water flow separated from the main inlet flow
(3) Directly to the water in the fish tank
(4) To an individual circuit where the water is supersaturated with oxygen.

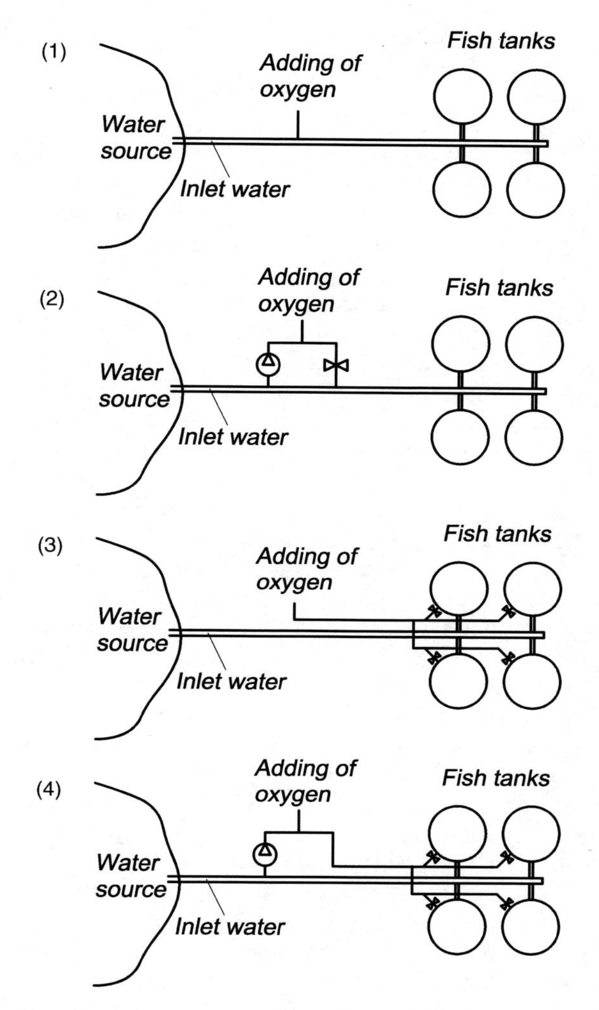

Figure 8.8 Oxygen can be injected: 1, into the main water inlet pipe to the farm; 2, into a part water flow separated from the main inlet flow; 3, directly into the water in the fish tank; 4, into an individual circuit until the water is supersaturated with oxygen.

When adding oxygen to the main flow, the gas is added directly into the main pipeline to the farm. It is important to be aware that the oxygen gas added to the water needs space so the water flow through the pipe will be reduced, and this must be taken into consideration when specifying the inlet pipes. With this method it is difficult to increase the pressure in the water significantly because the equipment required to increase the pressure in the entire water supply will be very large and costly.

Oxygenation of part of the water flow is carried out by diverting some of the water from the main pipeline into a smaller pipe. The pressure in this smaller pipe can and is normally increased, before oxygen gas is added by means of a pump. The pump is normally of the high-pressure centrifugal multi-stage type. As the pressure in the water is increased to 2–4 bar, a normal centrifugal pump is not powerful enough. After adding the oxygen gas, the pressure is reduced by a reduction valve; the water, which is now supersaturated with oxygen, is piped into the main flow.

In the third method oxygen is supplied directly to the fish tank either by direct addition via the inlet pipe to the tank or through diffusers at the bottom of the tank. The advantage with this method is that individual oxygenation of the separate fish tanks is possible; hence it is also possible to improve oxygen utilization because only the amount necessary for the fish in the tank is added. By having an oxygen meter in the outlet of the tank an acceptable value can be set (for instance 7 mg/l) and a simple control system can be used to add the necessary oxygen. If the temperature increases or the fish grow faster, more oxygen will be added to avoid the oxygen level in the outlet dropping. Another advantage achieved by using diffusers directly at the bottom of the fish tank is that saturation in the tank will always be below 100%. It will therefore be easier to get the supplied oxygen into the water, because it is not being supplied against a large oxygen gradient as is the case when oxygenating supersaturated water (refer to K_{LA} value).

When adding oxygen to a closed circuit, part of the inlet water to the farm is taken out and supersaturated with oxygen. This supersaturated water is then piped directly to each of the fish tanks on the farm through separate pipelines. The fish tank has two inlet pipes, one delivering aerated water at up to 100% oxygen saturation and another delivering water supersaturated with oxygen. Because it is possible to adjust the oxygen supply to the various tanks on the farm individually, good efficiency can be achieved. The major disadvantage of this method is the high cost of the additional piping needed.

Oxygenation may be carried out both in freshwater and seawater. It is simpler to add oxygen to seawater because of its higher ion concentration. Oxygen microbubbles will be smaller in seawater than in freshwater because the surface tension is reduced.[15] Small bubbles have a reduced rise velocity and so the contact time will increase. In addition, the relative surface area is higher for small bubbles than for large ones. Hence the efficiency of mixing, and thus the necessary addition pressure, varies with the salt concentration.

8.7.3 Evaluation of methods for injecting oxygen gas

Many methods are available to inject oxygen into water, using both low and high pressure. To be able to evaluate the performance, evaluation methods are available for aerators. As pure oxygen is much more expensive than air, its utilization is very important. This is given by the oxygen transfer efficiency (OTE), which is also known as the absorption efficiency[6] and is the amount of oxygen dissolved in water in relation to the amount of oxygen supplied:

$$\text{Absorption efficiency} = \frac{(C_{in} - C_{out})Q_w}{m_{O_2}}$$

where:

C_{in} = concentration of oxygen before injection of oxygen
C_{out} = concentration of oxygen after injection of oxygen
Q_w = water flow
m_{O_2} = mass flow of oxygen gas.

Performance values ranging from a few percentage points up to 100% have been reported.[6]

Another important parameter is oxygenation efficiency (OE) which is a measure of the amount of oxygen dissolved in relation to the power supplied. Normally excess pressure is used to achieve high oxygen concentrations in the water and for this power is necessary, for example to run a high-pressure pump. This can be described as follows:

$$\text{Oxygenation efficiency} = \frac{(C_{in} - C_{out})Q_w}{\text{Power supply}}$$

Units for OE are mg O_2/kW.

Most important of course, is the amount of oxygen dissolved in the water in relation to the investment and running costs for the injection equipment; this is measured in mg O_2 per unit cost.

It is important to remember that it is the amount of oxygen available for the fish in the production unit that is of interest. For example, the absorption efficiency directly after the injection unit, which may be very high, is of no concern; neither are values that are achieved in laboratories where all other factors are optimized in a way that is impossible to achieve under real farming conditions. Several design factors for the whole farm will have a great influence on the efficiency of the oxygenation system; these include the water transfer system and tank inlet design. No two farms will be exactly the same and this must be taken into account. Comparison of systems is therefore not straightforward; tests must be performed under exactly the same conditions and if possible on the site where the equipment is to be used.

8.7.4 Examples of oxygen injection system designs

A number of systems for injection of oxygen gas are available.[6] What distinguishes the methods from aeration is that the gas is injected under pressure, so the equipment is pressurized. Injection systems can be divided into high- and low-pressure systems, where the pressure in the former is above 1 bar.

Low pressure

Packed column: A simple way to add oxygen to the water is to substitute the air in a packed column aerator by pure oxygen, and aerate in a pure oxygen atmosphere (Fig. 8.9); there is no excess pressure. Under practical conditions, an oxygen saturation of two to three times the normal value is achievable by changing the atmosphere from air to pure oxygen. Theoretically the oxygen content should be up to five times as high as in air, but the time available for the gas to transfer into the water is normally a limitation. The air in a column aerator can easily be exchanged with pure oxygen by placing the lower end of the column in the water and feeding pure oxygen gas into the lower part of the column so that it flows up against the descending water. This results in good mixing of the water and oxygen. Eventually surplus oxygen that reaches the top of the column and is not dissolved in the water can be withdrawn through a special valve and recycled through the column inlet water to prevent any loss of pure oxygen gas. The amount of oxygen that can be transferred into the water and the efficiency of oxygen supply can be increased if the column is pressurized; here the column is closed at the top and bottom. The pressure in such a system normally varies from 0.2 to 0.5 bar excess pressure.

Diffuser: A diffuser may also be used to supply oxygen gas to the water, in the same way that air is supplied for aeration (Fig. 8.9). The efficiency depends on the depth at which the oxygen is added; a greater depth will result in increased efficiency (see Section 8.3.1, Dalton's law), a depth of at least 2 m being recommended. The diffusers can be either tubes or ceramic. It is important it that the oxygen gas leaves the diffuser as small bubbles, because then better mixing is achieved as a result of increased contact area and reduced rise velocity.[15,23] Usually diffusers are used to add oxygen to the production units in connection with transport or as security systems. This is normal procedure almost all fish farms and for all fish transport. When using diffusers, the oxygen is normally supplied from gas bottles because the amounts are quite small. Diffusers are also used in normal production; one method is to have some central oxygenation of the incoming water and treat each tank individually by having a diffuser on the tank bottom.

Diffuser placed directly in the inlet pipe: If oxygen is to be added to the fish tank directly through the inlet pipe, a diffuser is placed inside the inlet pipe after the regulating valve (Fig. 8.9). Such an oxygenation system requires a correct design of inlet pipe to the tank (Chapter 11). The water depth in the tank ought to be more than 1.5 m to give satisfactory pressure. The highest efficiency is seen when using this method in seawater because oxygen is more soluble in seawater than in freshwater. With a direct supply via the inlet pipe the greatest advantage is that single tanks can easily be supplied with oxygen individually and a better total utilization of the added oxygen achieved.

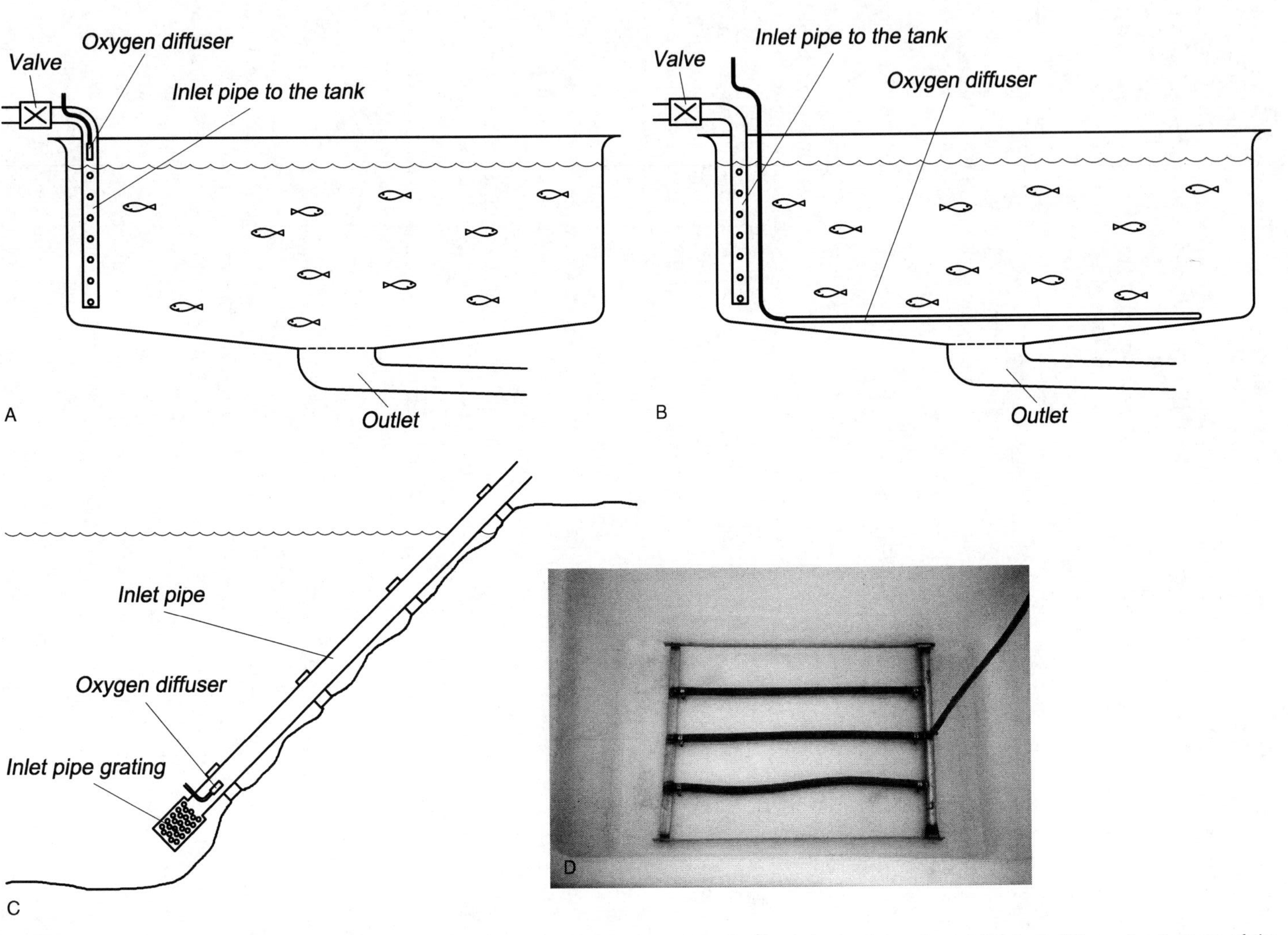

Figure 8.9(A–D) Methods for low pressure injection of pure oxygen gas: oxygen diffuser in the inlet pipe of the tank (A), on the bottom of the tank (B and D), in the inlet pipe to the farm (C).

High pressure

There are various methods utilized for injecting oxygen gas into the water under high pressure. Two of the most commonly used are the oxygen cone and deep wells, while others employ different types of injector.

Oxygen cone: In the oxygen cone the oxygen gas and water enter at the top through different pipes (Fig. 8.10). The oxygen will be pushed down in the cone because of the water flow and a large oxygen bubble, actually a layer of oxygen gas, will be created lower down in the cone. The water will flow in small drops through this layer, creating a situation with water drops flowing through a pure oxygen atmosphere. Since the pressure inside the cone can be up to 5 bar, good gas transfer is achieved and because of this quite good efficiency. After flowing through the oxygen layer, the oxygenated water will flow out, from the bottom of the cone. However, if the cone is 'overloaded' with water, the oxygen layer will be pushed down to the lower end of the cone, perhaps also through the cone outlet and efficiency is reduced. Because of the high pressure used inside the cone to transfer oxygen gas into the water, the pressure of the water must be increased before it enters the cone. To do this, part of the water flow is piped from the main pipeline, and a high-pressure centrifugal pump used to increase the pressure in this part flow before it enters the cone. After leaving the cone, having been supersaturated with oxygen, the water is returned to the main flow. However, the water pressure must be reduced by passage through a pressure reduction valve before entering the main flow again.

Use of oxygen wells: Oxygenation in wells is sometimes used and gives quite good efficiencies (70–90%) (Fig. 8.10). Well design varies, but the main purpose is to mix oxygen and water at quite large depths where the high pressure together with effective mixing ensures good efficiency. One method that utilizes this principle is the U tube. Oxygen gas in injected into the water before it flows into the U tube. At the bottom of the U the pressure will be increased depending on the depth of the U; 10–30 m is usual. The construction ensures effective mixing by creating turbulence between the oxygen gas and the water and the gas bubbles now dissolve. The same principle is used when oxygen is added directly into a deep-water intake. In land-based fish farms in Norway this system has been used with direct addition of oxygen to the inlet pipe at depths of 20–30 m. Here the pressure is high and good transfer of the oxygen gas into the water is achieved.

When using such methods it is important to be aware that the water flow rate through the pipe will be reduced because the oxygen gas takes up space and the real cross-sectional area where the water flows is reduced. If the water is sent to a pump after injection of oxygen, care must be taken because the conditions are now optimal for cavitation with gas bubbles that may implode as the water passes through the pump.

Oxygenation in sea cages: Recent research has shown that lack of oxygen may occur in sea cages, possibly as the result of algal blooms in the area where the cages lie. The oxygen concentration may be reduced to below 70% saturation at the end of dark nights because both fish and algae use oxygen during the night. The conditions can be so bad that the effects of current induced water exchange cannot meet the reduced oxygen concentration caused by the algal bloom.

Lack of oxygen may also be caused by high temperatures and high fish densities in sea cages. The oxygen consumption of the fish increases with temperature and at the same time the oxygen content of the water is reduced; in such cases the lack of oxygen will be local to the cage. However, if the water current increases, more oxygen-rich water can be transported to the cage; a propeller may therefore be used to create an artificial current to increase the supply of oxygen-rich water.

It is difficult to monitor the oxygen concentration in a sea cage in a representative way, because the cage can be large and there may be individual variations within the cage, mainly due to the depth. Several oxygen meters could therefore be advantageous.

When pure oxygen gas is to be added to the cage, oxygen diffusers can be lowered into the cage. It is very advantageous to create an equal pressure on the diffuser, so that an equal amount of oxygen gas is released all around the cage. As it is important to

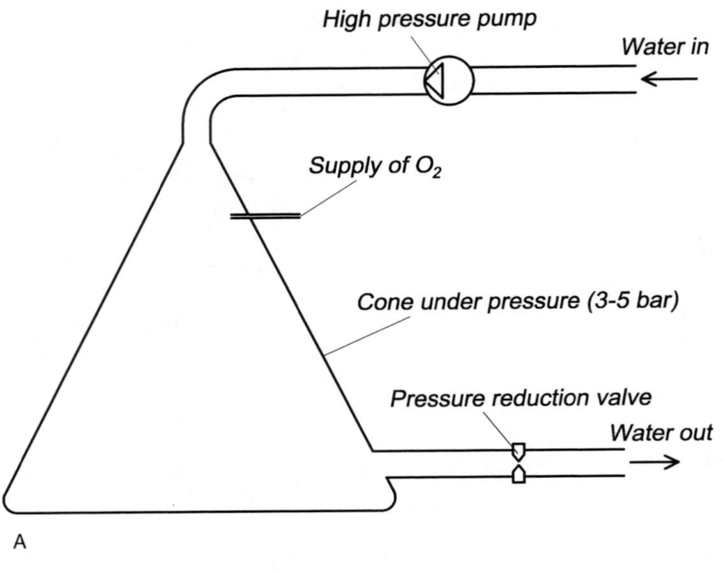

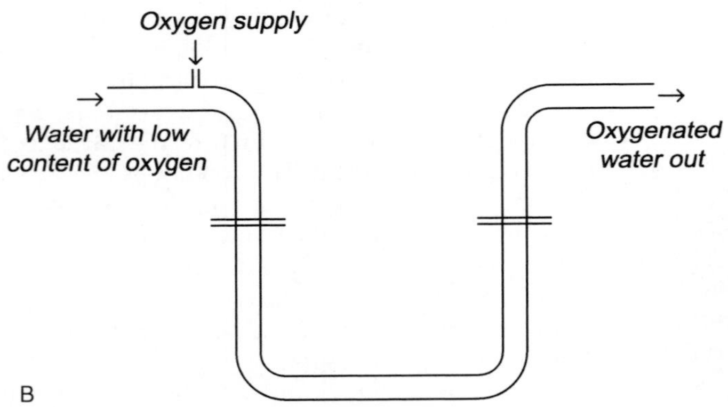

Figure 8.10 Oxygen cones and oxygen wells are typical systems for injecting oxygen gas into water under high pressure: (A) oxygen cone cross-section; (B) oxygen well cross-section. (C) cones *in situ*; (D) locked column.

get an even distribution of the added oxygen, specially designed diffusers are recommended.

8.8 Oxygen gas characteristics

Oxygen gas (O_2) has a boiling point of –183°C at normal atmospheric pressure. If the pressure is increased, the boiling point will also increase. At normal temperature O_2 is therefore only available in gas form. Liquid oxygen is light blue without odour; the density of the liquid is 1.15×10^3 kg/m^3 at boiling point, while the density of the gas is 1.36 kg/m^3 at 15°C and normal atmospheric pressure. One litre of liquid oxygen will therefore expand 820 times when changing phase to gas at 15°C.

Liquids with boiling point below –100°C are commonly called cryogenic liquids. Handling cryogenic liquids involves certain elements of risk:

- Extremely low temperature can result in frost injures on the skin when handling; the cold gas may also cause internal injuries
- Materials that come into contact with the liquid easily become brittle
- There is a danger of explosion when uncontrolled transformation from liquid to gas is allowed to occur in closed volumes.

By itself, O_2 is inflammable but oxygen supports a fire and is normally the limiting factor for fire development. If much oxygen is available in the atmosphere, only a small spark can create an explosive fire. Care should therefore be taken when handling oxygen gas.

8.9 Sources of oxygen

Oxygen gas may be delivered from oxygen producers, either as compressed gas in bottles or as liquid oxygen, which the farms transforms to gas under controlled conditions. This oxygen is produced in special factories that produce only oxygen. Whether gas or liquid, commercially produced oxygen is characterized by high purity (>99%) and the presence of a small amount of other gases.

The other alternative is on-site production of oxygen, which means that the oxygen is produced at the farm with air as the source.

8.9.1 Oxygen gas

Oxygen gas may be delivered as compressed gas in bottles. In Norway the bottles are blue and can be delivered as single bottles or in batteries, normally of 12 (Fig. 8.11). The usual pressure inside an oxygen gas bottle is as high as 200 bar; the bottle contains 50 l of compressed gas at this pressure and this gives 10 000 l gas at normal atmospheric pressure. On the top of the bottle there is a pressure reduction valve. A manometer, also on the top of the bottle shows the internal pressure; this also

Figure 8.11 Compressed oxygen gas in high pressure bottles, either supplied singly or in a battery of 12.

indicates the amount of gas remaining in the bottle, because the pressure inside the bottle will decrease as oxygen is released. The bottles are normally rented from the oxygen producers, one reason being that the producer will ensure the bottles are properly maintained. The price of oxygen gas depends on the distance from the oxygen production site amongst other things. The pressurized bottles represent a real danger, for instance in connection with a fire. It is therefore advisable not to have too many bottles on the farm, and if possible to place them outside; legal restrictions may limit the number of gas bottles that can be stored inside buildings.

8.9.2 Liquid oxygen

Liquid oxygen (LOX) is produced in special factories and transported to the farms by truck or boat in special tanks. On arrival, the liquid oxygen is transferred to the farm's specially designed tank for storage (Fig. 8.12). The usual tank size for storing

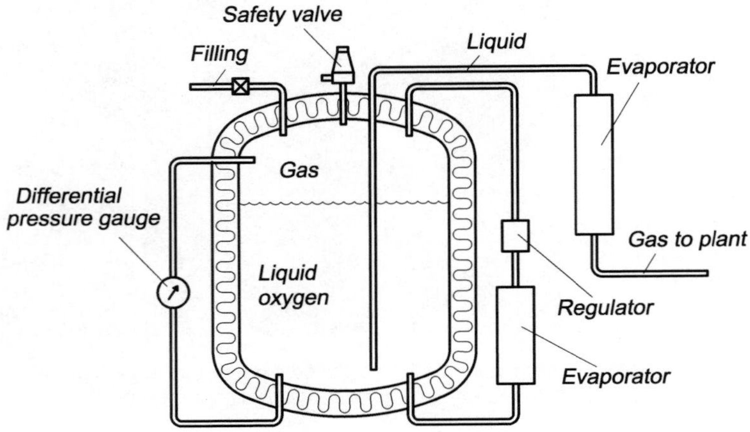

Figure 8.12 Oxygen can be stored as liquid in special tanks at the farm and a cold-gas evaporator ensures that the liquid oxygen is transformed to gas before entering the fish tanks. The schematic shows the principle; the photograph shows a tank *in situ*.

liquid oxygen is between 2 and 50 m^3. The oxygen is stored as liquid in the tank with a temperature below the boiling point and of a pressure of between 15 and 25 bar.

The pressure in the liquid oxygen tank will stay constant. An internal evaporator system, normally placed under the liquid tank, ensures this. Liquid oxygen is taken from the tank, goes through the evaporator where the liquid is transformed to gas, and back again as gas into the top of the tank. The expansion when oxygen goes from liquid to gas is utilized in this process and maintains the pressure in the tank. The tank is fitted with a safety valve to prevent it exploding, because even though the tank is constructed with two shells with space between them it is impossible to achieve 100% effective isolation of the tank. Therefore some liquid will always evaporate to gas inside the tank. With the large expansion in volume when changing phase from liquid to gas, the tank will easily blow if nothing is done, which is where the safety valve comes into play. If the daily use of oxygen from the tank is too low, oxygen gas will pass through the safety valve and into the atmosphere. Normally 0.1–0.6% of the tank volume must be used daily, depending on tank size, to avoid loss of oxygen through the safety valve to the atmosphere. This also shows the importance of choosing a tank of the correct size.

When the farm uses oxygen from the tank, it is drained as liquid from a pipe near the bottom of the tank. Because of the pressure in the tank, the liquid oxygen will be pressed out through the outlet pipe and into the evaporator when the outlet valve is opened. In the evaporator liquid oxygen changes phase to gas. The gas is transported from the evaporator through a pipe and into the farm's oxygen injection system.

The evaporator is known as a cold gas evaporator and must be designed according to the farm's oxygen consumption. It functions as a heat exchanger transferring heat from the air into the oxygen liquid (air–liquid exchange). It is important to have a large surface area to get a large contact area between the air and the oxygen. Usually at least two evaporators are installed, one runs while the other is switched off. The evaporator takes energy from the air and after a while is totally covered with ice; deicing is therefore necessary and is the reason for at least two evaporators being required.

The liquid oxygen tank should be placed outside on a concrete base and be fenced in. In addition, there should be a safety zone around the tank. Oxygen tanks are rented from the oxygen suppliers for safety reasons. With online measurement of the remaining oxygen content in the tank by the supplier, it is also possible for the supplier to automatically replenish the oxygen before the tank is empty. It is then not necessary for the fish farmer to keep control of this.

8.9.3 On-site oxygen production

The alternative to buying oxygen from a producer is to have on-site production facilities (Fig. 8.13). Air contains 78% nitrogen, 21% oxygen, 0.9% argon and 0.1% of other gases by volume (Table 8.1). In a specially designed adsorption unit it is possible to remove the nitrogen and some of the other gases from the air, leaving mainly oxygen. The absorption unit is fed by compressed air from a compressor; different names are used (oxygen generator or xorbox) according to the producer. Put simply, it is a unit that filters air and lets only oxygen pass. The process is known as pressure swing adsorption (PSA).

The compressed air passes into an absorption unit that is normally filled with clinoptilolite, a natural zeolite which is a clay mineral. To describe the construction and function in detail is beyond the scope of this book but a brief and simplified description is as follows. The crystals in the filter material form a latticework where the nitrogen molecules are trapped. Oxygen molecules, which are large, do not fit into the latticework and pass through; almost pure oxygen is produced. Provided that the filter medium is not contaminated with oil or water from the compressor, it is quite durable. An inlet air filter must be used and cleaned at regular intervals. A normal generator used for aquaculture can reach a purity of 95% oxygen, the rest being argon (4%) and nitrogen (1%). Special units may achieve 99% oxygen purity, but they are very expensive and not viable for on-site installation in fish farming. The efficiency of the filter gradually decreases. After a period the filter medium, will become saturated with nitrogen. Nitrogen gas will then follow the oxygen gas out of the generator and to the fish. For this reason two columns are always used in alternation: one is used and the other is regenerated. When a column is regenerated, air

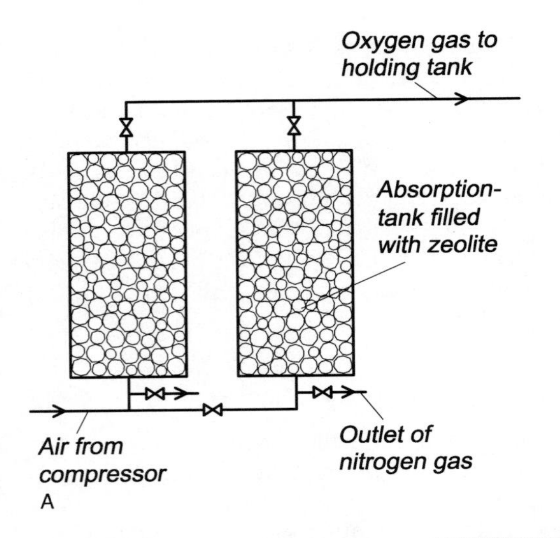

Figure 8.13 On-site production of oxygen from air using an oxygen generator. Principle (A), the air production unit and the two absorption units (B) in a cabinet shown with a bottle of compressed oxygen as security (C).

is 'driven' through the column in the opposite direction, and the nitrogen is blown out to the atmosphere.

An oxygen generator normally delivers oxygen gas with a maximum pressure of 4 bar. This must be taken into consideration because some of the equipment for injecting oxygen gas into the water may require higher pressure. If so, a specially designed oxygen compressor must be used after the oxygen generator. After the generator there must be a storage tank for the oxygen produced to ensure pressure equalization. From this tank the oxygen is tapped and used for injection into the water on the farm.

When a generator has been in use for some years (depending on type of equipment and running time) the purity of the delivered oxygen will fall to 80–85%. The amount of nitrogen gas will also increase, which can be toxic to the fish. If the oxygen gas is supplied under high pressure, there will be possibilities for supersaturation of nitrogen. Maintenance and replacement or cleaning of the filter medium inside the columns must then be carried out. It is therefore important to sample the oxygen gas from the generators at fixed intervals and measure its purity. An easy way to do this is to collect a sample of the oxygen gas in a plastic bag. The oxygen meter that is used to measure the content of oxygen in the water on the farm can then be used to measure the approximate oxygen purity. This is done by first adjusting the meter to show 21% purity in air; now when the oxygen probe is put into the plastic bag, it will show the purity of the oxygen gas sample directly, for instance 95%.

8.9.4 Selection of source

The choice of oxygen source depends on several factors. If the oxygen consumption is very low or if oxygen is only used for security purposes, it is normal to use gas bottles. Otherwise there is a choice between purchasing liquid oxygen and on-site production. If only the price of oxygen is considered, on-site production will in almost every case be cheaper. The reliability, maintenance costs and necessary back-up system, however, often cause farmers to buy liquid oxygen from commercial producers because they guarantee the oxygen supply. Other important factors when choosing are the distance to the oxygen production factory, infrastructure for transport of liquid oxygen, and the price of electricity. The amount of electric energy used for production of oxygen gas is size-dependent; larger generators are more efficient than smaller ones.

Appendix 8.1

Solubility of oxygen in water with different temperature and salinity with normal atmospheric pressure (1013 mbar)

	Salinity, parts per thousand (ppt)				
Temperature (°C)	0	10	20	30	40
0	14.6	13.6	12.7	11.9	11.1
2	13.8	12.9	12.1	11.3	10.5
4	13.1	12.3	11.5	10.7	10
6	12.4	11.6	10.9	10.2	9.6
8	11.8	11.1	10.4	9.8	9.1
10	11.3	10.6	9.9	9.3	8.7
12	10.8	10.1	9.5	8.9	8.4
14	10.3	9.7	9.1	8.6	8.0
16	9.9	9.3	8.7	8.2	7.7
18	9.6	8.9	8.4	7.9	7.4
20	9.1	8.6	8.1	7.6	7.2
22	8.7	8.2	7.8	7.3	6.9
24	8.4	7.9	7.5	7.1	6.7
26	8.1	7.7	7.2	6.8	6.5
28	7.8	7.4	7.0	6.6	6.3
30	7.6	7.2	6.8	6.4	6.1

Appendix 8.2

Bunsen's coefficient for oxygen as a function of temperature and salinity

	Salinity, parts per thousand (ppt)				
Temperature (°C)	0	10	20	30	40
0	0.049	0.046	0.043	0.040	0.037
2	0.047	0.043	0.041	0.038	0.036
4	0.044	0.041	0.039	0.036	0.034
6	0.042	0.039	0.037	0.034	0.032
8	0.040	0.037	0.035	0.033	0.031
10	0.038	0.036	0.034	0.032	0.030
12	0.036	0.034	0.032	0.030	0.028
14	0.035	0.033	0.031	0.029	0.027
16	0.034	0.032	0.030	0.028	0.026
18	0.032	0.030	0.029	0.027	0.025
20	0.031	0.029	0.028	0.026	0.025
22	0.030	0.028	0.027	0.025	0.024
24	0.029	0.027	0.026	0.024	0.023
26	0.028	0.026	0.025	0.024	0.022
28	0.027	0.026	0.024	0.023	0.022
30	0.026	0.025	0.024	0.022	0.021

Appendices based on:
Benson, B.B., Krause, D. (1984) The concentration and isotopic fraction of oxygen dissolved in freshwater and seawater in equilibrium with the atmosphere. *Limnology and Oceanography*, 29: 620–632.
Colt. J. (1984) *Computation of dissolved gas concentration in water as a function of temperature, salinity and pressure*. American Fishery Society. Special Publication No 14.
Weiss. R.F. (1970) The solubility of nitrogen, oxygen and argon in water and seawater. *Deep-sea Research*, 17: 721–735.

References

1. Colt, J., Orwicz, K. (1991) Aeration in intensive aquaculture. In: *Aquaculture and water quality* (eds D.E. Brune & J.R. Thomasso). World Aquaculture Society, Louisiana State University.
2. Fivelstad, S., Haavik, H., Løvik, G., Olsen, A.B. (1997) Sublethal effects and safe levels of carbon dioxide in seawater for Atlantic salmon postsmolts (*Salmo salar* L.): ion regulation and growth. *Aquaculture*, 160: 305–316.
3. Tucker, J.W. (1998) *Marine fish culture.* Kluwer Academic Publishers.
4. Bergheim, A., Gausen, M., Næss, A., Krogedal, P., Hølland, P., Crampton, V. (2006) A newly developed oxygen injection system for cage farms. *Aquacultural Engineering*, 34: 40–46.
5. Colt, J. (1986) Gas supersaturation – Impact on the design and operation of aquatic systems. *Aquacultural Engineering*, 5: 49–85.
6. Colt, J., Watten, B. (1988) Application of pure oxygen in fish culture. *Aquacultural Engineering*, 7: 397–441.
7. Colt, J. (1984) *Computation of dissolved gas concentration in water as a function of temperature, salinity and pressure.* American Fisheries Society Special Publication 14.
8. Gebauer, R., Eggen, G., Hansen, E., og Eikebrokk, B. (1992) *Oppdrettsteknologi – vannkvalitet og vannbehandling i lukkede oppdrettsanlegg*. Tapir forlag (in Norwegian).
9. Mackay, D., Shiu, W.U. (1984) Physical-chemical phenomena and molecular properties. In: *Gas transfer at water surfaces*. (eds W. Brutsaert, G.H. Jirka), D. Reidel.
10. Lincoff, A.H., Gossett, J.M. (1984) The determination of Henry's constant for volatile organics by equilibrium partitioning in closed systems. In: *Gas transfer at water surfaces* (eds W. Brutsaert, G.H. Jirka), D. Reidel.
11. Tchobanoglous, G., Burton, F.L., Stensel, D.H. (2002) *Wastewater engineering*. McGraw-Hill.
12. Boyd, C.E., Watten, B.J. (1989) Aeration systems in aquaculture. *Reviews in Aquatic Science*, 1: 425–472.
13. Colt, J. (2000) Aeration systems. In: *Encyclopedia of Aquaculture* (ed. R.R. Stickney). John Wiley & Sons.
14. Stenstrom, M.K., Gilbert, R.G. (1981) Effects of alpha, beta and theta factor upon the design, specification and operation of aeration systems. *Water Research*, 15: 643–654.
15. Boyd, C.E., Tucker, C.S. (1998) *Pond aquaculture water quality management*. Kluwer Academic Publishers.
16. Shelton, J.L., Boyd, C.E. (1983) Correction factors for calculation of oxygen-transfer rates of pond aerators. *Transaction of the American Fishery Society*, 112: 120–122.
17. Lawsons, T.B. (2002) *Fundamentals of aquacultural engineering*. Kluwer Academic Publishers.
18. Wheaton, F.W. (1977) *Aquacultural enginering*. R. Krieger Publishing Company.
19. Huguenin, J.E., Colt, J. (2002) *Design and operating guide for aquaculture seawater systems*. Elsevier Science.
20. Bouck, G.R., King, R.E., Bouck-Scmidt, G. (1984) Comparative removal of gas supersaturation by plunges, screens and packed columns. *Aquacultural Engineering*, 3: 159–176.
21. Summerfelt, S.T. (2000) Carbon dioxide. In: *Encyclopedia of aquaculture* (ed. R.R. Stickney). John Wiley & Sons.
22. Summerfelt, S.T., Davidson, J. Waldorp, T. (2003) Evaluation of full-scale carbon dioxide stripping columns in coldwater recirculation systems. *Aquacultural Engineering*, 28: 155–169.
23. Lunde, T. (1987) Oksygenering. In: *Vannrensing ved akvakulturanlegg*. NIF kurs (in Norwegian).
24. Boyd, C.E. (1998) Pond water aeration systems. *Aquacultural Engineering*, 18: 9–40.
25. Cancino, B., Rothe, P., Reuss, M. (2004) Design of high efficiency surface aerators. *Aquacultural Engineering*, 31: 83–98.

9
Ammonia Removal

9.1 Introduction

In aquaculture it is often necessary to reduce the concentration of ammonia in the water, because it is toxic for the fish.[1–4] It is particularly important when re-using water (Chapter 10) or when transporting fish long distances without changing the water, since the fish produce ammonium compounds as a metabolic waste product.

In water there is an equilibrium between the concentrations of ammonium ion (NH_4^+) and ammonia (NH_3) (Fig. 9.1):

$$NH_3 + H^+ \Leftrightarrow NH_4^+$$

This equilibrium depends on pH. The sum of NH_3 and NH_4^+ is known as the total ammonia nitrogen (TAN). Because of the equilibrium between NH_3 and NH_4^+, reducing one of them automatically reduces the other. NH_3 is the more toxic substance for fish, and therefore is the substance of interest.

This chapter gives an overview of the most common methods for ammonia nitrogen removal in fish farming, including biological and chemical processes. There are also a number of other possible methods for removal of ammonia, such as ammonia stripping, break-point chlorination, membrane filtration and addition of ozone, but these methods are not commonly used in aquaculture. Much information is available about this subject in the general literature for municipal wastewater treatment.[5,6] Removal of nitrogen from municipal wastewater has become more and more important during the past few years. In aquaculture the subject has also become more important due to the increase in water re-use systems.[7–9]

9.2 Biological removal of ammonium ion

When using a biological filter, bacteria are used to oxidize ammonium to nitrite and nitrate, and perhaps further to molecular nitrogen. The bacteria are grown in a biofilm. Three processes are included in the biological removal of ammonium:

- Transfer of NH_4^+ (ammonium ion) to NO_2^- (nitrite)
- Transfer of NO_2^- (nitrite) to NO_3^- (nitrate)
- Transfer of NO_3^- to N_2 (molecular nitrogen)

The two first processes are carried out simultaneously and are known as nitrification; the process is performed in a nitrification filter. The third process is denitrification and is performed in a denitrification filter. The two first are aerobic, so air must be added. The last process is anaerobic so air must be removed from the water. Two different filters involving different bacteria are therefore used. In most cases only the nitrification process will be needed for aquaculture purposes, because fish have a higher tolerance for nitrate than for ammonia. Very high degrees of water re-use and high fish densities might require a denitrification filter, but knowledge of denitrifiaction filter function and optimization for fish farming is scant.

9.3 Nitrification

The nitrification process is carried out in two steps and is performed by bacteria which oxidize ammonia. These bacteria are autotrophic and use O_2 as oxidizing agent and CO_2 or HCO_3^- as a carbon source for growth. NH_4^+ is transformed

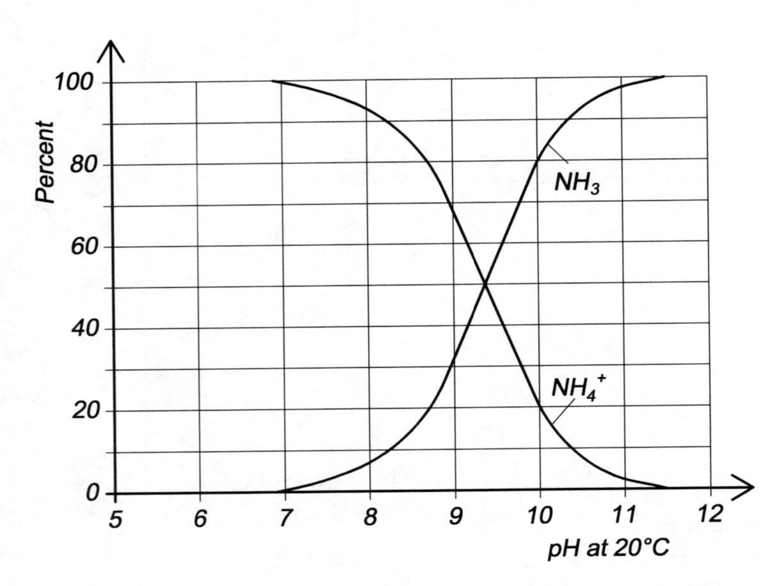

Figure 9.1 The equilibrium between the concentrations of ammonium (NH_4^+) and ammonia (NH_3) is pH dependent.

to NO_2^- by *Nitrosomonas* bacteria and then to NO_3^- by *Nitrobacter bacteria.* Both of these processes require energy that is supplied by the substrate. The chemical processes involved are as follows:

$$NH_4^+ + \frac{3}{2}O_2 \rightarrow NO_2^- + 2H^+ + H_2O$$

$$NO_2^- + \frac{1}{2}O_2 \rightarrow NO_3^-$$

The bacteria grow in the biofilm on the filter medium. Nitrification takes place in this film so the biofilm must be established for the nitrification filter to function. The process creates more biofilm as the bacteria grow and divide, and the cell mass increases. This cell mass can be described as $C_5H_7NO_2$. The processes including creation of cell mass can then be described as follows.[10]

Step 1 using *Nitrosomonas* bacteria:

$$55NH_4^+ + 5CO_2 + 76O_2 \rightarrow C_5H_7NO_2 + 54NO_2^- + 109H^+ + 52H_2O$$

Step 2 using *Nitrobacter* bacteria:

$$400NO_2 + 5CO_2 + 76O_2 \rightarrow C_5H_7NO_2 + 400NO_3^- + H^+$$

The effectiveness of a nitrification filter can be described by the nitrification rate, defined as the amount of ammonium oxidized per unit biofilm surface area and unit time (mg $NH_4^+/(m^2$ min)). The efficiency of the nitrification process and the establishment of the biofilm process depend on several factors.[11] It is important that the bacteria grow as optimally as possible.

The following important factors regulates the growth of the bacterial culture:

- Concentration of ammonia
- Temperature
- Oxygen concentration
- pH
- Salinity
- Organic substances
- Toxic substances.

One of the main factors affecting bacterial growth is the amount of ammonia in the water. In fish farming this concentration is normally too low for maximal growth, compared to municipal wastewater, for instance, where the same processes is used. Generally values above 3 mg ammonium nitrogen per litre are recommended for maximal growth.[12] Values that are too high may inhibit bacterial growth.

Since these bacterial cultures requires oxygen to grow, and in order to have transformation of ammonia to nitrate, it is important that sufficient oxygen is available throughout the entire nitrification process. Experiments have demonstrated a reduction in *Nitrosomonas* activity with oxygen levels below 4 mg/l water, while the corresponding value for *Nitrobacter* is 2 mg/l.[10]

Bacterial growth rate depends on the temperature.[13–15] Bacterial activity occurs from 0 to 30°C and increases with temperature: the optimal range is around 30°C. Temperatures that are too high will result in mortality. However, the bacteria may acclimate to lower temperatures over time, so high bacterial activity can also be achieved at lower temperatures. If the temperature reaches values below 5°C growth will be slow and it will be difficult and time-consuming to establish new bacterial cultures. It is difficult to give general values for the nitrification rate in relation to temperature, due to both acclimation and the number of other factors affecting bacterial metabolism (Table 9.1).

Nitrification also depends on the pH of the water: optimal values are between 8 and 9.[18] It was shown experimentally that the nitrification rate decreased by 90% when the pH was reduced from 7 to 6.[12] It is important to remember that hydrogen ions (H^+) are produced in the nitrification process and, depending on the buffering capacity of the water, may reduce the pH. If the loads on the biofilter are high, systems for adjusting the pH are necessary; addition of lime is an example.

The presence of organic matter can inhibit the functioning of the nitrification filter.[19–21] Other bacterial cultures may start to grow inside the filters in competition with nitrification bacteria. Heterotrophic aerobic bacteria may use the organic matter as a carbon source and replace the nitrification bacteria because they have a faster growth rate.[22] Nitrification will be reduced by increasing the carbon/nitrogen ratio;[23,24] a 60–70% reduction in nitrification rate was observed when increasing the chemical oxygen demand/nitrogen (COD/N) ratio from 0 to 3 for a substrate containing 10 mg TAN.[1,25]

Table 9.1 Large variations in nitrification rates are observed at different temperatures and in different experiments. The average of values from refs 13, 16 and 17 are shown.

Temperature (°C)	Nitrification rate (g NH_4/(m^2 day))
5	0.3
10	0.5
15	0.8
20	1.0

Some substances may be toxic to nitrification bacteria and may kill or inhibit the bacterial culture in the biofilter. These can include metal ions, organic substances or medicaments such as formaldehyde.[26,27]

Salinity affects the nitrification rate because the chloride ions inhibit bacterial growth. For this reason the nitrification is faster in freshwater than in seawater.[28,29]

Nitrification filters need to be shielded from light, because it may reduce nitrification.[11,30] Predators may also have a negative influence.

The problem about giving reliable values for the nitrification rate is, of course, that all the factors affect each other. For this reason it is also normal to incorporate high safeguards when designing nitrification filters.

9.4 Construction of nitrification filters

The main purpose when constructing a nitrification filter is to create a surface for optimal growth of the biofilm. Depending on the construction and the filter medium on which the biofilm is established, it is possible to distinguish four types of biological filter:

- Flow-through system
- Bioreactor
- Fluid bed/active sludge
- Granular filters/bead filters.

9.4.1 Flow-through system

The flow-through system may again be divided into three types depending on how the water flows through the filter medium (Fig. 9.2):

- Trickling filter
- Submerged up-flowing system
- Submerged down-flowing system.

In a trickling filter, the water trickles through the filter medium where the biofilm is established.[31–33] The filter medium is located above the surface of the water and it is very similar to a column aerator. Advantages of this filter type are its simple construction, that good natural aeration for the process is achieved and that it is impossible to block. The

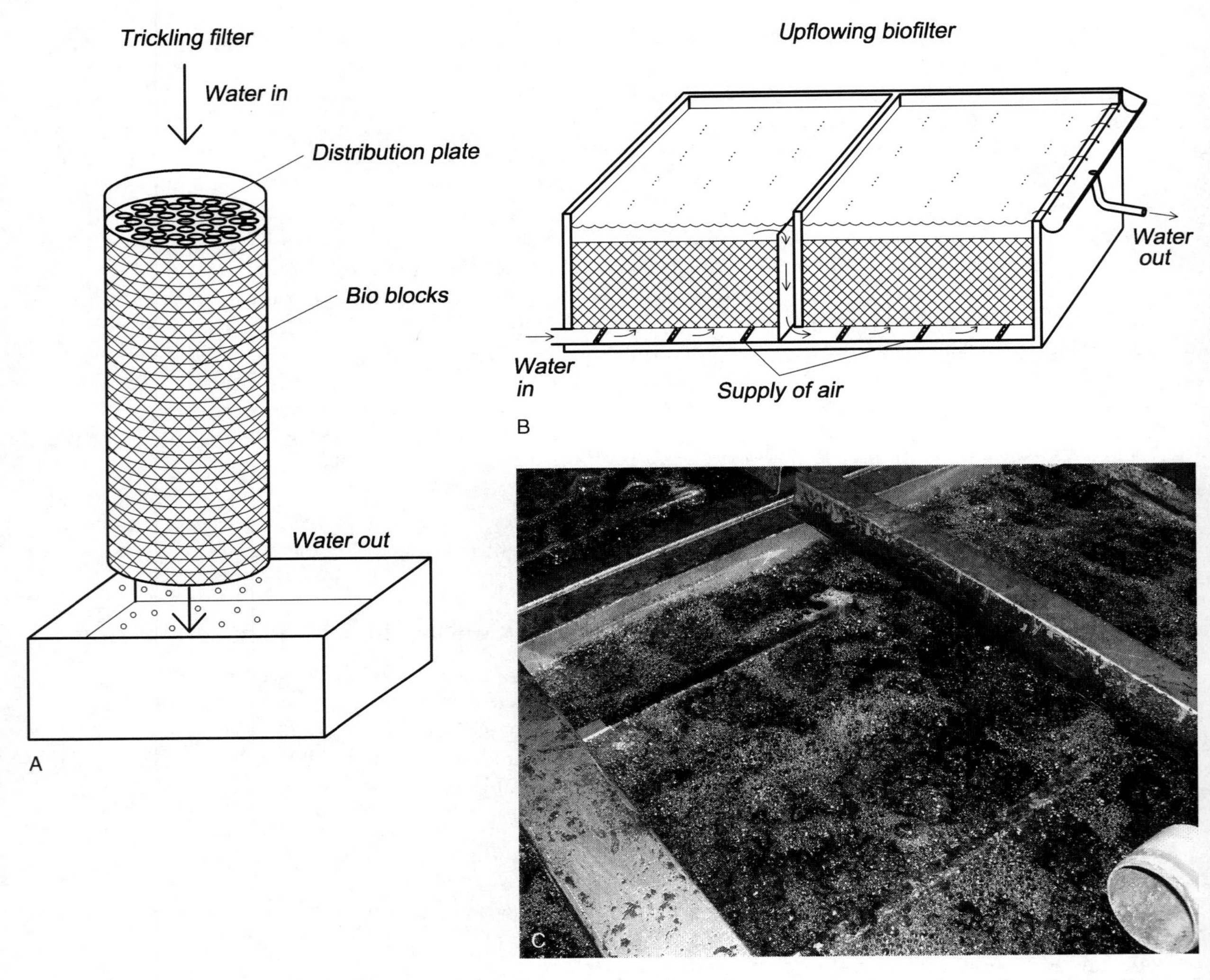

Figure 9.2 Different designs of biofilter with a fixed medium: (A) trickling filter and (B) submerged up-flowing filter; (C) up-flowing filter in use.

disadvantage is that it has a low nitrification capacity compared to other filter types.

In an up-flowing filter the water is supplied under the filter medium, which is submerged. The incoming water is forced through the filter medium by the flow and head of the water. There are several advantages with this system: it provides good medium structure, good distribution of the water in the total medium volume is possible, and there is good contact between the water and the filter medium where the biofilm grows. The disadvantage is that it is necessary to add oxygen or air with positive pressure, for example by using an air blower, below the filter medium to ensure that the bacteria in the biofilm get enough oxygen.

A down-flowing filter uses a similar technique to an up-flowing filter; here also the filter medium is submerged. The water is supplied at the top of the filter and is distributed through it. There is a 'water lock' at the bottom of the filter and this ensures the filter medium is always submerged. An advantage with this filter type is that the water flows in the opposite direction to the air bubbles, so quite good mixing of the air/oxygen and water is achieved. The disadvantage is the same as that for the down-flowing filter: it is necessary to supply oxygen or air; however, in addition, it is difficult to obtain the same good distribution of water in the total volume of the filter medium as in an up-flowing filter.

9.4.2 The filter medium in the biofilter

It is important that the surface of the filter material on which the biofilm grows is optimal. In fish farming where the ammonia concentration and concentration of organic matter are low compared to levels in municipal wastewater, it is most effective to use systems where a biofilm is established on an artificial surface.[17] The following requirements must be met by the filter medium:

- Have large specific surface areas where biofilm can be established (surface area per unit volume, m^2/m^3)
- Ensure proper contact between the water and the surface of the medium
- Create low head loss
- Be difficult to clog
- Ensure an even distribution of water in the entire filter medium
- Be simple to clean.

In practical farming conditions it is difficult to fulfill all these requirements. There is a conflict between having a nitrification rate as high as possible and the need for a simply operated and maintained system.

In the past gravel and sand were much used as filter media in biological filters. Today different types of plastic filter media have replaced sand and gravel in many applications[34] (Fig. 9.3). This is because they have a have large specific surface areas where biofilm can be established, and they are not so easily clogged. Leca (lightweight clay aggregate) is a material with very large specific surface are that has been used to create surface for biofilm with good results.[35] However, there are problems with clogging when the grain size is small and water flow large.

9.4.3 Rotating biofilter (biodrum)

A rotating biofilter, also known as a biodrum or rotating biological contactor (RBC), utilizes the same basic principle as the submerged filter (Fig. 9.4). The filter medium where the biofilm is established rotates at 2–3 rpm, partially above the water surface and partly submerged. The oxygen necessary for nitrification is supplied when the medium is above the water surface. Two different designs of rotating biofilter are used: (1) a cylinder filled with biobodies (filter medium in small elements); (2) parallel discs made of plates where the biofilm grows on the surfaces. In both cases the biofilter is mounted on a rotating shaft.

The advantage of this type of filter is that as a result of the rotation aeration occurs when the filter medium is above the water surface, so nitrification can occur. In such systems the efficiency is quite low and it may be necessary to add extra air/oxygen to the tank where the drum is rotating.

Figure 9.3 Different filter media used in a biofilter.

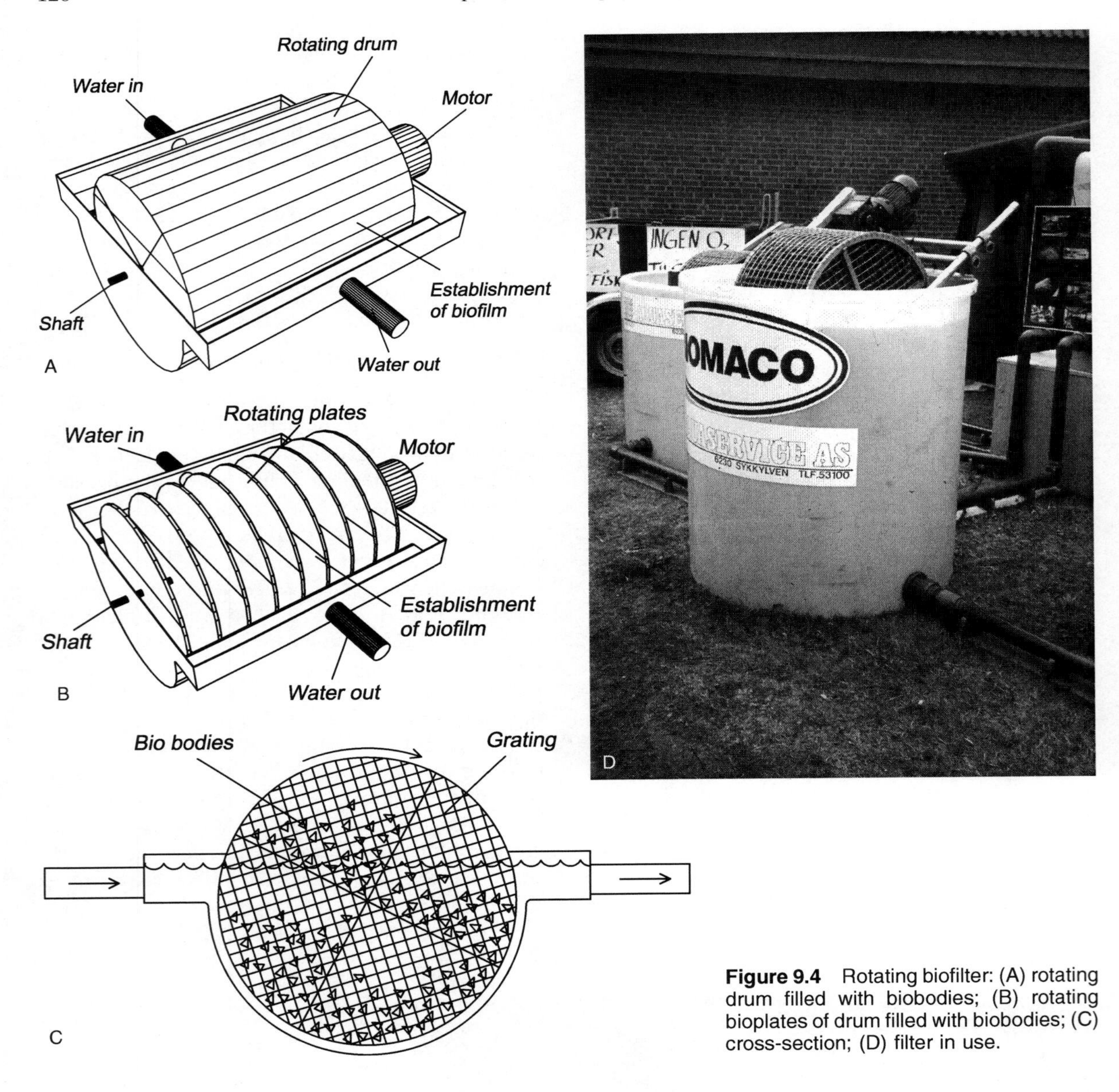

Figure 9.4 Rotating biofilter: (A) rotating drum filled with biobodies; (B) rotating bioplates of drum filled with biobodies; (C) cross-section; (D) filter in use.

9.4.4 Fluid bed/active sludge

In an active sludge reactor the bacteria are attached to suspended sludge material and kept floating in the water column. The nitrification process occurs on the surface of the suspended solids. Because of the relatively low content of ammonia, suspended solids and organic substances, this is not an appropriate method for use in aquaculture. The method is, however, widely used for treatment of municipal wastewater.

A fairly new method is to utilize a so called fluidized bed reactor, which has much in common with the active sludge method (Fig. 9.5). Here an artificial filter medium, normally plastic, floats in an upflowing current of water and air bubbles (supplied

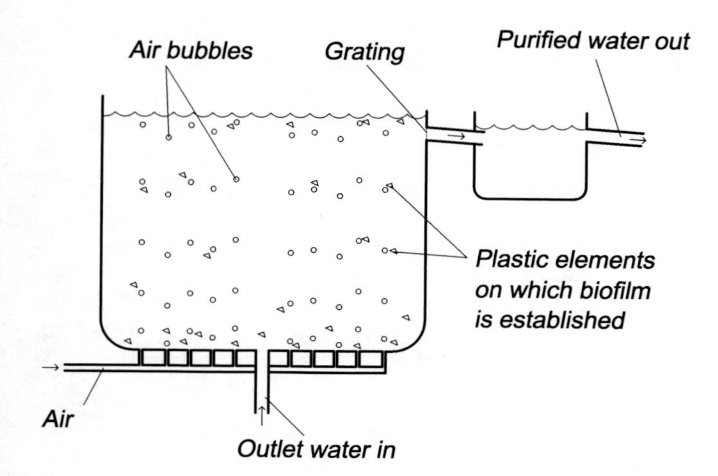

Figure 9.5 A fluidized bed reactor: biofilm grows on artificial medium floating in the water.

by diffusers at the bottom). The medium is maintained in suspension by the current, like a fountain.[36] Specially designed filter medium is used for this purpose, with a density slightly below that of water, so it is quite easy to keep fluidized. In addition, the medium is in small elements, so has a large specific surface area and will float in rotating movements in the up-flowing water. A filter grating is necessary on the outlet to prevent the elements following the water out of the reactor.

Biological nitrification takes place on the surface of the elements where the biofilm is established. It is very important to supply enough oxygen/air in this system to provide optimal growth conditions for the biofilm on the elements. Correct construction of the reactor where the elements are floating is important if high nitrification rates are to be achieved. One advantage with this system is that it is difficult to clog. A quite stable water flow is, however, important to obtain optimum functioning. The efficiency of such filters is also quite good because old biofilm is constantly removed by the air bubbles, the water flow and the rotation of the elements, which all erode to the biofilm. Only new thin fresh biofilm is therefore established on the filter medium elements. Thin new biofilm has been shown to have the highest nitrification rate. The nitrification process is reduced in thick film because it is difficult to get enough oxygen and nutrients to the deeper layers. Plastic media have been shown to produce good results in several water re-use plants.

9.4.5 Granular filters/bead filters

A granular filter used for removal of ammonia is the same as that used for particle removal. Hence two operations can be carried out by one filter. This type of filter will, however, require frequent back-washing to remove particles.

A bead filter is a type of granular filter, which is commonly filled with spherical plastic pellets (polyethylene (PE)/or polypropylene (PP)) with slightly positive buoyancy.[37,38] The water flows upwards through the layer of plastic pellets and biofilm is created on the pellet surface. Back-washing of the filter medium to avoid blockage can be performed with air bubbles, circulating water or mechanically by paddles or propellers. Experiments have shown reduced nitrification rates in such filters compared to other flow-through systems, even if on a per volume basis they are quite effective as a result of their the large surface area.[39,40]

Traditional up-flowing sand filters may also be used as biofilters, but the nitrification rate is low and requirements for maintenance and back-washing are high.

9.5 Management of biological filters

A biofilter will always need a certain start-up time before it becomes functional[41,42] (Fig. 9.6) because it takes some times to establish the culture of nitrifying bacteria on the filter medium (the biofilm). Of course there must be nitrificants in the water and the environment must be suitable, for instance with

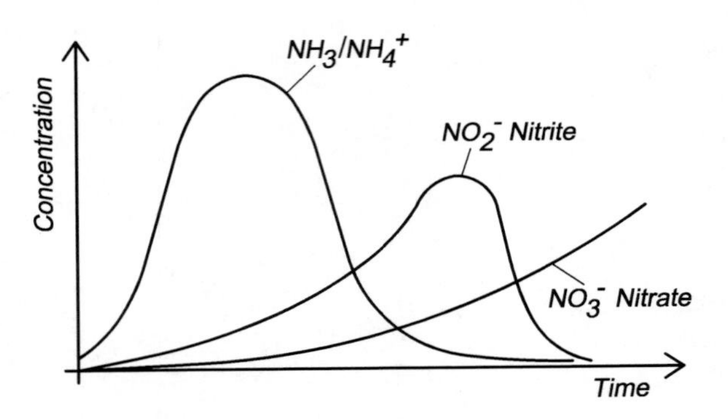

Figure 9.6 A biofilter will require a certain start-up time before being fully functional.

a water temperature above 7–8°C, dissolved oxygen content above 4 mg/l and biological oxygen demand at 7°C (BOD_7) values below 10 mg/l.[7]

If a biofilter is put into service before the bacterial culture has been established, there will be no or low efficiency. Normally, the start-up time for the filter is between 20 and 40 days, depending upon, among other factors, temperature and pH.[28,35] The start-up time can be reduced by inoculation or placing old bacterial cultures in the filter. It is normal to only clean a part of the biofilter each time so that there will always be some biofilm left on a parts of the filter. Old filter surfaces will always have some film left, so the start-up time for a previously used filter surface will be reduced compared to completely new ones.

In the start-up period, some nitrite may be supplied to the fish with the water that has passed through the biofilter because this is the end product of the first process that takes place in the filter. This is therefore a critical period, because small concentrations of nitrite can be toxic to the fish and care must therefore be taken at this stage. The best course of action is to have as few fish as possible in the start-up period or to inoculate the filter with bacteria.

9.6 Example of biofilter design

In a fish tank the consumption of feed is 100 kg/day. The amount of TAN produced by the fish, depends on growth rate, protein content in the feed and in the fish, amount of nitrogen in the protein, and the protein uptake and digestion. A secretion of TAN to be between 30 and 40 g per kg feed can be used for further calculation. The water temperature is 20°C and the pH is 7. The design and size of the biofilter can be calculated as follows.

The secretion of TAN will be 3–4 kg per day. At pH 7 and a water temperature of 20°C only 0.4% is NH_3 (see Fig. 9.1). The TAN concentration is therefore almost equal to the the concentration of NH_4^+.

Under normal conditions the nitrification rate mainly depends on water temperature. From Table 9.1 it can be ascertained that at 20°C a value of 1.0 g NH_4^+ per m^2 biofilter surface area per day can be used. The biofilter medium must therefore have a surface area (B_a) of:

$$B_a = (3000 \text{ to } 4000)/1$$
$$= 3000\,m^2 \text{ to } 4000\,m^2$$

It is normal to have a security factor when designing the biofilter: a value of 2 is quite commonly used. The necessary area is therefore 8000 m^2.

The next steep is to choose a filter medium to fit the necessary volume. If choosing a filter with a surface area of 300 m^2 per m^3 filter medium, the total volume required will be 8000/300 m^3, which is close to 27 m^3. Actual designs use submerged up-flowing filters or trickling filters. A combination is also quite commonly used. The recommended ratio if both are used is from 2:1 to 1:1. This means that the filter can have an 18 m^3 submerged filter and a 9 m^3 trickling filter.

9.7 Denitrification

In a biological filter ammonium (NH_4^+) is transformed into nitrate (NO_3^-). With a high degree of water re-use (see Chapter 10) and large loads on the filter (high fish biomass and rapid growth) the amount of NO_3^- may exceed the tolerance of the fish, and result in mortality. To inhibit this, it may be necessary to add a new filter with a denitrification step. A denitrification filter transforms nitrate to nitrite (NO_2^-) and further on to nitrogen gas (N_2), which is degassed from the water. Denitrification is for instance used in some eel farms that have a very high degree of water re-use.

Biofilm-hosting bacteria may also carry out denitrification. This is an anaerobic two-step process. In the first step nitrate is transformed to nitrite and in the second step nitrite to nitrogen gas. The bacteria that carry out this process need a supply of organic carbon for growth, not oxygen. Normally there is not enough easily available organic carbon in the outlet water so carbon must be supplied. Carbon is commonly added as methanol, ethanol or sugar in liquid form before the denitrification filter. When using methanol the following equations apply:

Step 1

$$CH_3OH + 3NO_3^- \rightarrow 3NO_2^- + CO_2 + 2H_2O$$

Step 2

$$CH_3OH + 2NO_2^- \rightarrow N_2 + CO_2 + H_2O + 2OH^-$$

If the created cell mass is described empirically by the formula $C_6H_7O_2N$, the total equation, including steps 1 and 2, is as follows:[4]

$$CH_3OH + 0.92NO_3^- + 0.92H^+ \rightarrow 0.06C_6H_7O_2N + 0.43N_2 + 0.7CO_2 + 2.25H_2O$$

Since denitrification is an anaerobic process it will not happen in the presence of oxygen. Therefore it is necessary to remove the oxygen from the water to get denitrifaction to start. Methanol and ethanol may also be used as a means of removing the oxygen; after addition of methanol or ethanol there will be no free oxygen left in the water (i.e. the concentration of dissolved oxygen will be close of 0 mg/l).

The design of the denitrification filters is the same as that of nitrification filters, requiring a large area where the bacteria can grow. An established biofilm is also necessary. However, the filter medium must be submerged to avoid ingress of oxygen. Trickling filters and drum filters are not therefore used.

9.8 Chemical removal of ammonia

9.8.1 Principle

Ion exchangers have sometimes been used in aquaculture to remove ammonia.[43,44] Ion exchangers utilize the fact that different ions have different electrical charges. Some substances have the ability to attract specific ions in the water because of their charge and exchange them with ions attached less strongly. Ion exchangers may be divided into cation and anion exchangers. For example, cation exchangers are used to remove positively charged calcium and magnesium ions from hard water, while anion exchangers can be used to remove negatively charged nitrate ions. With ammonia, it is the ammonium ion (NH_4^+) that is removed from the water, so a cation exchanger is required. Removal of ammonium ions ensures a reduction of ammonia levels because of the equilibrium between ammonia and ammonium.

9.8.2 Construction

An ion exchanger comprises a column filled with gravel-like ion exchange substance (Fig. 9.7). The water to be purified, for example by removal of NH_4^+, flows through this column. The choice of substance to be loaded into the column depends on the ions to be exchanged. To remove NH_4^+ a clay mineral (zeolite) called clinoptilolite if often used. This mineral occurs naturally in the ground in certain places. The ion exchange substance is delivered as a granulate, normally in sizes up to 5 mm. When using clinoptilolite, Na^+ is used as the exchange ion and NH_4^+ is attracted to the ion exchange substance (R^-). The following equation then applies:

$$NH_4^+ + R^- Na^+ \Leftrightarrow Na^+ + R^- NH_4^+$$

When water with high ammonia content enters the ion exchanger, NH_4^+ will be attracted to the ion exchange substance and Na^+ will be released. After

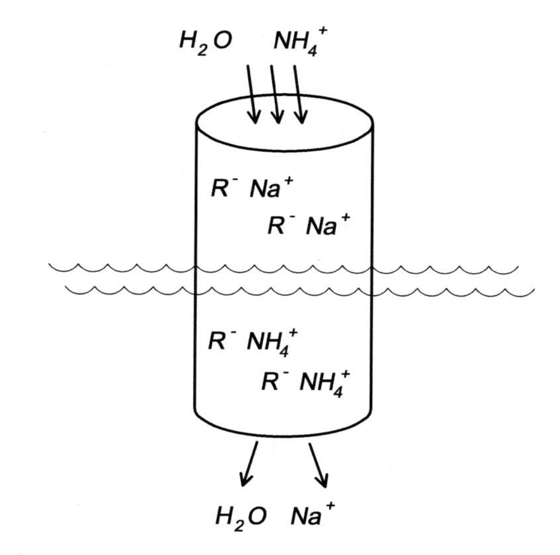

Figure 9.7 Principle of an ion exchanger.

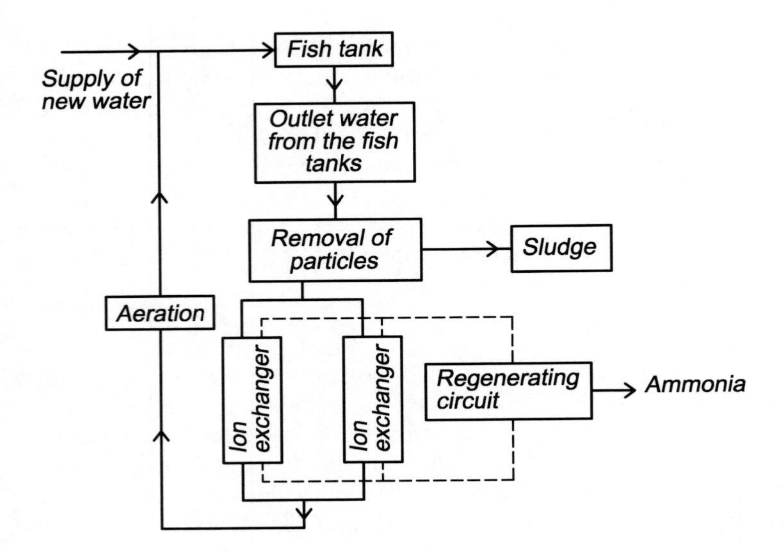

Figure 9.8 Use of an ion exchanger to remove ammonia chemically.

a period of time all the Na^+ in the ion exchange substance will have been released and the exchange process will stop; all the NH_4^+ in the water will flow directly through the column, and the outlet concentration will be the same as the inlet concentration. The ion exchanger will then have to be regenerated by removing all the NH_4^+ bound to the ion exchange substance. This can be carried out by allowing a regenerating solution containing a very high concentration of sodium ions, for example a salt solution, to flow through the ion exchanger (Fig. 9.8). Often the regenerating solution is reused; in this case a cleaning circuit for the salt solution is necessary. A stripping process may be used to clean the regenerating solution for NH_4^+.

To function optimally the water entering the exchanger must be free of particles. A fine particle filter must therefore be installed before the exchanger. Furthermore, the substance in the ion exchanger must be disinfected at specific intervals to avoid the exchanger starting to function as a biological filter. This is because biofilm will be established on the surface of the exchange substance. The capacity as a biofilter will be very low compared to the exchange capacity. Clogging and blockage will also occur rapidly.

Even if the ion exchanger can remove up to 95% of the ammonia from freshwater, the economic yield of the system is normally negative. The system requires several columns with separate regenerating circuits so that maintenance can be carried out, and to avoid breakdown of the whole system if a column is out of order. Because of this, very few fish farms use ion exchanger technology today. The advantage of an ion exchanger compared to a biofilter for nitrification, is that there is no start-up time and no waste product such as nitrate. There have been examples where this technology has been used in relation to long distance fish transport (Chapter 17).

References

1. SECL (1983) Summary of water quality criteria for salmonid hatcheries. SECL 8067, Report to Department of Fisheries and Oceans, Vancouver, BC.
2. Mead, J.W. (1985) Allowable ammonia for fish culture. *Progress in Fish Culture*, 47: 135–147.
3. Knoph, M.B. (1995) *Toxicity of ammonia to Atlantic salmon (*Salmo salar *L.).* PhD thesis. Department of Fisheries and Marine Biology, University of Bergen.
4. Russo, R.C., Thurtston, R.V. (1991) Toxicity of ammonia, nitrite and nitrates to fishes. In: *Aquaculture and Water Quality. Advances in World Aquaculture, vol. 3* (eds D.E. Brune, Thomasso, J.R.). World Aquaculture Society.
5. Montgomery, J.M. (1985) *Water treatment: principles and design.* John Wiley & Sons.
6. Tchobanoglous, G., Burton, F.L., Stensel, D.H. (2002) *Wastewater engineering.* McGraw-Hill.
7. Gebauer, R., Eggen, G., Hansen, E. og Eikebrokk, B. (1992) *Oppdrettsteknologi – vannkvalitet og vannbehandling i lukkede oppdrettsanlegg.* Tapir Forlag (in Norwegian).

8. Van Rijn, J. (1996) The potential for integrated biological treatment systems in recirculating fish culture – a review. *Aquaculture*, 3–4: 181–210.
9. Timmons, M.B., Losordo, T.M. (1997) *Aquaculture water reuse systems: engineering design and management*, 2nd ed. Elsevier Science.
10. Haug, R.T., McCarty, P.L. (1971) Nitrification with submerged filters. *Water Pollution Control Federation*, 44: 2086–2102.
11. Wheaton, F., Hochheimer, J., Kaiser, G.E. (1991) Fixed film nitrification filters for aquaculture. In: *Aquaculture and Water Quality. Advances in World Aquaculture, vol. 3* (eds D.E. Brune, Thomasso, J.R.). World Aquaculture Society.
12. Odegaard, H. (1992) *Wastewater treatment*. Tapir Forlag (in Norwegian).
13. Wortman, B., Wheaton, F. (1991) Temperature effects on biodrum nitrification. *Aquacultural Engineering*, 10: 183–205.
14. Fdez-Polanco, F., Villaverder, S., Garcia, P.A. (1994) Temperature effect on nitrifying bacteria activity in biofilters – Activation and free ammonia inhibition. *Water Science and Technology*, 30: 121–130.
15. Zhu, S., Chen, S. (2002) The impact of temperature on nitrification rate in fixed film biofilters. *Aquacultural Engineering*, 26: 221–237.
16. Speece, R.E. (1973) Trout metabolism characteristics and rational design of nitrification facilities for water reuse in hatcheries. *Transactions of the American Fish Society*, 2: 323–333.
17. Rusten, B. (1986) Ammoniakkfjerning i resirkuleringsanlegg for fiskeoppdrett. In: *Vannbehandling i akvakultur* (ed. H. Odegaard). Tapir Forlag (in Norwegian).
18. Henze, M., Harremoës, P. (1990) Chemical–biological nutrient removal: the hypro concept. In: *Chemical water and wastewater treatment* (eds H.H. Hann, R. Klute). Springer Verlag.
19. Hem, L.J, Rusten, B., Ødegaard, H (1994) Nitrification in a moving bed biofilm reactor. *Water Research*, 28: 1425–1433.
20. van Loosdrecht, M.C.M, Tijhuis, L, Wijdieks, A.M.S., Heijnen, J.J. (1995) Population distribution in aerobic biofilms on small suspended particles, *Water Science and Technology*, 31: 163–171.
21. Bovendeur, J., Zwaga, B.A., Lobee, B.G.J., Blom, J.H. (1990) Fixed-biofilm reactors in aquacultural water recycle system: effect of organic matter elimination on nitrification kinetics. *Water Research*, 24: 207–213.
22. Grady, C.P.L., Lime, H.C. (1980) *Biological wastewater treatment – Theory and applications.* Marcel Dekker.
23. Zhu, S., Chen, S. (2001) Effect of organic carbon on nitrification rate in fixed film biofilters. *Aquacultural Engineering*, 25: 1–11.
24. Carrera, J., Vicent, T., Lafuente, J. (2004) Effect of influent COD/N ratio on biological nitrogen removal (BNR) from high strength ammonia industrial wastewater. *Process Biochemical*, 39: 2035–2041.
25. Ling, J., Chen, S. (2005) Impact of organic carbon on nitrification performance of different biofilters. *Aquacultural Engineering*, 33: 150–162.
26. Levine, G., Meade, T.L. (1975) *The effects of disease treatment on nitrification in closed systems aquaculture.* University of Rhode Island.
27. Kaiser, G.E., Wheaton, F.W. (1983) Nitrification filters for aquatic culture systems: state of the art. *Journal of the World Mariculture Society*, 14: 302–324.
28. Bovendeur, J. (1989) *Fixed-biofilm reactors applied to waste water treatment and aquacultural water recirculating system.* PhD thesis, Agricultural University, Wageningen.
29. Nijhof, M., Bovendeur, J. (1990) Fixed film nitrification characteristic in seawater recirculating fish culture systems. *Aquaculture*, 87: 133–143.
30. Alleman, J.E., Preston, K. (1991) Behaviour and physiology of nitrifying bacteria. In: *Commercial aquaculture using water recirculating* (ed. L. Swann) Illinois State University.
31. Thorn, M., Mattsson, A., Sorensson, F. (1996) Biofilm development in a nitrifying trickling filter. *Water Science and Technology*, 34: 89–83.
32. Wik, T., Mattsson, A., Hansson, E., Niklasson, C. (1996) Nitrification in a tertiary trickling filter at high hydraulic loads – Pilot plant operation and mathematical modeling. *Water Science and Technology*, 32: 185–192.
33. Parker, D.S., Jacobs, T., Bower, E., Stowe, D.W., Farmer, G. (1997) Maximising trickling filter nitrification rates through biofilm control: research review and full scale application. *Water Science and Technology*, 36: 255–262.
34. Kruner, G., Rosenthal, H. (1983) Efficiency of nitrification in trickling filters using different substrates. *Aquacultural Engineering*, 2: 49–67.
35. Lekang, O.I., Kleppe, H. (2000) Efficiency of nitrification in trickling filters using different filter media. *Aquacultural Engineering*, 21: 181–201.
36. Rusten, B., Hem, L.J., Odegaard, H. (1995) Nitrification of municipal waste water in moving-bed biofilm reactors. *Water Environmental Research*, 67: 75–86.
37. Huguenin, J.E., Colt, J. (2002) *Design and operating guide for aquaculture seawater systems.* Elsevier Science.
38. Malone, R.F., Beecher, L.E. (2000) Use of float bead filters to recondition recircualting waters in warm water aquaculture production systems. *Aquacultural Engineering*, 22: 57–73.
39. Singh, S., Ebling, J., Wheaton, F. (1999) Water quality trials in four recirculating aquaculture systems. *Aquacultural Engineering*, 20: 75–84.
40. Greiner, A.B., Timmons, M.B. (1998.) Evaluation of the nitrification rates of microbead and trickling filters. *Aquacultural Engineering*, 18: 189–200.
41. Muir, J.F. (1982) Recirculating water systems in aquaculture. In: *Recent advances in aquaculture* (eds. J.F. Muir, R. Roberts). Westview Press.
42. Skjølstrup, J., Nielsen, P.H., Frier, J.O., Mclean, E. (1997) Biofilters in recirculating aquaculture systems

– state of the art. In: *Technical solutions in the management of environmental effects of aquaculture* (ed. J. Makkonen). Scandinavian Association of Agricultural Scientists, Seminar 258, pp. 33–42.

43. Dryden, H.T., Weatherly, L.R. (1989) Aquaculture water treatment by ion exchange with clinoptilolite. *Aquaculture Engineering*, 8: 109–126.

44. Rosental, H. (1993) The history of recycling technology. A lesson learned from past experience? In: *Fish farming technology* (eds H. Reinertsen, L.A. Dahle, L. Jørgensen, K. Tvinnereim). Proceedings of the first international conference on fish farming technology. A.A. Balkema.

10
Recirculation and Water Re-use Systems

10.1 Introduction

In water re-use or recirculation systems, the outlet water from the fish tanks is re-used instead of being released into a recipient water body (Fig. 10.1). The outlet water is cleaned and used again, which means that the amount of added new water can be reduced. Theoretically, all the outlet water can be re-used, as in aquaria where no new water is required except for that which is lost through evaporation. This is only theoretical, however, because in most cases the cost of removing all contaminants from the outlet water is very high, but this is of course dependent on the water quality requirements of the fish species being farmed. Usually, if maximum growth is required the water quality needs to be high.

A water re-use system includes the fish tank(s) or units for the aquaculture species, an adapted water treatment system and a pump to transport the water around the system. The pump and the water treatment system are the items that make the system distinct from traditional flow-through systems. The water treatment system, which is the heart of the re-use system, may include physical, chemical and biological processes to improve the water quality to acceptable levels.

The aim of this chapter is to describe important factors about re-use systems, definitions and equipment. Specialized literature is available in this field (ref.1 is recommended for an overview).

10.2 Advantages and disadvantages of re-use systems

10.2.1 Advantages

Reduction of water flow

Re-use of water will reduce the amount of new water required for the fish farm. Therefore farms can be established on sites where the amount of water is a limiting factor, or established farms can increase production without increasing the amount of new water required.

Limited resourses of freshwater are today a serious problem in the world. Water consumption has shown exponential growth during the past few years. Warning signs are evident, such as lowered groundwater tables, reduced size of lakes and disappearance of marshland. This is indicative of more competition for freshwater resources in the future, which will of course affect the fish farming industry because of its huge consumption of freshwater. All methods that reduce water consumption in fish farming, such as re-using water, are therefore of general interest.

Re-use of energy

As shown in previous chapters, the heating of water requires much energy and for this reason is expensive. By reducing the amount of new water

Flow-through system

Water in
Fish tank
Water out

Reuse/recycling of water

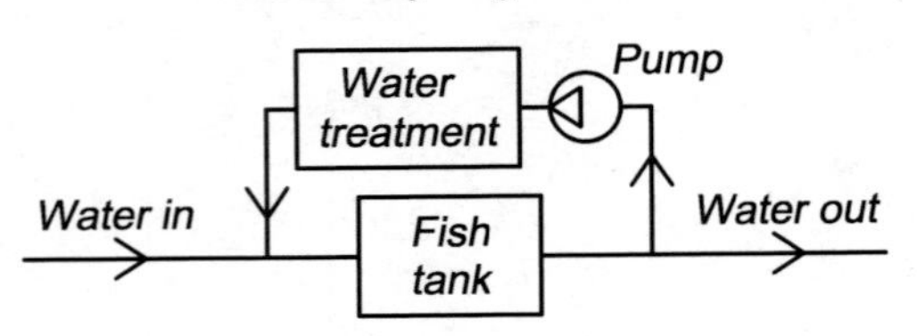

Figure 10.1 Re-use system compared to a traditionally flow through system, remember the pump.

supplied, the energy requirements for water heating will also be reduced; again this will reduce the total heating costs.

By using a water re-use system, it is possible to farm fish species that have higher temperature requirements than the natural temperature in the area, for instance to grow warm-water species at high latitudes in the northern hemisphere.

Simpler cleaning

If there are stringent requirements for cleaning the effluent water, re-use systems will assist the process because the amount of water to be treated is reduced.

Poor water sources

If the water supply to the farm is of poor quality, the requirements for improvement will be increased. Re-use systems will be of interest in such cases because the amount of new incoming water, where the quality must be improved, is reduced together with the treatment costs.

If the inlet water has to be pumped to a higher level to get it to the farm, the costs can be considerable; this may favour a re-use system. The same is the case if new water supplied to the farm is metered.

10.2.2 Disadvantages of re-use systems

Although re-use systems have advantages, they also have several disadvantages; these must therefore be weighed against each other. In most cases the disadvantages are greater than the advantages. It is best therefore, to have a site with enough good quality water of the correct temperature to suit the species grown, and low costs associated with transferring the inlet water from the source to the farm.

The two main disadvantages of re-use systems are the investment and operating costs. Because the number and size of the components for water treatment is higher than for a flow-through farm, the investment costs are also higher. In systems with a high degree of re-use (>95%) the investments can be significantly greater than for traditional flow-through farms, several times as high per unit farming volume.

In a normal re-use system there is continuous transport of water performed by some kind of pump, which results in constant running costs for the pump(s). Since so much technology is usually involved in purification systems, a re-use system will also be more exposed to technical faults. To ensure a functioning system, the requirements for monitoring water quality and water flow are greater than in traditional flow-through systems which translates to larger monitoring systems and more back-up systems. Furthermore, the time limit/reaction time is reduced when a fault occurs, e.g. pump failure, filter blockage, which increases the requirements for having operators on stand-by.

Some of the equipment used in the re-use system also requires a high level of technological and biological knowledge to operate; this imposes extra requirements for the competence of farm operators. The need for maintenance of the equipment is much higher, which also represents a significant cost.

10.3 Definitions

To describe re-use or recycling systems a number of parameters are required.

10.3.1 Degree of re-use

Normally the degree of re-use (R) is used to give the percentage of the new incoming water (Q_N) in relation to the total amount of water (Q_T) flowing into the fish tank. This is described by the following relationship (Fig. 10.2):

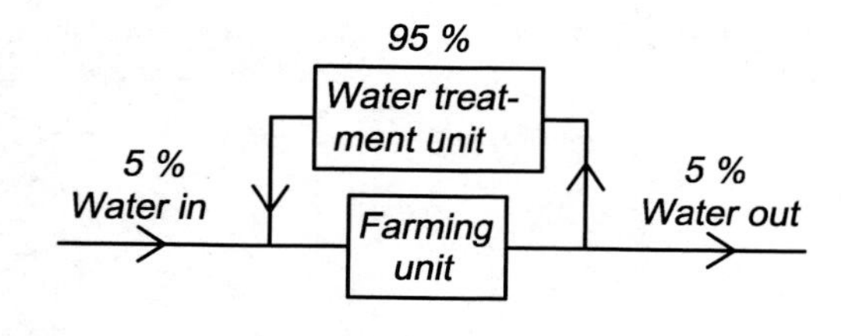

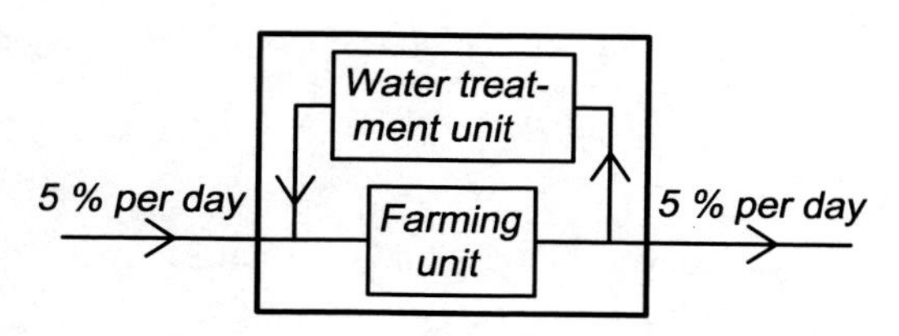

Figure 10.2 The degree of re-use can be defined in several ways. There can either be a continuous supply of new water, or there can be a batch exchange of water, for instance once a day.

$$R = (1 - (Q_N/Q_T)) \times 100$$

where:

R = degree of re-use
Q_N = new incoming water
Q_T = total water supply to the fish tank
= new incoming water + re-used water.

Example
If the supply of new water to a farm is 150 l/min and the total water flow is 1000 l/min calculate the degree of re-use.

$$R = (1 - (Q_N/Q_T)) \times 100$$
$$= (1 - (150/1000)) \times 100$$
$$= 85\%$$

This, however, requires a continuous supply of new water, which also represents the normal way to operate a re-use farm.

Another way to operate a re-use system is to change the water in batches. This is the same as is done in aquaria, so information and knowledge from the operation of aquaria can be transferred to fish farming. However, in fish farming the emphasis is on optimal growth and this requires optimal environmental conditions; fish densities are also much higher. If the aim is for the fish to survive with some spare capacity, the requirement for water quality is reduced. For a re-use system that changes the water in batches, a given amount of water can be changed once a day or once a week, for instance. In this case the water quality will gradually decrease, until a new batch is exchanged and the quality returns to top level. In this case the degree of re-use can be expressed either as the percentage of water exchanged in relation to the total flow in the system throughout a day and night, or it can be given as the amount of water that is exchanged in relation to the total volume of water in the system (Fig. 10.2):

$$R = (1 - (Q_B/Q_T)) \times 100$$

where:

R = degree of re-use
Q_B = size of batch of new incoming water
Q_T = total amount of water flowing in the system between the exchanged batches of water.

$$R = (1 - (Q_B/Q_T)) \times 100$$

where:

R = degree of re-use
Q_B = size of batch of new incoming water
Q_T = total amount of water in the re-use system.

Example
The total amount of water in the system including the fish tank(s) and re-use circuit is 1000 l. The internal water flow in the system is 10 l/min. The amount of water in the exchange batch is 250 l and it is exchanged once a day. Calculate the degree of recycling based on both definitions.

$$R = (1 - (Q_B/Q_T)) \times 100$$
$$= (1 - (250/1000)) \times 100$$
$$= 75\%$$

Total water flow in the system during a day: 10 l/min × 60 min × 24 h = 14400 l.

$$R = (1 - (Q_N/Q_T)) \times 100$$
$$= (1 - (250/14400) \times 100$$
$$= 98.3\%$$

This example shows the importance of knowing which definition of re-use is used. Systems with continuous exchange of water and not batch systems are most commonly used.

10.3.2 Water exchange in relation to amount of fish

The degree of re-use does not, however, give enough information about the performance of the system and is not sufficient to describe a recirculation system properly. Neither does it take into account the amount of fish in the system. This can be illustrated in the following way: there are few problems in having only one fish in a large re-use system, compared to having a high density of fish in the same system. The degree of re-use for the systems can, however, be equal. To describe the re-use system it will also be necessary to know the amount of new water added per kg fish (litres new water per kg fish).

Example
A re-use system with a total tank volume of 100 m³ has a total circulating water flow of 2000 l/min. The re-use degree is 95%, meaning that the amount of added new water is 100 l/min. The fish density is in case 1, 10 kg/m³ and in case 2, 100 kg/m³. This represents a total amount of fish of respectively, 1000 kg and 10 000 kg, so the amount of new water in these cases is 0.1 l/(min kg fish) and 0.01 l/(min kg fish).

Even with this information it is difficult to compare re-use systems, because factors such as species, size and growth rate will have effects. For easily evaluating a separate re-use system, the growth rate of the fish is the best indicator. If the growth rate is optimal it can either be compared to growth tables or to growth in a flow-through system.

10.3.3 Degree of purification

Another important factor in re-use systems is the degree of cleaning C_P of the water treatment system. This factor indicates how effectively the cleaning plant in the re-use circuit removes unwanted substances. It can be described as follows (Fig. 10.3):

$$C_P = (C_{in} - C_{out})/C_{in} \times 100$$

where:

C_P = degree of purification
C_{in} = concentration of actual substance entering the cleaning unit

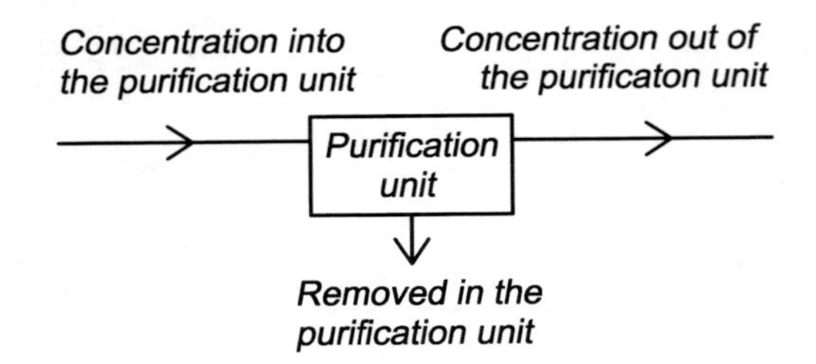

Figure 10.3 The filters in the re-use circuit removes substances according to their efficiency.

C_{out} = concentration of actual substance leaving the cleaning unit.

Example
The concentration of suspended solids entering the particle filter is 25 mg/l; after the filter the concentration is measured as 10 mg/l. Find the effectiveness of the filter.

$$\begin{aligned} C_P &= (C_{in} - C_{out})/C_{in} \times 100 \\ &= (25\,\text{mg/l} - 10\,\text{mg/l})/25\,\text{mg/l} \times 100 \\ &= 60\% \end{aligned}$$

10.4 Theoretical models for construction of re-use systems

10.4.1 Mass flow in the system

To understand more about what is happening in a re-use system, a theoretical approach can be taken.[1–4] In re-use systems is it important to have control over the different water flows (Q), amount of substances added and removed from the system (M) and, based on this, the concentration of substances at different stages in the system (C):

$$C = M/Q$$

where:

Q = water flow (l/min)
M = mass flow of substances (mg/min)
C = concentration of substances (mg/l).

Based on this, a total picture of the re-use system, including the fish tanks, the water treatment system and water transport system can be drawn up as shown in (Fig. 10.4). Here the following indices are used:

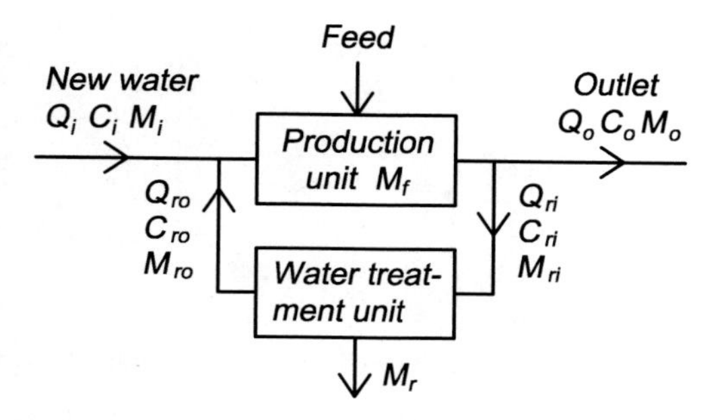

Figure 10.4 Sketch showing the water flow, concentration of substances and mass flow of substances in a re-use system.

i = new water into the tank
o = out from the tank to outlet (= $Q_i + Q_{ri}$)
ri = out from the tank and into the re-use circuit
ro = out of the re-use circuit and into the tank (Q_{ri})
f = amount of substances produced or consumed by the fish
r = amount of substances removed from the re-use system.

10.4.2 Water requirements of the system

The necessary amount of flowing water in the system, and the separation of new and re-used water depends on a number of factors including:

- Amount of water to satisfy the oxygen requirements of the fish
- Amount of water to dilute and remove waste products to acceptable levels
- Amount of water to ensure self-cleaning of the tanks (see Chapter 13)
- Degree of re-use
- Effectiveness of the water treatment system.

Water flow to satisfy oxygen requirements of the fish

The addition of water to a fish tank to satisfy the oxygen requirements depends on the oxygen consumption of the fish, the oxygen concentration in the inlet water and the lowest acceptable concentration in the outlet water to achieve optimal growth for the fish species. The specific water requirements can be calculated from:

$$Q_{in} = M_f/(C_{in} - C_o)$$

where:

Q_{in} = specific water flow per kg fish
M_f = specific oxygen consumption of the fish (mg O_2/(min kg fish))
C_i = concentration of oxygen in the inlet water to the tank (mg/l)
C_o = concentration of oxygen in the outlet water from the tank (mg/l).

Example
The fish size is 2000 g, the water temperature 12°C and the specific oxygen consumption is 3.63 mg O_2/(min kg fish). The oxygen concentration in fully saturated water is 10.8 mg/l (from tables). The acceptable concentration in the outlet is set to 7 mg/l. Calculate Q_{in}.

$$\begin{aligned} Q_{in} &= M_f/(C_{in} - C_o) \\ &= 3.63/(10.8 - 7) \\ &= 0.96\,\text{l/(min kg fish)} \end{aligned}$$

Here is it also important to remember the requirements to dilute for other substances and those for self-cleaning.

By supersaturating the inlet water with pure oxygen the water requirements can, of course, be reduced. This means that the C_{in} is increased.

Example
The same numbers as in the previous example are used, but the inlet water has a supersaturation of oxygen of 150%, meaning that the concentration is 16.2 mg/l. Calculate the new Q_{in}.

$$\begin{aligned} Q_{in} &= M_f/(C_{in} - C_o) \\ &= 3.63/(16.2 - 7) \\ &= 0.39\,\text{l/(min kg fish)} \end{aligned}$$

Water flow to dilute waste products to acceptable concentrations

The amount of water required to dilute and remove substances produced by the fish (suspended solids (SS), CO_2 and total ammonia nitrogen (TAN)) to acceptable concentrations can be calculated based on mass balance equations for the single substances:

$$M_{in} + M_{ro} + M_f = M_o$$

where:

M_{in} = mass of substances in new incoming water

M_{ro} = mass of substances from water entering from the re-use circuit
M_f = mass of substances produced by the fish in the tank
M_o = Mass of substances in the outlet from the tank.

If M_o and the water flow out of the tank Q_o are known, the concentration of a substance in the outlet from the tank (C_o) can be calculated; this must not exceed a value that is acceptable for the fish. Based on this, can the lowest acceptable outlet water flow can be calculated from the following equation:

$$Q_o \geq M_o/C_{o\text{-acc}}$$

where:

$C_{o\text{-acc}}$ = acceptable concentration of the substance in the outlet to avoid reduction in growth.

Example
A fish of size 50 g has a specific growth during one day of 31 g resulting in waste production measured as SS = 6.2 g. The acceptable level of SS in the outlet is set to 25 mg/l. Calculate the necessary water flow out (Q_o).

$$\text{SS produced per minute} = 6200\,\text{mg}/(24\,\text{h} \times 60) = 4.3\,\text{mg/min}$$

$$Q_o = 4.3\,\text{mg/min}/25\,\text{mg/l} = M_o/C_{o\text{-acc}} = 0.17\,\text{l/min}$$

This means that the water flow into the tank must be higher than 0.17 l/min to dilute the concentration of SS to acceptable levels or less.

10.4.3 Connection between outlet concentration, degree of re-use and effectiveness of the water treatment system

When starting up a re-use system the concentration of substances in the system will gradually increase until it is stabilized at a given level.

Example
A simple re-use system uses a degree of re-use of 50%; a filter with 50% efficiency is installed in the re-use circuit (Fig. 10.5). Show how many times the water must circulate in the re-use system before the system is in balance (this condition is assumed as ideal) regarding the concentration level of metabolic products, presented as parts of M.

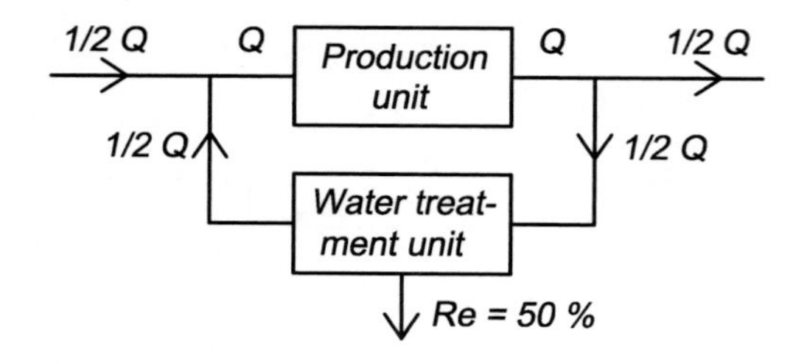

Figure 10.5 The system used in the previous example.

No. circuits	*1 (all water is new water)*	*2*	*3*	*4*	*5*
Concentration of metabolic products (M) out of the production unit	*M*	$\frac{5}{4}M$ *(M is added to the incoming concentration* $\frac{1}{4}M$*)*	$\frac{21}{16}M$	$\frac{4}{3}M$	$\frac{4}{3}M$
Concentration into re-use circuit	$\frac{1}{2}M$ *(half of the water flow)*	$\frac{5}{8}M$	$\frac{21}{32}M$	$\frac{2}{3}M$	$\frac{2}{3}M$
Concentration out of re-use circuit	$\frac{1}{4}M$ *(50% is removed) by the filter*	$\frac{5}{16}M$	$\frac{21}{64}M$	$\frac{1}{3}M$	$\frac{1}{3}M$

As shown, the system will be stabilized when the water has completed four circuits.

An equation to determine the concentration in the tank outlet in a re-use system (C) compared to the outlet concentration in a flow-through tank has been developed.[2] This is based on the degree of re-use (R) and the removal efficiency (re) of the filter system in the circuit and is as follows:

$$C = 1/(1 - R + (R\,\text{re}))$$

Example
A system has a degree of re-use of 96%, while the removal efficiency is 50%. Calculate the concentration in the outlet of the system compared to a traditional flow-through system.

$$C = 1/(1 - 0.96 + (0.96 \times 0.5)) = 1.92$$

This means that the concentration of substances is 1.92 times those from a flow-through system. For instance, if the flow-through system had an SS concentration of 20 mg/l in the outlet, the concentration in the re-use system will be 20 × 1.92 = 38.4 mg/l. If the maximum concentration that the fish tolerate without growth reduction is 25 mg/l, the re-use system is not useful because the SS concentration is too high. Either a better filter must be installed or the degree of re-use must be lowered, which means greater dilution by adding more new water.

Based on this, it is possible to calculate how much new water has to be added to have a system that functions. First the maximum allowed SS concentration C_{max}, *is found:*

$$C_{max} = 25/20 = 1.25\,mg/l$$

Then this value substituted in the formula and in the equation solved for the degree of re-use (R):

$$C = 1/(1 - R + R\mathrm{re})$$

$$1.25 = 1/(1 - R + 0.5R)$$

$$R = 0.4$$

This means that the maximum degree of re-use that can be used is 40% and 60% of new water must be added. In practice, however, a better filter unit will be installed instead of adding so much new water.

The general formula is as follows:

$$C = (1/(1 - R + R\mathrm{re}))M_f/Q_{out}$$

where M_f/Q_{out} represents the outlet concentration in a flow-through system. By rearranging this equation it can be used to find the necessary efficiency of a filter system, based on acceptable outlet concentrations and degree of re-use. The equation may also be rearranged and solved to find the acceptable degree of re-use when a filter system has been chosen.

If the inlet water contains the substances in question, their concentrations must also be added to the equation. The mass of substances in the new inlet water (M_i) must be calculated using the mass balance equation and be added to M_f. This can be expressed as $C_{in}Q_i$, which gives the following equation:

$$C = (1/(1 - R + R\mathrm{re}))(M_f + (C_{in}Q_i))/Q_{out}$$

10.5 Components in a re-use system

Fish require oxygen for respiration and produce faecal waste, urine and dissolved substances released over the gills (Fig. 10.6). The reasons for treating the water are therefore either to add new substances (oxygen) or to dilute waste products. The components required for water treatment, either for addition or removal of substances in the re-use circuit, can be calculated based on the equation given earlier in the chapter, and depend on species, size and growth rate. The usual order of necessary water treatment efforts is shown below, and compared to what happens in a tank were the inlet water flow is reduced (Fig. 10.7). To reduce the amount of incoming water to the tank is actually the same as using a re-use system without any purification units such as particle or ammonia removal filters.

If the water is re-used without any treatment, the effect is the same as reducing the water inlet volume. The first problem is that the concentration of oxygen in the outlet water will be below recommended values; the main reason for adding water to the tank is to have an outlet oxygen level high enough to achieve maximum growth of the fish. If this water is re-used directly, the oxygen concentration will be too low and will result in growth reduction. If the reduction is excessive, mortality will occur; values are species-dependent. There are numerous ways to increase the concentration of oxygen in the water (see Chapter 8), including the usual method of adding an aerator before the inlet to the tank system. In addition, after the water has been aerated up to near 100% saturation pure oxygen can be added to increase the concentration further (supersaturation). Both aeration and oxygenation can also be done directly into the

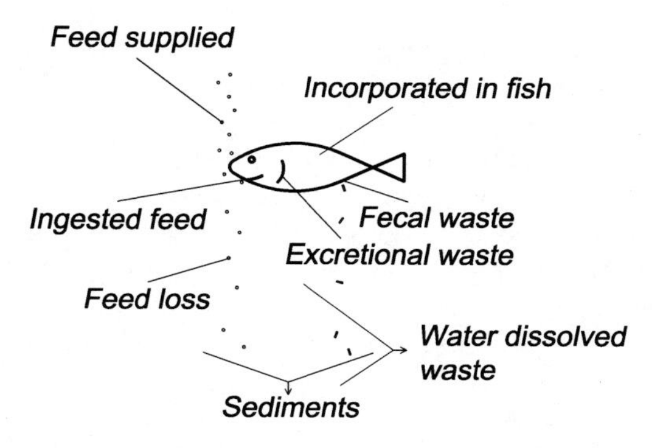

Figure 10.6 Fish in a system will produce faecal waste and soluble waste.

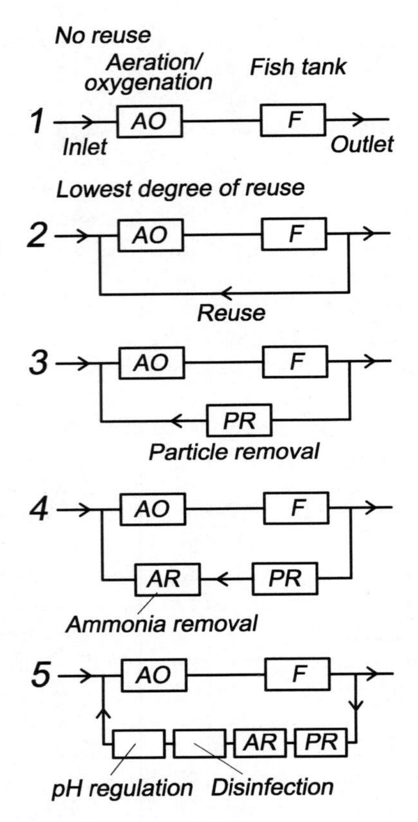

Figure 10.7 Diagrams of plants suitable for different degrees of water re-use.

tank which will also reduce the requirement for additional water.

If aeration/oxygenation is carried out, the amount of water added to the system can be reduced. This new value is normally set by the acceptable SS concentration, but may vary with organism and species. The water flow-through the tanks must now be sufficiently high to dilute the SS concentration to avoid reduction in growth. Methods of achieving this are given in Chapter 5, but usually the outlet water is sent over some kind of micro-strainer where particles are removed before the water is recycled. Another way to deal with this is to use a dual drain outlet with a separate particle outlet from which the SS are drawn off in 1–5% of the total water flow (see Chapter 5) enabling the rest of the water to be re-used without any other treatment.

If the amount of new water is reduced significantly, the next problem that may occur is accumulation of CO_2 in the tank to critical levels, because the fish release CO_2 through the gills as a waste product of metabolism and the concentration in the water increases. Tank aeration or piping the water through an aerator are two possible solutions to this problem; however, normally this has already been done to increase the concentration of oxygen in the water. If continuously high concentrations of CO_2 are experienced a vacuum aerator may be used because this is more effective for removing CO_2 than a traditional aerator designed to increase the oxygen concentration (see Chapter 8).

Reduction of the water flow into the tank can lead to excessive concentrations of NH_3 (actually TAN, because there is a connection between NH_3 and NH_4^+ levels which depends on the pH). The normal method of reducing the NH_3 concentration in the system is to use a biofilter in the re-use circuit that transforms ammonia to less harmful nitrate (NO_3).

The normal units included in a re-use system are described above. However, other problems may occur when adding a small amount of new water or with a high degree of re-use. This problem mainly concerns the components in the re-use circuit: the complexity of the system means that a good knowledge of water quality and the way in which different water quality parameters influence each other is very important.

There will be a drop in the pH in the system with high degrees of re-use. The reasons for this are that the biofilter process release H^+ ions to the water, and that CO_2 is released by the fish. Therefore pH regulation must be included in the water treatment for the system, for instance addition of lime.

When having high percentages of water re-use (>99%), high fish densities and a biofilter in the system for ammonia removal, the concentration of nitrate (NO_3) can reach values that can be toxic to the fish if it is not removed. Usually a denitrification filter is added to the re-use circuit. Here nitrate is transformed to nitrogen gas (N_2) that can be removed by aeration.

High degrees of water re-use will normally also increase the total number of bacteria, some of which might be pathogenic. It is therefore normal to include disinfection in the circuit, for instance UV irradiation.

10.6 Design of a re-use system

A re-use plant may be established either with continuous addition of new water or batch exchange; the former is most common because it maintains stable water quality.

Two different principles are used for construction of re-use plants (Figs. 10.8 and 10.9) regardless of whether there is continuous or batch exchange of water:

- Centralized re-use system for handling water from several fish tanks

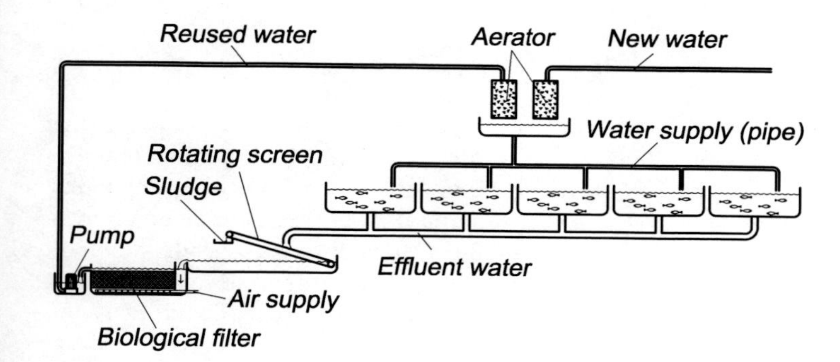

Figure 10.8 A centralized re-use system serving several fish tanks.

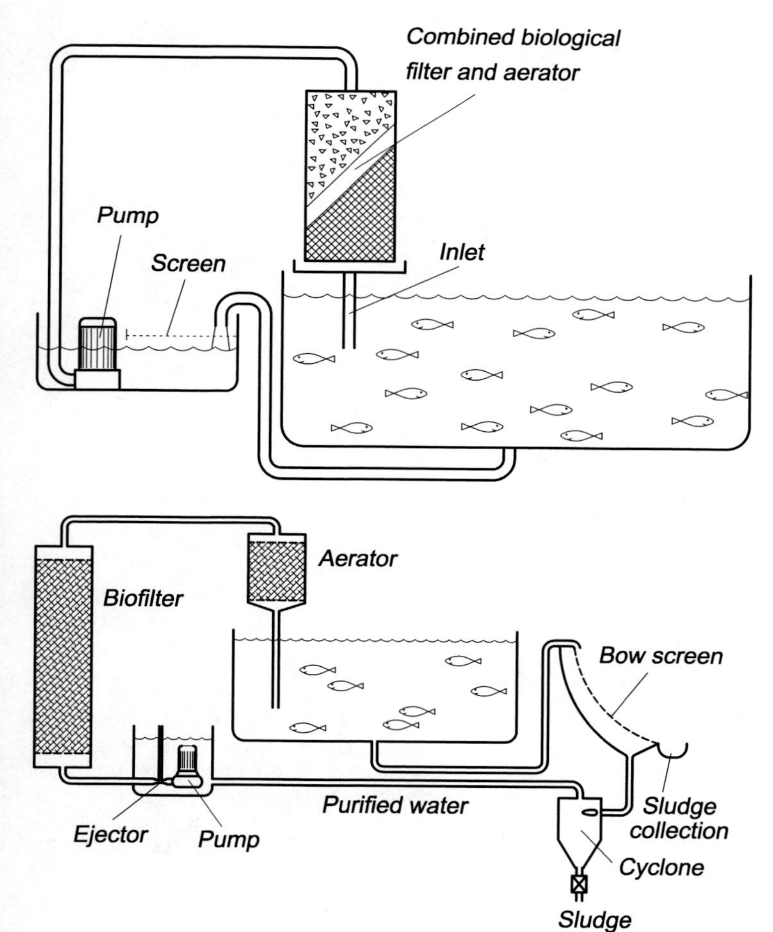

Figure 10.9 Two designs of tank internal re-use system serving only one tank.

- Re-use system placed in a single fish tank, also known as a tank internal re-use system.

Most current re-use systems are based on the centralized principle. Outlet water from all the fish tanks is colleted in a common pipeline that leads directly to one centrally placed water treatment system which includes all the necessary water treatment components. After treatment the water is returned to the tanks via a common inlet pipe (Fig. 10.10). Addition of the new water and removal of old water is also performed here, either on a continuous or batch basis.

The advantage of a centralized system is that more investment can be put in the water treatment components because they handle more tanks and greater weight of fish. However, one disadvantage with this system is that if infection occurs in one tank it will be transferred via the water to all other tanks in the re-use circuit, although this can be eliminated or reduced by installing a disinfection plant in the circuit. Another disadvantage with a centralized water treatment system is that it is more difficult to gradually increase the size of the farm/system.

In a single tank re-use system, the outlet water from the tank is lead directly into a water treatment system before it is returned to the same tank. Thus every tank has its own water treatment system and there is no mixing of water from different tanks. The water treatment system can either be an integral part of the fish tank, partly inside the tank volume, or it can be a separate external unit attached to the fish tank. Great flexibility is the

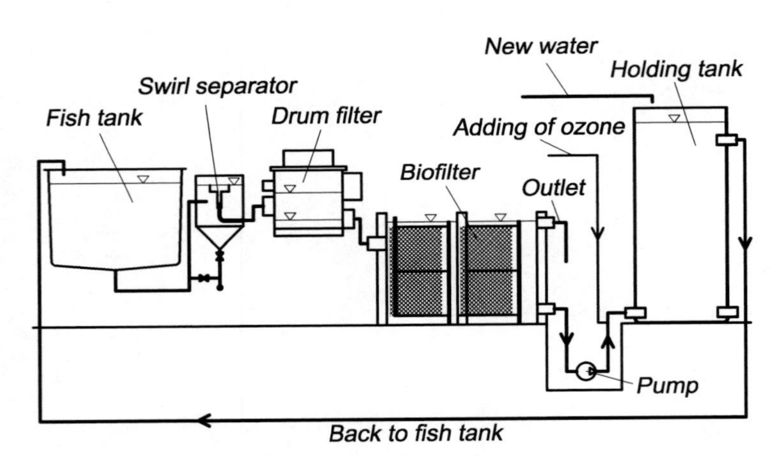

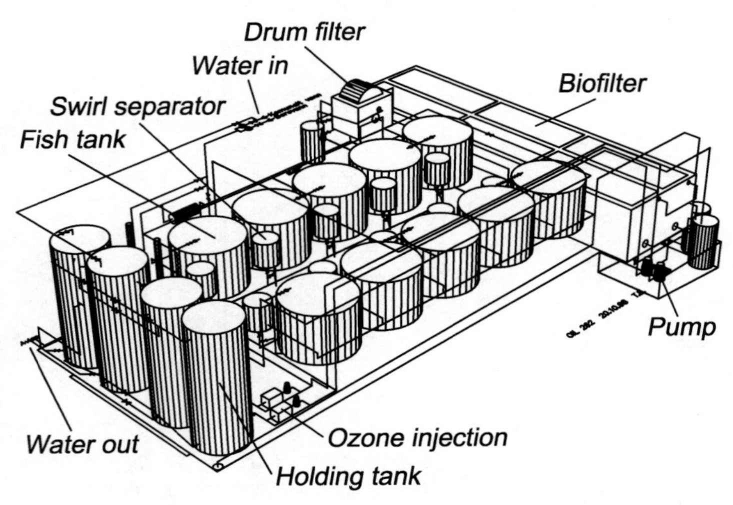

Figure 10.10 The re-use plant at the Norwegian University of Life Sciences is centralized and the water treatment system consists of particle removal facilities (swirl separators and drum filter), submerged biofilter filled with bioblocks, dry-placed pumps and addition of ozone for combined disinfection and oxygenation of the water.

major advantage with this system. It is easier successively to expand a fish farm. In addition, there are possibilities for better adaptation to individual loads, meaning that the degree of re-use can vary from tank to tank, and the single systems can in this way be operated optimally. The risk of spreading pathogens between tanks is also eliminated because there is no water connection between the tanks.

The disadvantage is, however, the price, which inhibits development of the model. Each tank needs a separate water treatment system; this only allows use of low cost simple systems, but even then it is difficult to compete with centralized systems. The management cost of such systems is also increased because there are several units that have to be maintained and controlled; for this a larger monitoring system is needed. Such systems are not favoured when having high fish densities or very high percentage of re-use (>99.5%) because this normally includes several high cost steps, such as pH regulation, denitrification and disinfection. What is generally important when constructing a water re-use plant is that the components are compatible with each other and of the correct size.

In addition to the water treatment components, it is necessary to establish a water flow in the re-use circuit; this is done by some kind of pump. Types used include airlift pumps, propeller pumps or centrifugal pumps. When using airlift pumps it is possible to combine aeration with the transport of water and by this eliminate the need for traditional pumps. In tank internal systems with a low degree of re-use this might be done to create a low cost system. In larger systems, traditional centrifugal pumps are most commonly used, because system efficiency is increased. The pump is either dry-placed or submerged. When using submerged pumps some heat will be transferred to the water from the pump, because the pump creates heat when running (see Chapter 2); when having a high degree of re-use this can contribute an important part of the total heating needs of warm-water species. In addition, the fish in the system will create heat from their metabolism, so the amount of heat that must be added is reduced even more.

References

1. Timmons, M.B., Ebeling, J.M., Wheaton, F.W., Summerfelt, S.T., Vinci, B.J. (2002) *Recirculating aquaculture systems*. Cayuga Aqua Ventures.
2. Liao, P.B., Mayo, R.D. (1972) Salmon hatchery water reuse systems. *Aquaculture*, 1: 317–355.
3. Gebauer, R., Eggen, G., Hansen, E., Eikebrokk, B. (1992) *Fish farming technology – water quality amd water treatment in closed production farms*. Tapir Forlag (in Norwegian).
4. Langvik-Hansen (1995) *Recircualtion of water in fish farming*. Master thesis. Norwegian University of Life Science (in Norwegian).

11 Production Units: a Classification

11.1 Introduction

The aim of the production unit is to create a restricted area where aquatic organisms can be reared under the best possible growth conditions. A habitat must be created for the aquatic organisms within a body of water. In the unit the organisms (apart from eggs) need to have access to food and oxygen; in addition, the waste products must be removed. Optimal performance of the production units is of major importance because they constitute the production system on the farm.

When starting to develop production units, the aim is to create an environment that resembles natural conditions as much as possible. Over a period of time after the species have gone through breeding programmes and become more adapted to farming conditions, the production units may also be developed to increase cost effectiveness by achieving greater production coupled with reduced investment and operating costs.

The design of the units depends on the organisms, for instance whether fish or shellfish are being farmed, and will also vary with the species. For instance, requirements for flatfish are different from those of pelagic fish, the former requiring a larger bottom area. Requirements may also change with the development stage and will be different for eggs and on-growing fish.

During the past few years, animal welfare considerations have been introduced into fish farming. Therefore the production units should be designed to allow the fish to behave as naturally as possible.

In the production unit oxygen and eventually feed are consumed by the organism, and waste products are released. The amount of oxygen consumed depends on the fish density, amount of feed supplied and the growth rate.

The following chapters give more information about the most commonly used production units, commencing with those for storing and hatching eggs, continuing with tanks and ponds, and concluding with cages.

11.2 Classification of production units

There are a number of different designs of production units (Fig. 11.1). Several classifications systems can therefore be used, either directly related to the production unit or the production method, which again influences the design of the production unit.

11.2.1 Intensive/extensive

One classification of production systems described in Chapter 1, is intensive, semi-intensive and extensive. The same can be used for the production units. In extensive units the biomass is lower than is normally the case for cultivation. An example of a very extensive rearing unit is the utilization of small lakes for fish farming purposes. Before use the lakes should be cleared of natural predators. In use, fry, for instance, can be released and harvesting can be done with a seine net when the fish have reach the required size. In dry periods such systems may be run without any water or oxygen supply, the only available oxygen being that produced by photosynthesis occurring in the lake. No food is supplied, the only food is that naturally produced. More advanced and intensive is the use of artificially created ponds (excavated or

Figure 11.1 A number of different designs of closed production units are in use: tanks (A), closed sea cages (B), sea cages (C), ponds (E), raceway (F), tidal basin (G).

Figure 11.1 *Continued.*

embankment ponds). At low density such systems may also be run without any supply of water, the ecosystem in the pond ensuring proper water quality in the pond. However, when artificial feeding of the fish in the pond is commenced, it is necessary to increase the oxygen supply and remove the waste products, thus gradually progressing to more intensive systems. Tanks allowing high fish density represent intensive production units.

11.2.2 Fully controlled/semi-controlled

Another way to create simple and quite extensive rearing units is to fence in a water volume, either in lakes, rivers or the sea, and so create a restricted volume (a pen) where fish or other aquatic organisms can be reared. Normally the fence is made of net or wire, but the use of electric barriers has also been proposed. If using a net, the fence may be established by means of a post planted in the ground. The area that is fenced in will vary with the geographical conditions, amongst other things. For example, by restricting an area in the sea between two narrow necks of water may create a very large farming volume. Dams of natural earth masses or concrete may also be used to restrict small natural bays and give a low cost production volume.

Classification can also be based on the possibilities for controlling the environment inside the production unit. If the bottom or walls are made of fixed materials such as concrete, steel or plastic, control over the water environment may be possible, as in a closed production unit. In addition, a light-tight insulated superstructure will give full control of the environment. Cages floating in seawater represent an open production unit where full control of environmental factors is impossible; the only factor that is controlled is that the fish are colleted in a restricted area. However, if the cage is closed with a tarpaulin, for example, more control will be possible (Fig 11.2).

11.2.3 Land based/tidal based/sea based

Units can also be classified depending of where they are placed: 1, on land; 2, in the tidal zone; 3, in the water (sea or freshwater) (Fig. 11.2). On land the units will be closed. The water supply and exchange is either gravitational or pumped.

Units in the tidal zone are normally closed, but may also be open. In the second case the water level in the unit wills vary according to the tide. If the unit is closed the water may either be pumped or the tide can be used to ensure water supply and exchange. If using the tide there can be a batch exchange, meaning exchange only occurs when the tide is high. Specially designed valves may also be used either taking in only surface water or only taking in bottom water (Fig. 11.3).

Sea-based farms float in the water, normally on the surface, but submerged units may also be used.

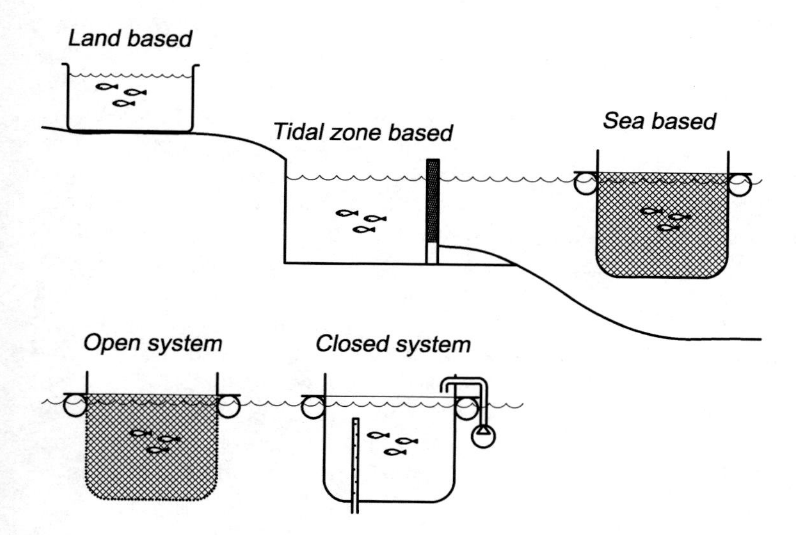

Figure 11.2 Production units can be open or closed, and placed onshore, in the tidal zone or in the sea.

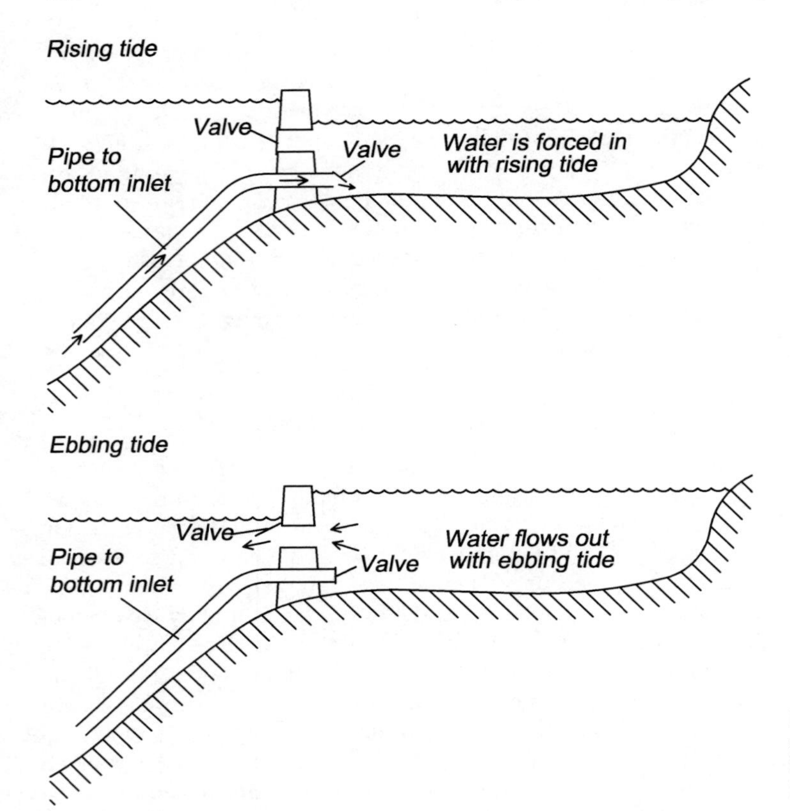

Figure 11.3 Use of a tidal basin where special valves control the water supply and exchange so only bottom water is taken into the basin.

Normally they have an open construction, like sea cages, where water supply and exchange is ensured by the natural currents. However, the units can also be closed with sealed walls and bottom (Fig. 11.2), in which case a pump must be used to ensure water supply and exchange. The advantage with such systems compared to land-based units is that the pumping head is reduced.

11.2.4 Other

A number of other categories are also possible. One is based on the means of water supply and exchange, whether continuous or batch. Another is based on the investment cost per unit farming volume or per unit farming area. Sea cages and ponds represent relatively low-cost systems, while circular concrete tanks represent high-cost systems.

If freshwater is fed into a seawater pond, because of the density difference, it will form a layer on the top of the seawater for a period of time. This can be utilized in specially designed production units for Artic charr farming during the winter because they do not then tolerate full salinity. A freshwater/ brackish layer can be kept in a sea cage by having a tarpaulin skirt in the upper part of the net bag walls with an open lower netted part and sending freshwater in through a pipe (Fig. 11.4).

Another way that this density gradient can be utilized is by sending freshwater into a closed lagoon or basin of seawater. Because of the density difference the freshwater will float on top in a separate layer and a 'greenhouse effect' can be achieved. Heat radiation will go down to the seawater, but the reflection is reduced because of the border layer between the saltwater and freshwater so the temperature of the seawater in the basin will increase more than that of the freshwater on the surface. This principle can be used for oyster spat farming in the northern hemisphere, for example (Fig 11.4).

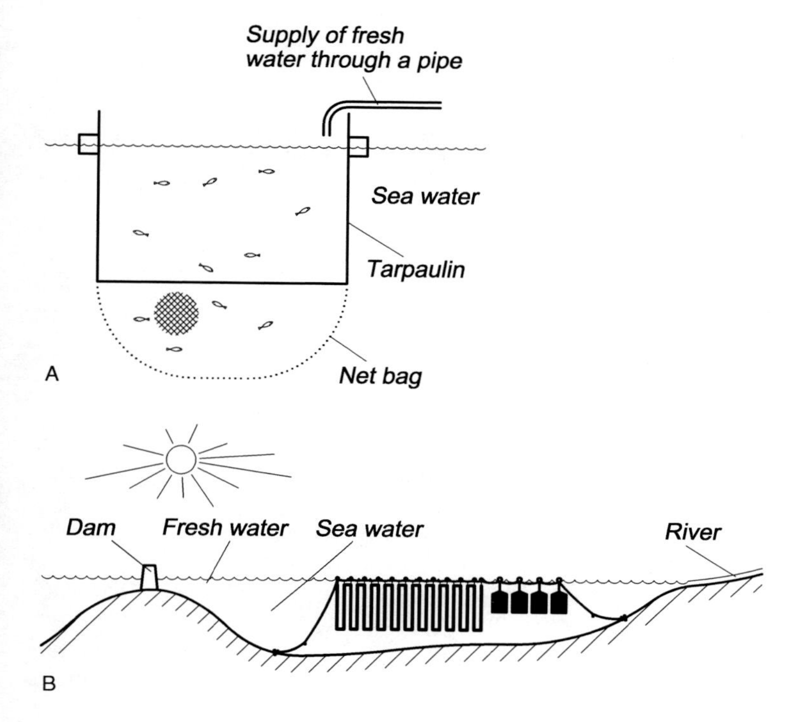

Figure 11.4 Production units utilize the fact that freshwater has a lower density than salt water, either to create a volume of freshwater in a sea cage (A) or to heat the water in a basin (B).

11.3 Possibilities for controlling environmental impact

It will become increasingly important to control the environmental impacts of aquaculture, mainly the discharge of nutrients, organic substances and micro-organisms, as well as escape of fish. The possibilities for control will depend very much on the design and construction of the production unit. For closed units, control is normally possible by collecting and treating the outlet water from the unit. Closed production units may also float in the sea; here treatment of the outlet water is rather more difficult; it is best to have a small height difference between the production unit and the treatment plant. Floating production units may also be damaged by waves, so shallow sites must be chosen for such installations.

For open production units like sea cages the possibilities for controlling the environmental impact are limited. It is therefore important to use sites that tolerate discharge of nutrients and organic substances without local accumulation.

In open production units in the sea there will always be possibilities for fish to escape as a result of construction failure. The weather is unpredictable and large waves can result in breakage of the production unit. Necessary precautions must therefore be taken, including optimal design of the unit and units adapted to the site. A typical open production unit is the sea cage for which correct mooring is important. In addition, it is important that the net bags are strong enough for the site and are regularly inspected. Double net bags have also been used; as have nets around the farm area, all to prevent fish escaping.

Fish escape from closed production units may also occur, especially at the fry stage, even if this theoretically can be avoided. The fry follow the outlet pipes out of the unit. An absolutely secure screen ought therefore to be set in the outlet system, its sole purpose being to prevent escape of fish.

12
Egg Storage and Hatching Equipment

12.1 Introduction

The main purpose of units for storage and/or hatching of eggs is to create a restricted area where the eggs can grow under optimal conditions. Separate units can be used for storage and hatching, or the same unit can be used for both purposes. It is also possible to combine the hatching equipment with later holding of fry and eventually also for first feeding. Units for the incubation of eggs are often called incubators.

In the storage unit, the eggs must be supplied with sufficient new water to meet their oxygen requirements and remove metabolic waste products. Continuous addition of new water will also achieve the necessary water exchange that is essential to inhibit the growth of fungus which may increase mortality. In addition, the quality of the water is of great importance at the egg stage. As the oxygen requirement is low at the egg stage, it is possible to not have a continuous supply of water, but to exchange water in batches, provided that there is no fungus problem or an antifungal agent is added to the water. Addition of air through diffusers may also be used to supply oxygen.

The design and function of the units depends on how the eggs need to be stored and the intensity of production. Eggs from different species have different storage requirements. Some prefer to lie on the bottom or on/in a bottom substrate; others prefers to stay pelagic in the free water mass, while others again are stored inside the females: under wild conditions, for instance, *Tilapia* stores the eggs in the mouth and initially the wolf fish stores them inside the 'belly' for fertilisation before laying demersal eggs. Salmonids and catfish are representatives of the first group where eggs prefer to stay on the bottom, while several marine species such as cod and halibut and several freshwater species belong to the second group. Under farming conditions, however, one species may be adapted to use a rearing system different to that used in the wild, such as holding pelagic eggs on a bottom substrate. Normally, systems in which the eggs lie on the bottom or on a bottom substrate are easiest to build and control. There are also differences in how the egg lies, because some species have single eggs, while in others the eggs lie together in a matrix or with a 'cover' around the egg batch. Egg size varies significantly between species and this is also of major importance when designing storage and hatching units with the required water supply; the task is more difficult with smaller eggs.

Egg production can be separated into intensive and extensive farming, and this also influences the design of the equipment. For more extensive farming, ponds, net pens or cages may be used, but of course the production per volume unit will be reduced. If using extensive systems, the eggs can be collected and put into intensive hatching systems or the hatching can be performed in the extensive system, depending on production strategy. Egg production is normally based on artificial spawning, but egg production can also be based on collecting wild eggs that are introduced into extensive or intensive farming systems.

In this chapter the focus is on intensive farming systems. Because of the great differences between species, it is difficult to give a general overview of the units for storage and hatching of eggs. Therefore information is given concerning the two basic methods: 1, systems where the eggs stay pelagic; 2, systems where the eggs lie on the bottom or on a bottom substrate. The basic principles of the two categories are however, the same and also used for other species. Many textbooks are available with more detailed information regarding the requirements of the various species, as this is of great importance for egg storage equipment (see, for example, refs 1–5).

12.2 Systems where the eggs stay pelagic

Two production methods are commonly used for pelagic eggs, which may influence the design of the incubator. Either the eggs stay in the same unit until hatching is finished, or there is a two-step process using two separate units where the eggs are removed before hatching. In both units the eggs stay pelagic (Fig. 12.1). The normal difference is the size of the incubators; for the second step large units with volume of several cubic metres may be used. The system chosen is also species dependent, and there will always be species-dependent adjustments to be made.

If the eggs are to stay pelagic in the water column, it is important to create an environment that makes this possible. A suitable incubator has both a water inlet and a water outlet. Inside the incubator an environment must be created that is as close to natural conditions as possible for pelagic eggs; at the same time oxygen must be supplied and waste products removed. Most usually a small up-flowing (up-welling) current is created so that the eggs will stay in suspension in the incubator, as in a fluidized bed.

12.2.1 The incubator

The traditional shape of an incubator for storage and hatching of pelagic eggs is a cylinder with a conical bottom (Fig. 12.2). With this shape it is easier to achieve a good flow pattern in the unit. Funnel-shaped units, however, can also be used (Fig. 12.2). Traditional tanks (see Chapter 13) can be used, but are not ideal.

Having a conical bottom makes it possible to remove the dead eggs easily by tapping them out

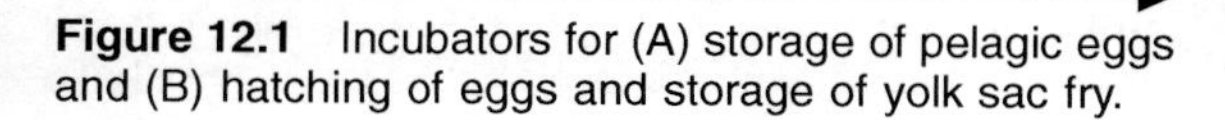

Figure 12.1 Incubators for (A) storage of pelagic eggs and (B) hatching of eggs and storage of yolk sac fry.

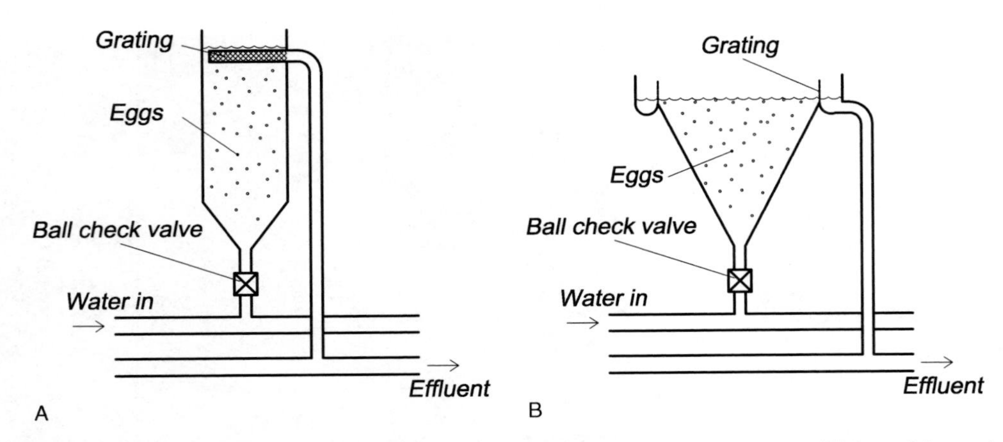

Figure 12.2 Designs of incubator for pelagic eggs: (A) cylinder with a conical bottom; (B) funnel-formed.

through an outlet in the bottom of the cone. In practice, several methods are used: for instance, stopping the water flow or adding a plug with higher salinity and thereafter stopping the water flow.[6,7] In both systems dead eggs will sink to the bottom, while live eggs will remain in suspension due to their greater buoyancy.

Incubators are normally constructed of fibreglass or polyethylene (PE). Glass jars may also be used, but this occurs more typically in smaller units. The normal size of incubators varies from some litres to several thousand litres.

12.2.2 Water inlet and water flow

To create an up-flowing current in the incubator, the inlet is normally placed at the bottom and the outlet at the top. Various flow patterns can be created in the up-flowing water.[8] It is important that the up-flowing water forms a 'plug' that equals the complete cross-sectional area of the incubator. The up-flow velocity must be the same at every point over the cross-sectional area of the incubator; this will enable the eggs to remain in suspension. Although it is common to place the inlet at the bottom of the incubator, vertical spray inlets may also be used provided that the outlet is placed at the top of the water column. An up-flow will also be created in this way. Air bubbling through diffusers on the bottom of the tank may also be used to create an up-flowing current, and this can used to reduce the necessary water flow. Air bubbling may also be used to prevent aggregation of the eggs on the water surface.

The amount of water to be added is dependent on the size of the incubator, the number of eggs and the species. For example, in a 700 l incubator for cod eggs, the water supply must be at least 3 l/min,[7] while the recommended water flow for a 250 l incubator for halibut eggs is 3–4 l/min.[6] It is important that the vertical flow velocity is not too high to avoid lifting the eggs to the surface. This can be monitored if transparent materials are used for the incubator or there are inspection windows.

12.2.3 Water outlet

It is important to avoid having too high a flow velocity through the outlet screen otherwise the eggs are dragged towards the outlet screen where they may get stacked (a large surface area of the outlet screen is therefore important). Several solutions are used to avoid this (see, for instance, Refs. 9, 10). Outlet systems can either have a screen around the total circumference or part of the circumference, or the outlet can be within the incubator; the last may, for example, be an open pipe covered with a screen (Fig. 12.3) or what might be called a banjo screen. A banjo screen is a banjo-shaped screen placed inside the incubator. When using an internal screen it could be place some distance below the surface to ensure pressure on the outlet screen.

The holes in the screen must be small so the eggs are not lost. Typically, a plankton cloth with suitable mesh size is used. For halibut, a mesh size of 250 μm can be used.[6] To avoid blockage

Figure 12.3 Outlet screen in an incubator for halibut eggs (seen here as white on the surface).

of the outlet screen, air bubbles may be blown against it.

12.3 Systems where the eggs lie on the bottom

There is much experience concerning eggs lying on the bottom and several systems have been developed for intensive fish farming, especially for use in salmonid farming. It is possible to divide these into three different systems:

- System where the eggs remain in the same unit for the whole process from spawning up to fry ready for first feeding
- System where the eggs lie in thick layers and must be moved before hatching
- System where storage, hatching and first feeding are carried out in the same unit.

The system chosen depends on the management strategy for the hatchery, whether the farm produces eggs for its own use or for sale. If the farm is to produce eggs for sale they may lie in thick layers, because these will be sold before hatching. This system is suitable for salmonid eggs that are sold as eyed eggs, because they tolerate a lot of handling.

12.3.1 Systems where the eggs lie in the same unit from spawning to fry ready for start feeding

Hatching troughs

A common unit is the hatching trough with trays inside; this is also known as the California system (Fig. 12.4). Trays or boxes are placed beside each other lengthwise, in the trough. The trays have a perforated bottom and one of the sidewalls is also perforated. Water is supplied at one end of the trough and leaves from the opposite end. A level outlet controls the water level in the trough.

Inside the trough the tray is installed so that an undercurrent of water is forced to flow up through the perforated bottom, through the layers of eggs lying in the tray and then out through the perforated side of the tray. The water is then forced down to the bottom of the trough and up through the perforated bottom of the next tray. In this way an undercurrent is generated in all the trays in the trough.

After hatching there are large amounts of eggshells to be removed to prevent obstruction of the outlet grating. To increase the grating area an L-shaped outlet grating can be installed in the tray during hatching.

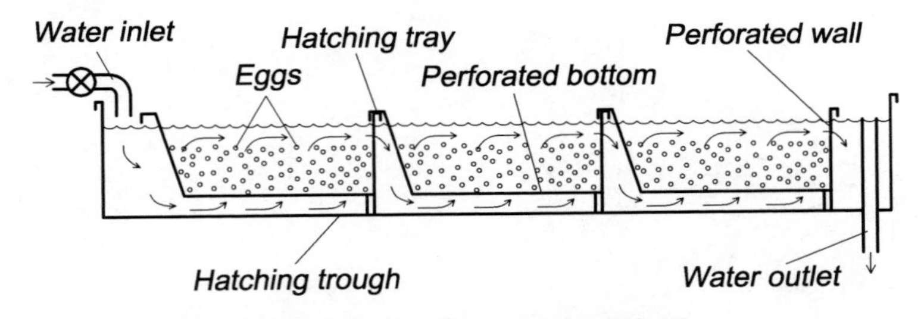

Figure 12.4 Hatching troughs with trays inside are much used in salmonid farming where the eggs are lying on the bottom.

The typical size of one type of hatching tray is 40 cm × 40 cm × 15 cm. In each trough there may be 4–7 trays. The number of eggs in each tray is species dependent; normally 1–2 l or two layers of eggs is recommended for salmon. The recommended water flow to each trough is 7 l/min and 12 l/min, for troughs with four and seven trays, respectively. If the water supply is too large, the undercurrent may lift the eggs in the trays; increased mortality may occur if the eggs are moved during critical phases in the incubation period. The troughs and trays are usually made of glass-reinforced plastic.

There are also simpler hatching troughs where the eggs are not distributed in individual trays, but placed along the whole surface of the trough. The water enters on one side and leaves from the opposite side. Such a system has no up-flowing water through the egg layers, but a horizontal flow. The capacity per area unit is therefore lower. In addition, the hatching results are normally reduced, and the work requirement for production is increased.

A special type of hatching trough, or small raceway, can be used in channel catfish production.[1,2] Here the eggs are put into stiff cloth baskets. Between each basket there is a space where a small paddle wheel rotates (Fig. 12.5) to simulate the fanning action of the adult male. As catfish eggs are deposited in an adhesive yellow mass, it may need to be broken into smaller pieces before being placed in the baskets.

Artificial substrate

Artificial substrate can be placed in the bottom of the trays to improve the results (Fig. 12.6). The eggs are laid on top of a perforated plate in the hatching trough, and when hatching occurs the yolk sac fry (alevins) will move down through the perforations. The artificial hatching substrate is located below the perforated plate. This substrate creates small spaces where the yolk sac fry can stay in an upright position. In this way only a small percentage of the yolk sac is used for swimming and maintaining balance. Instead this energy is used for growth. The substrate can be designed in many ways, from squared cells to mats made of artificial grass (for example, AstroTurf™). The mats are usually made of plastic. Increased growth and reduced mortality have been achieved for salmonids with the use of artificial substrate. It is

Figure 12.5 System for rearing eggs of channel catfish.

Figure 12.6 Artificial substrate is placed in the hatching tray to improve the development of the yolk sac fry.

important that the artificial substrate does not release toxic substances into the water.

Hatching cabinet

In the hatching cabinet the eggs are placed in drawers or on racks (low boxes) on top of each other. There are two different designs of hatching cabinet, either water drops fall from the top, or there is an individual water inlet and outlet in each drawer (Fig. 12.7).

The second design is the most used; the construction includes an individual inlet and outlet to each drawer, with the outlet keeping the water level constant. The drawer has a perforated bottom where the water flows up through the layers of eggs. After passing through one drawer the water is sent into the drawer below. This system has the advantage that it maximizes the space utilization in relation to hatching trough and trays, but it is more difficult to control and reduced production may result.

Systems using the first design comprise a number of drawers made of perforated plates where the eggs are distributed. Water is supplied from the top and flows down through the layers. However, the eggs must be removed before hatching in this system.

12.3.2 Systems where the eggs must be removed before hatching

In egg-rearing cylinders the eggs are layered on top of each other (Fig. 12.8), so this system cannot be used through to hatching, and for salmonids only to

Figure 12.7 In hatching cabinets the eggs are placed in drawers or racks above each other.

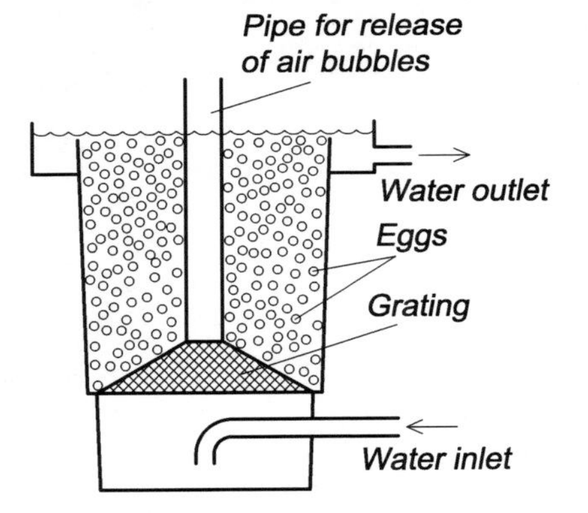

Figure 12.8 In the egg-rearing cylinder layers of eggs are stacked on top of each other, but must be removed before hatching. During incubation the eggs must lie completely still.

the eyed egg stage. In these cylinders the water is taken in at the bottom and a distribution plate ensures that it is distributed evenly through the layers of eggs via an underflow. The water will then flow up through the layers of eggs and over the top edge of the cylinder to the outlet. To avoid air bubbles or clogging, the distribution plate must be set at an angle and there is an aeration pipe in the centre of the cylinder. From here air bubbles can go to the surface without passing through the layers of eggs in the cylinder. If the air bubbles must pass through the egg layers they may increase mortality by moving the eggs during a critical phase of development.

A commonly used egg-rearing cylinder contains approximately 30 l of eggs; the necessary water supply is 5–7 l/min. The great advantage of using these cylinders is that more eggs can be stored in a restricted area; up to 120 l/m^2 floor area. The cylinders are often made of polyethylene (PE); typically 80 cm high and 50 cm diameter. Larger tanks of up to 200 l, may also be used with an upflow of water, but here control of conditions is reduced.

12.3.3 System where storing, hatching and first feeding are carried out in the same unit

There is also special hatching equipment that may be placed directly into the first feeding units (Fig. 12.9). The space, water supply and water outlet to the first feeding units can then also be used for storing and hatching of the eggs. In one arrangement the hatching installation consists of a perforated inner bottom and a fixed exterior bottom where the hatching substrate can be attached. It is placed inside the tank on legs and a special larvae outlet is directed into the tank outlet. The advantage with this system is that the same unit is used for egg storage, hatching and first feeding. Therefore less space is needed and no separate hatchery is necessary. Problems with disease, due to the lack of a special isolated area, can be a disadvantage with the system; neither is it as suitable for specialized egg production. The system is most commonly used in smaller farms.

Figure 12.9 Tank for combined egg storage, hatching and start feeding.

References

1. Tucker, C.S., Robinson, E.H. (1990) *Channel catfish farming handbook*. Van Nostrand Reinhold.
2. Stickney, R.R. (1992) *Culture of non salmonid freshwater fish*. CRC Press.
3. Stickney, R.R. (1994) *Principles of aquaculture*. John Wiley & Sons.
4. Tucker, J.W. (1998) *Marine fish culture*. Kluwer Academic Publishers.
5. Lucas, J.S., Southgate, P.C. (2003) *Aquaculture, farming aquatic animals and plants*. Fishing News Books, Blackwell Publishing.
6. Mangor-Jensen, T. (2001) *Klekkeridrift. I kveitemanualen*. Havforskningsinstituttet, Bergen (in Norwegian).
7. Van der Meeren, T. (2002) *Gyting, innsamling, innkubering og klekking av egg. I havbruksrapport 2002*. Havforskningsinstituttet, Bergen (in Norwegian).
8. Danielberg, A., Berg, A., Lunde, T. (1993) Design of inlets for yolk sac larvae of Atlantic halibut (*Hippoglossus hippoglossus* L.). In: *Fish farming technology. Proceedings of the first international conference on fish farming technology* (eds H. Reinertsen, L.A. Dahle, L. Jørgensen, K. Tvinnereim). A.A. Balkema.
9. Myre, P., Danielberg, A., Berg, L. (1993) Experiments on upwelling tank systems for halibut larvae (*Hippoglossus hippoglossus* L.). In: *Fish farming technology. Proceedings of the first international conference on fish farming technology* (eds H. Reinertsen, L.A. Dahle, L. Jørgensen, K. Tvinnereim). A.A. Balkema.
10. Reitan, K.I., Evjemo, J.O., Olsen, Y., Salvesen, I., Skjermo, J., Vadsetin, O., Øye, G., Danielsberg, A. (1993) Comparison of incubator conceps for yolk sac larvae of Atlantic halibut (*Hippoglossus hippoglossus* L.). In: *Fish farming technology. Proceedings of the first international conference on fish farming technology* (eds H. Reinertsen, L.A. Dahle, L. Jørgensen, K. Tvinnereim). A.A. Balkema.

13 Tanks, Basins and Other Closed Production Units

13.1 Introduction

The main purpose of a closed production unit is to create a restricted volume where the fish or other aquatic organisms can be fed in a good water environment. For fish, the units may be used from first feeding of fry up to on-growing. Closed production units are used for both freshwater and seawater species. A unit includes the water inlet, the production unit and the water outlet (Fig. 13.1).

To get a closed production unit to function as optimally as possible, a number of requirements can be set in addition to the main requirement that the unit should produce as much fish as possible at low cost. These are as follows:

- The supplied water, including the oxygen dissolved in the water, should be distributed evenly throughout the entire production unit
- The fish should be evenly distributed in the total production volume
- The fish must be transported effectively in and out of the unit
- The unit should require a minimum of manual cleaning
- The inner surface should be smooth and require little cleaning and maintenance
- The unit should be easy to operate, which means it should be easy to clean, to remove dead fish and to perform other handling fish tasks
- The unit should have low investment costs per unit effective farming volume.

This list shows that there are a number of requirements that need to be fulfilled before a production unit functions optimally. However, compromises have to be made because it is difficult to fulfil all requirements. For example, some species prefer to lie on the bottom and do not utilize the entire water column (non-pelagic species). This chapter gives a survey of the design and construction of closed production units. The focus is on tanks with a circular flow and species that utilize the entire water column. However, there is a wealth of general information for all types of closed production units.

13.2 Types of closed production units

The flow pattern of the water in the unit can be used to classify closed production units (Fig. 13.2):

- Production units with a circulating water flow
- Production units with a one-way water flow.

Production units with a circulating water flow may again be separated into tanks with a circular flow pattern (as is most common), or oval tanks that have an oval flow pattern, of which there are several types;[1,2] for example, Foster Lucas tanks and Burrow tanks, as well as different ovals and pipe connections. Among the traditional tanks with circular water flow, one type may be defined as a farming silo. This is a circular tank of greater height than diameter, i.e. a tower. It is normally difficult to achieve satisfactory water exchange and self-cleaning in silos (see Section 13.7).

Earth ponds belong to the group of production units with one-way water flow and represent the oldest type of closed production units used for fish production. They are mainly used for extensive fish farming; i.e. there is a low production per unit farming volume. Earth ponds are described separately in Chapter 14 because they have so little in

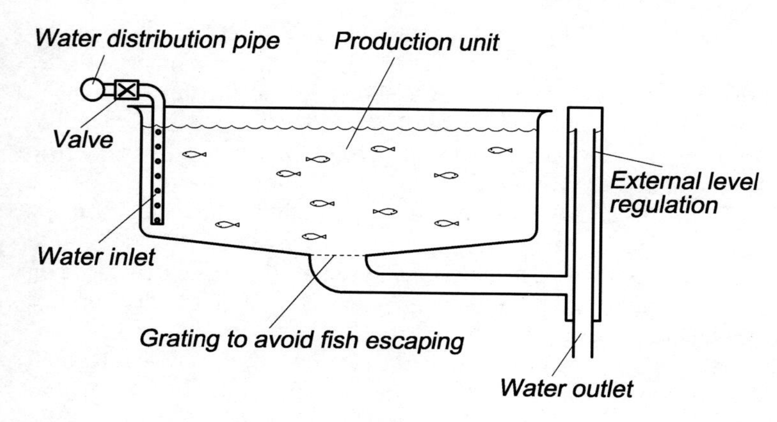

Figure 13.1 The main components in a closed production unit include the water inlet, the storage unit and the water outlet. A circular tank is used as an example.

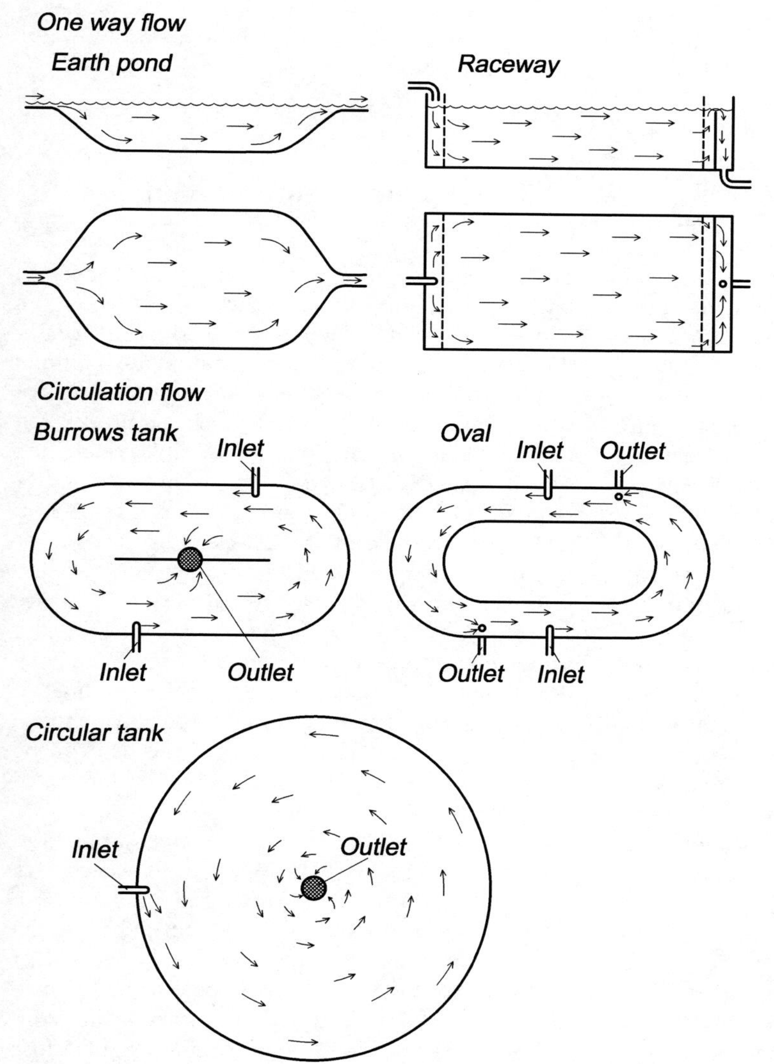

Figure 13.2 Production units can be separated into units with a circulating water flow and those with a one-way water flow.

common with other closed production units such as tanks. According to the requirements for closed production units, the main interest in using ponds is the low initial cost per unit farming volume. In addition, normally a natural ecosystem that can be utilized is created inside the ponds. In some countries such as Norway, no new permits for earth ponds are given unless they are dried once a year. This is because it is quite difficult to control disease as pathogenic micro-organisms may survive in the earth. If the ponds are dried during the winter season these micro-organisms will probably be killed. A layer of lime which increases the pH may also be used as a disinfectant. If the ponds are covered with a plastic tarpaulin (polyvinyl chloride (PVC) or polyethylene (PE)) the problems are avoided because such ponds may be cleaned inside, but this is not a normal pond design, and such installations are more like a traditional tank with no ecosystem inside.

A further development of the earth pond is the raceway, which also uses the one-way flow pattern. This is a fixed construction often of concrete, built as a long rectangle. The water is supplied at one end and the outlet is located at the opposite end. Raceways are quite commonly used for various species throughout the world. However, raceways require quite large amounts of water to have effective hydraulic self-cleaning of their total volume,[3,4] and even then it is very difficult to get good cleaning results. Normally, some kind of mechanical equipment is necessary for additional cleaning of the raceway. It is therefore important to create a good flow pattern inside the unit, with a correctly designed flow inlet and outlet, to ensure uniform water flow through the entire cross-sectional area and length of the raceway to reduce the requirement for manual cleaning. During the past few years, a specially designed raceway with a very low water level (10–50 cm) has been developed.[5] The unit is specially designed for fish species that need a bottom to lie on and do not utilize the entire water column, for example halibut and wolf fish. The unit is constructed so that it can be installed in tiers, one above the other.

Closed floating cages are also a type of closed production unit. In this case, both circulating water flow and one-way flow systems have been tried in different variants that have been constructed using different materials. In one variant the traditional net bag in a sea cage has been substituted by plastic sheeting. Water is pumped into the bag tangential to the edge and the outlet is placed in the centre of the unit. This creates water circulation inside the bag (see, for example, Ref. 6). The advantage of this type of unit is that it lies on the water surface and there is only a small head to overcome to pump the water into the cage, compared with, for instance, closed production units placed on shore and where seawater is normally pumped several metres.

As shown, a closed productions unit can be built in several ways and can have different water flow patterns. The design of the production units, however, depends on the type of aquatic organism to be grown and its requirements regarding water distribution and bottom conditions.

13.3 How much water should be supplied?

The reason for adding water to a closed production unit is to give the fish or shellfish access to oxygen and remove waste products excreted by the fish. In this way, a water environment that creates optimal conditions for growth is established. The amount of water that must be supplied to a closed production unit depends on a number of variables, including species, fish density, growth stage and rate, water temperature, whether adding pure oxygen or not, and whether hydraulic self-cleaning occurs or manual cleaning is required.

Tables showing the supply of water necessary to satisfy the oxygen requirements of fish and shellfish at different water temperatures have been developed for many species. The amount of water that must be added to the production unit to cover the oxygen consumption of the fish can be calculated based on the amount of fish and water temperature.

Provided the quality of the water supplied is acceptable, one way to regulate the necessary supply of new water to a closed production unit can be to monitor the oxygen concentration in the outlet. This must be at an acceptable level for optimal growth of the actual species, and is normally around 7 mg/l. By using electronically controlled actuator valves for controlling the inlet flow, it is possible to have fully automatic flow control based on the oxygen concentration in the outlet water.

The amount of inlet water can be reduced by adding pure oxygen to the incoming water or inside the production unit. With high fish density and rapid growth, consideration must, however, be given to the concentrations of CO_2, suspended solids (SS) and NH_3 that might become excessive when adding oxygen. If this is the case, these substances need to be removed; this requires water treatment as is carried out in a water re-use system (see Chapter 10).

The hydraulic forces in the water supplied to the tank may also be used to clean it; this is known as hydraulic self-cleaning. Extra requirements then apply to the amount of added water; these are described in Section 13.7. This gives another method for calculating the necessary supply of new water. If this water supply does not fulfil the requirements for oxygen, pure oxygen gas must be added to make up the shortfall.

13.4 Water exchange rate

The water exchange rate indicates how quickly the water in closed units is exchanged. This can be defined as the period for which a specific water molecule stays in the unit before leaving via the outlet (Fig. 13.3). As the new incoming water will be mixed with the 'old' water in the tank, the outlet water will always contain both new and old water. It is important to realize this, and means that if one tank volume is run into a tank full of old water, only part of the old water is exchanged, not all. To describe this, the term ideal water exchange is used. When expecting ideal water exchange and adding 1 l of new water to 100 l of old water, the new and old water will be mixed immediately; for example, the addition of 1 l red water to 100 l of clear water instantly results in pink water. This is a simplification, but it helps us to understand better what is happening with the water exchange, and does not require difficult equations.

Mathematically, the water exchange rate can be calculated as follows (based on developing a differential equation):

$$F = (1 - e^{-t/th}) \times 100$$

where:

t = time after start of filling water into the unit
th = time necessary to fill one tank volume at the actual water flow rate; also known as the theoretical retention time
F = water exchange rate (proportion of the water volume in the unit that is exchanged after time t)

Example
Fifty litres of new water is added to a tank that contains 100 l water, described as old water, over a period of 5 minutes (i.e. 10 l/min). The same amount of old water flows out through the outlet because the water volume and level are constant. How much of the water volume is exchanged after 5 and 10 min, respectively?

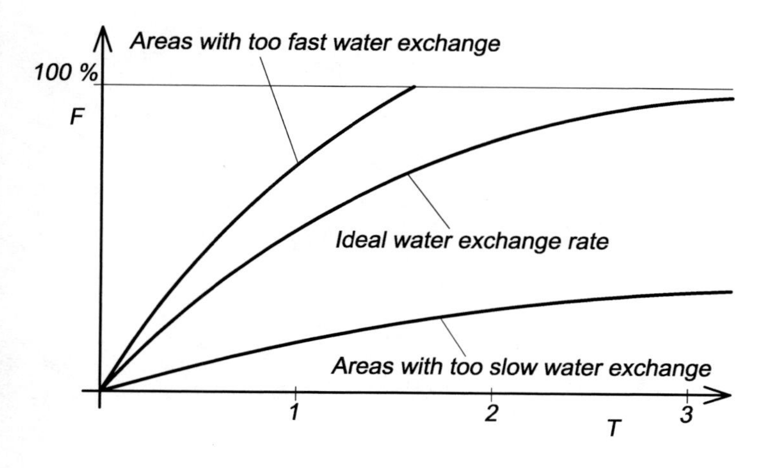

Figure 13.3 Amount of water exchanged in a tank in relation to time.

Setting $t = 5\,min$

$$th = 100\,l/(10\,l/min) = 10\,min$$

$$F = (1 - e^{-5/10}) \times 100 = (1 - 0.605) \times 100 = 39.5\%$$

Setting $t = 10\,min$

$$F = 63.2\%$$

This means that by adding a water volume equal to the tank volume only 63.2% of the water is exchanged, not all as might be expected, the reason being that new and old water are mixed.

13.5 Ideal or non-ideal mixing and water exchange

The aim in designing a closed production unit including the inlet and outlet, is to achieve the most effective mixing as possible of the new incoming water to the old water together with a good exchange of water in the entire farming volume. No new water must go directly to the outlet, by what could be called a short cut; neither must there be areas or zones in the unit were there is small or no exchange of water, so called 'dead zones' (Fig. 13.4). Since water exchange does not occur in the completely dead zones, no new water or oxygen is supplied, so the fish will prefer not to stay there and the effective farming volume will be reduced below the real tank volume. If there are short cuts, part of the inlet water will go directly to the outlet without having been properly utilized by the fish; this results is non-ideal mixing. Short cuts will also give zones in the unit where the water flows much more slowly and than elsewhere water exchange will therefore not be satisfactory.

A picture of the velocity gradients within the production unit can be found by using a specially designed small propeller – a velocity meter – to measure the flow rate at different points in the tank both horizontally and vertically (different depths). Eventually, dead zones and zones where the water flow is too fast will also be identified.

A number of factors, including design of the tank, the inlet and the outlet, will affect water exchange. Before starting to use a new tank design or inlet or outlet system, it is advantageous to test the flow pattern in the tank and find the velocity gradients.

13.6 Tank design

Several designs of tanks with circular water flow are in use. What is important when choosing is that the new water is uniformly distributed throughout the entire tank volume. Round or polygonal (6–8 edges) tanks with a circulating flow pattern are suitable because they have no dead zones provided that the inlet and outlet are correctly designed. Square tanks will, however, have dead zones in each corner and the effective farming volume is therefore not so large; for this reason square tanks are not recommended. Square and rectangular tanks with cut corners have, however, been shown to be quite good; experience with tanks having a total side

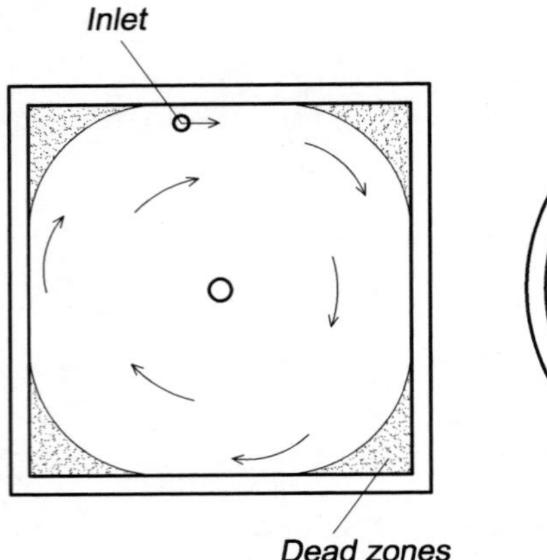

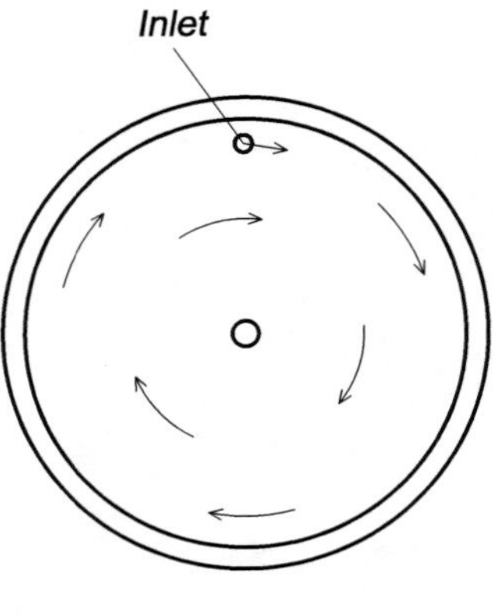

Figure 13.4 Water exchange rate in tanks with dead zones.

length of a and corners of size $a/5$ shows they are well suited[7] (Fig. 13.5). For other tank designs such as raceways or earth ponds, it is far more difficult to avoid dead zones, and the effective fish production volume is normally less than the actual tank volume.

When selecting a tank design, it is also important to take into account the utilization of the area; square tanks with cut corners utilize this well, achieving several m^3 of farming volume per m^2 surface area. Raceways will also utilize the area satisfactorily.

Utilization of the tank construction material is another factor that must be considered when deciding the shape of the tanks. Circular tanks will have the best utilization of the construction materials. The pressure of the water is equally distributed all around the circumference of the tank and therefore a thinner wall may be used than for square tanks.

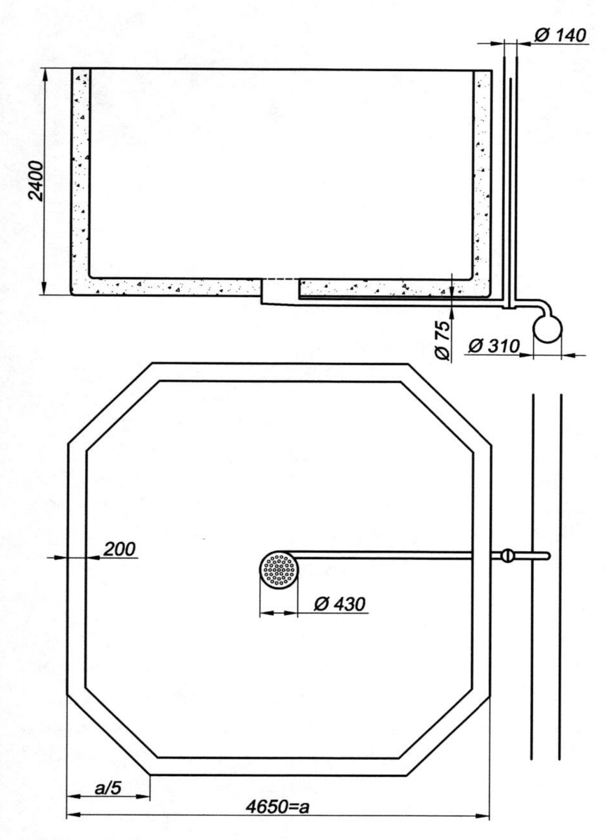

Figure 13.5 Example of an octagonal tank (a square tank with cut corners).

In square tanks the forces are greatest in the middle of the sides, and there is an accumulation of forces in the corners. The height of the tank will also be important because the pressure on the tank walls and hence the necessary thickness will increase.

The bottom of the tank could be horizontal or have a small slope towards the outlet which is usually in the centre of the tank; however, a part outlet might be in the tank wall, see Section 13.10. There is, however, little benefit from sloping the bottom towards the grating and outlet of the tank when having a correct flow pattern inside the tank. This is because the most important mechanisms for transport of the settled solids (faeces, feed loss) are the water flow and its hydraulic force, not gravity. Even a small upward slope (2–5%) to a centrally placed outlet has been used by the author with good results in tanks with a circulating flow pattern. This also confirms that the most important factor for transport of settled solids to the outlet is the force created by the water flow, rather than the bottom slope and force of gravity. When having non-self-cleaning flow conditions, for instance because fry production requires water flow of low velocity, it is important to have a slope to the outlet grating to be able to utilize graviditational forces. However, this slope must be quite large to really get an effect of gravity. In a filter unit the angle is recommended to be above 55° to utilize the force of gravity to get the settled solids to slide; this is because the density of aquaculture solids is low (1.05–1.2) and almost equal to that of water.[8,9]

The height of the tank compared to the diameter will also affect the water exchange. For tanks with a circular flow pattern, a tank diameter: height ratio of between 2 and 5 has been successfully used. If the tank diameter is 10 m, the height could therefore be between 2 and 5 m. For tanks that do not fall within these ratios, special attention must be given to the design and placement of the inlet and outlet. If the ratio is lower, the inlet should be placed some distance away from the tank wall, closer to the centre of the tank. Tanks where the height is greater than the diameter are often called silos; by using such tanks high production can be achieved per unit area. It is, however, difficult to get a proper water exchange throughout the entire volume in such constructions.

Various materials are used to fabricate tanks (Fig. 13.6). It is important that there is a smooth

Figure 13.6 Tanks of different materials: concrete (A), tarpaulin with a framework of wood plates (B), fiberglass (C), stainless steel (D), coated steel plates (E).

surface inside the tank to reduce problems with fouling, and that the material does not release any toxic substances into the farming water. Glass-reinforced plastic is a commonly used material for tanks, because it can be produced with a very smooth surface; small tanks are also light and easy to move. The tanks are either delivered completely finished or as elements that are screwed together on site. Plastic (polyethylene, PE) may also be used; this is also a light and cheap material. New material has a very smooth surface; however, is it more prone to ageing and the surface gradually becomes less smooth. The surface also scratches more easily. The tanks are either made of plates welded together into tanks or the tanks are rotation cast (a special casting process). Concrete is much used for larger tanks, where the price is competitive; concrete may also be mixed on site or prefabricated elements can be joined together. The other material that has been used to some extent is metal; for example steel plates (stainless, acid-proof or coated) or aluminium (special quality). Tanks of tarpaulin with a frame of steel or wood represent a low cost easily movable construction.

13.7 Flow pattern and self-cleaning

A flow pattern will be created inside a production unit having a water inlet and outlet. It is important that this flow pattern encompasses the entire unit so that all the fish can come into contact with flowing water. The flow pattern depends on the design of the production unit.

In a tank with a circulating water flow and correctly designed inlet and outlet, two flow patterns will occur: the primary flow and the secondary flow (Fig. 13.7). The primary flow causes even distribution of the water in the horizontal plane, while the secondary flow will clean the tank walls and bottom.

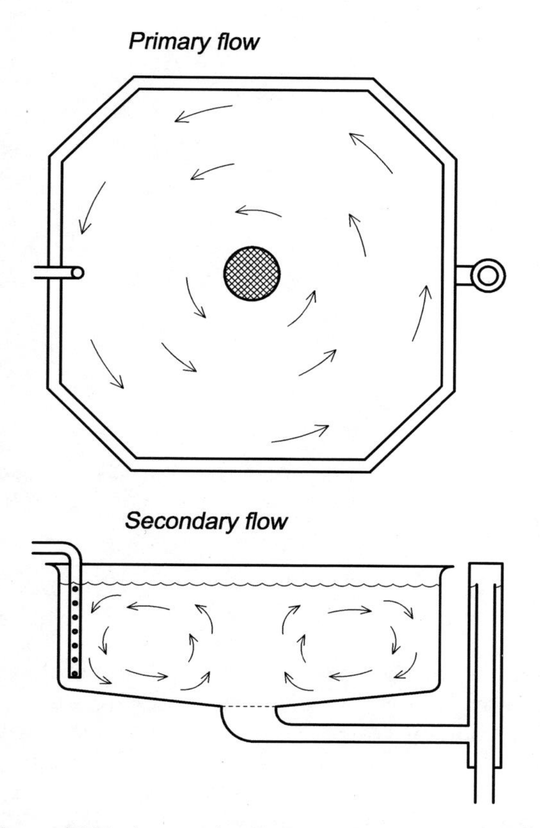

Figure 13.7 In a tank with a correctly designed inlet and outlet, both a primary and secondary flow will be generated.

In well-designed tanks with correctly designed and constructed inlet and outlet, the incoming water may therefore be used to clean the tank walls and bottom. This process is known as hydraulic self-cleaning. To achieve self-cleaning in a tank, a certain amount of water has to be added; the amount depends on the tank construction. The water velocity at the bottom of the tank must be so high that the settled solids are removed. To ensure transport of settled solids in circular tanks, the recommended bottom velocity to ensure self-cleaning is above 6–8 cm/s.[7] This will also remove algal growth from the tank sides. Inside the tank there will be a velocity profile equal to that in a channel, where the lowest velocity occurs near the bottom due to friction (Fig. 13.8). Bottom water velocities of between 6 and 8 cm/s normally represent a water velocity in the free water mass of between 12 and 15 cm/s.[7] Practical experience has also shown that high fish density promotes self-cleaning of the bottom. A lower velocity could therefore be accepted when the fish density is increased because the movement of the fish results in resuspension of settled solids, so the secondary flow pattern could more easily transport the particles to the drain.

In a correctly designed flow through tank with inlet and outlet, and a circular flow pattern, the water retention time should be between 30 and 100 min for satisfactory self-cleaning.[10] This means a flow through of between 10 and 33 l/m^3 farming volume. A retention time of less than 30 min may create a vortex around the centre drain. The peripheral velocity in the tank may also be so high that the fish will have problems staying there. When using low retention times, a specially designed inlet and outlet are necessary. With retention times above 100 min the self-cleaning effect is decreased and additional cleaning is necessary.

To attain hydraulic self-cleaning, a high volume of water is needed to create a high water velocity inside the tank. Even if the water velocity has yielded improved growth results,[11–13] there is a maximum velocity that not must be exceeded.[14] This will vary according to species and growth stage. Examples here are fry of marine or freshwater species, where only very low velocity is tolerated; to maintain satisfactory water quality for

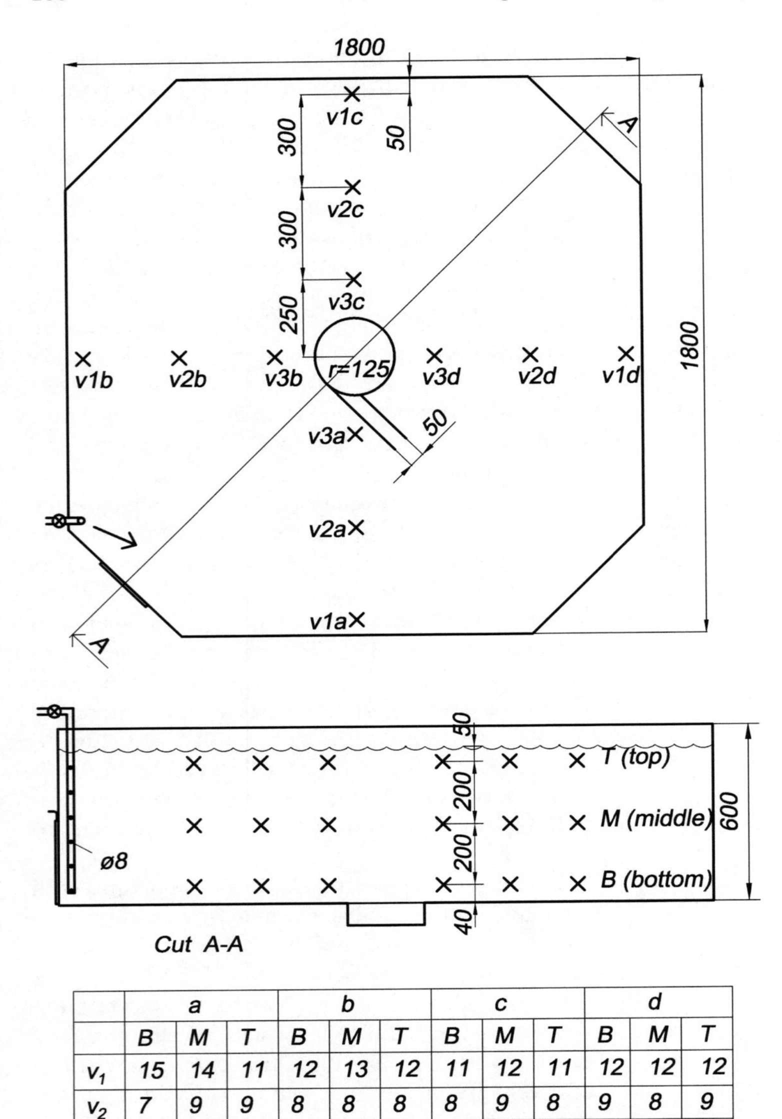

	a			b			c			d		
	B	M	T	B	M	T	B	M	T	B	M	T
v_1	15	14	11	12	13	12	11	12	11	12	12	12
v_2	7	9	9	8	8	8	8	9	8	9	8	9
v_3	7	8	8	6	7	7	7	7	7	7	7	7

Figure 13.8 An idealised velocity profile matrix across A–A in a 1800 × 600 mm tank.

these species is therefore a challenge. Settled particles and fouling on the tank bottom and sides will create a sub-optimal environment, and be a good substrate for unwanted bacterial growth. Regular removing of fouling is therefore absolutely essential. If this is done manually, it is labour intensive, and therefore commercially available automatic systems are preferred. Rotating brushes on the bottom of the tank, powered either by electric motors or by the pressure of the incoming water are one solution. Other solutions include a small turtle-like unit moving around on the tank bottom or installing a washing arm half the pipe diameter in length. Addition of chemicals that remove the fouling, such as oxidizing agents, has also been tried by the author and colleagues.[15]

13.8 Water inlet design

Correct design of the inlet flow arrangement to the tank is necessary to ensure even distribution and mixing of the new incoming water and if self-cleaning is to be attained. This requires the inlet water pipe to enter below the water surface in the tank and the water must pass through a narrow nozzle. The force of the inlet water can then be utilized to create a flow pattern inside the tank (Fig. 13.9). It is also important to spread the incoming water throughout the water column. This can be achieved by the use of several holes, splits or nozzles in the inlet pipe below the water surface. The force of the incoming water will now be distributed throughout the water column, not just in one place. Improved distribution of the new oxygen-rich incoming water is also achieved.

The impulse, as the force caused by the inlet water is called, depends on the water flow and water velocity; it can be expressed as follows for tanks with a circular flow pattern:

$$F = \rho Q\ (v_2 - v_1)$$

where:

F = impulse
ρ = water density
Q = water flow
v_1 = velocity of the water in the tank
v_2 = velocity out of the holes in the inlet pipe.

This equation shows that by increasing Q, the impulse will increase. The same result is achieved by increasing the velocity of the water emerging from the holes in the inlet pipe. Decreasing the cross-sectional area of the holes will increase the velocity of the water. However, the increased velocity will increase the turbulence and hence the head loss. Recommended values are below 1.5 m/s in the inlet pipe, while the velocity in the hole (or split or nozzle) should be below 1.2 m/s.[7]

The inlet pipe can be arranged in several ways depending on the tank design (Fig. 13.10). In tanks with a circular flow, a horizontal spray inlet has the advantage of creating good water distribution (primary flow), but the secondary flow is not optimal. The vertical spray inlet creates both good primary and secondary flow, and is therefore preferred. It is also possible to use a combined vertical and horizontal inlet with good results.

Normally a vertical spray inlet will be placed about one fish width away from the tank wall, so

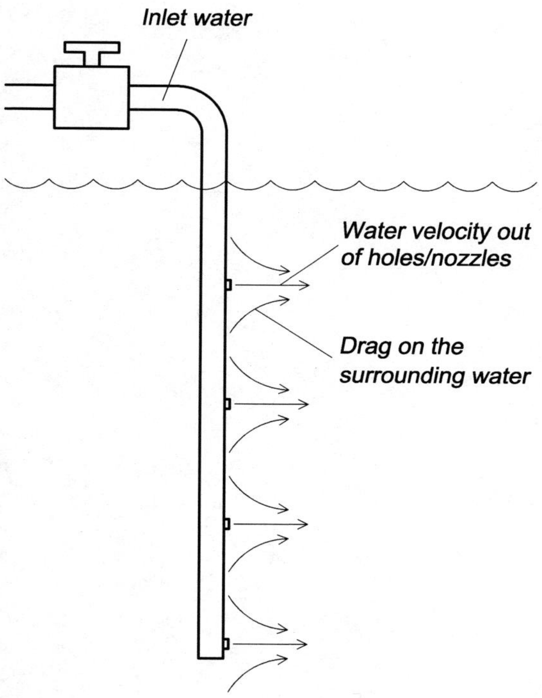

Inlet water

Water velocity out of holes/nozzles

Drag on the surrounding water

Figure 13.9 Having the tank water inlet below the water surface and using nozzles in the inlet pipe will create velocity gradients, and mixing of the water in the tank is achieved.

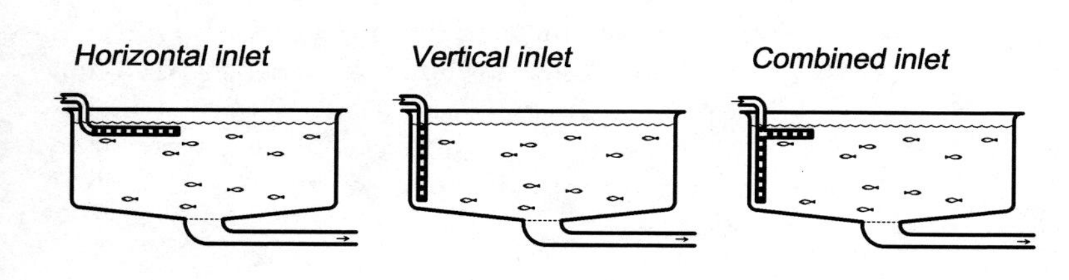

Figure 13.10 Various designs of inlet pipe in tanks with a circular flow pattern.

that the fish can pass behind, and to avoid too much friction from the tank walls. If it is too close to the wall, friction against the wall will reduce the impulse. However, in a low tank designed with a diameter: depth ratio of less than 0.2, the inlet has to be placed further into the tank closer to the drain, to create a good flow pattern. In silos is it especially difficult to get good inlets and effective transfer of the impulse, and hence effective water exchange throughout the water volume. However it may be possible to use several water inlets in the tanks to improve the flow pattern; testing of the velocity profile is recommended in such cases. Depending on the current velocity from the holes in the inlet pipe, the current velocity in tanks with circular flow is normally in the range 0.15–0.25 m/s.[7]

In raceways it has proved to be difficult to create an inlet that distributes the water in a uniform way throughout the entire cross-sectional area and total length. It is important that the impulse is distributed over as large a part of the cross-sectional area as possible. Because of the continuous reduction in water flow velocity close to the bottom due to friction, there have been experiments in which water was added at several places over the length of the raceway to improve the velocity over the total area; this, however, increases the costs.

The velocity of the inlet water out of the holes or nozzles in the inlet pipe (V_2) depends on the design of the nozzle (hole), the area and the amount of water that has to pass. It can be expressed as follows (based on the continuity equation; see Chapter 2):

$$V_2 = \frac{Q}{\sum A}$$

where:

V_2 = velocity out of the nozzles
Q = water flow out of the nozzles
ΣA = total cross-sectional area of all the nozzles.

The relation between the water velocity in the inlet pipe and the velocity out of the nozzles will be as follows:

$$V_2 = \frac{V_0 A_0}{\sum A}$$

where:

V_0 = velocity in the inlet pipe
A_0 = area of the inlet pipe.

Example
An inlet pipe to a circular tank is designed for a water flow (Q) of 50 l/min. Suggest an appropriate pipe diameter and area of the nozzles (holes).

First, calculate the area of the inlet pipe:

$$A = Q/V$$

Transform the units so that they correspond and take a maximum water velocity of 1.5 m/s.

$$50\,\text{l/min} = 0.00083\,\text{m}^3/\text{s}$$

$$A = \frac{0.00083}{1.5} = 0.000553\,\text{m}^2 = 5.53\,\text{cm}^2$$

Calculate the diameter:

$$A = \pi d^2/4$$

$$d = \sqrt{\frac{4A}{\pi}} = \sqrt{\frac{4 \times 5.53}{3.1415}} = 2.65\,\text{cm}$$

In practice the nearest standard dimension will be used.

Calculate the total nozzle/area (cross-sectional area) using a velocity of 1.2 m/s.

$$\sum A = Q/V = \frac{0.00083\,\text{m}^3/\text{s}}{1.2\,\text{m/s}} = 0.00069\,\text{m}^2 = 6.9\,\text{cm}^2$$

This must then be divided by the number of holes used in the total water column.

To force the water through the inlet pipe results in a head loss. Because of this, a minimum head (available water pressure) is necessary to get the water to flow through the nozzles/holes. Different shapes of the nozzles/holes will result in different head loss because of different degrees of turbulence created in the nozzles. This must be taken into consideration when constructing the inlet pipe.

Example
Head loss in the inlet pipe

The inlet pipe has a diameter of 63 mm and a water flow of 300 l/min is used. This gives a water velocity of about 1.6 m/s. Find the head loss when increasing the water velocity from 1 to 2.5 m/s ($f = 0.024$).

Water velocity (m/s)	*Head loss per metre of pipe (mH_2O)*
1.0	*0.019*
1.6	*0.05*
2.0	*0.08*
2.5	*0.12*

With a water velocity of 2.5 m/s in the inlet pipe, it is necessary to have a pressure in the pipe of 0.1 mH_2O per metre of pipeline to achieve the necessary water flow.

13.9 Water outlet or drain

The tank outlet or drain normally has two functions: 1, to remove the waste from the tank as quickly as possible, before the leakage of nutrient starts; 2, to maintain the correct water level inside the tank. Correct design of the outlet is also important for the water exchange rate and to ensure effective self-cleaning. Incorrect specification of the outlet system or the outlet pipe may result in settling of uneaten feed particles and faeces, and the outlet system will function as a settling basin. This situation can be rectified by shock draining the tank, by removing the device that controls the tank water level. Shock draining will increase the water velocity through the outlet above normal and eventually settled solids will be dragged out. If there are large number of particles coming out, there may be places in the outlet pipe system where settling of solids occurs if the design of the outlet system is sub-optimal. It should not be necessary to shock drain a well-designed outlet system to remove particles that have settled in the outlet. The outlet pipe should be designed for water velocities above 0.3 m/s to ensure no settling of solids. Velocities above 1.5 m/s in the outlet will result in rough treatment of the particles which may break up. This makes later filtration of the outlet water more difficult.

It is important to treat the outlet water as gently as possible to avoid increased particle breakage. Elbows and other pipe parts that create additional turbulence should therefore be avoided. It is important to use long bends and small bend angles in the piping. It is also important to understand that the inlet and outlet of a tank are designed for a given flow rate, but with some latitude outside which the inlet and outlet will not function optimally because the velocity is either too low or too high.

The head loss in the outlet system can be seen by comparing the water level inside the tank with the top level of the water in the outlet system. A large difference means a high head loss.

Example
The supply of water to a fish tank with a farming volume of 6 m^3 should be between 60 and 200 l/min to ensure sufficient self-cleaning. Find a suitable diameter for the outlet pipe.

Choose a velocity of 0.5 m/s and calculate for the smallest amount of water.

$$A = Q/V$$

$$Q = 60\,\text{l/min} = 0.001\,\text{m}^3/\text{s}$$

$$A = \frac{0.001\,\text{m}^3/\text{s}}{0.5\,\text{m/s}} = 0.002\,\text{m}^2$$

$$A = \pi r^2$$

$$r = 0.025\,\text{m} = 2.5\,\text{cm}$$

The next task is to calculate the water velocity in the pipe with the largest amount of water when the radius of the pipe is 2.5 cm.

$$V = Q/A$$

$$Q = 200\,\text{l/min} = 0.0033\,\text{m}^3/\text{s}$$

$$V = \frac{0.0033\,\text{m}^3/\text{s}}{0.002\,\text{m}^2} = 1.67\,\text{m/s}$$

A larger dimension for the outlet pipe is recommended because this velocity is rather high. This also shows that the outlet system is not designed optimally if there is too large a variation in the water flow.

When designing the outlet system, it is important that the outlet creates a drag on the water in the tank. Particles lying close to the outlet will be forced against the outlet by the reduced pressure there.

The design of the outlet can be either of the flat or tower type (Fig. 13.11). The flat outlet contains a horizontal screen inside the fish tank, adapted to the fish size. This normally covers an outlet pot. From the outlet pot the outlet pipe goes to a vertical standing pipe where the water level inside the tank is controlled. It is important to increase the speed through the holes in the outlet screen to avoid fouling and blockage. Velocities above 0.3–0.4 m/s are recommended. The outlet screen is a perforated sheet of stainless steel, aluminium or plastic. Good results have been achieved by using oblong slots instead of holes in the outlet screen.[14,16] Slots do not clog as quickly and are easier to clean. The holes or slots are recommended to be as large as possible, but of course not so large that fish escape. It may also be possible to run the tank without any outlet screen, because the fish prefer not to go down in the outlet pipe. However, this is species dependent, and some species that have a crowding behaviour, such as eel, will go down. For salmonids, the author's experience is that it is possible to run without any outlet screen. If this is done it is necessary to install a trap in the outlet system to collect the dead fish (see Section 13.11).

The outlet pot, if used in the tank, has been shown to be a critical part of the construction (Fig. 13.12). Settling of solids is very common here. It is therefore important not to reduce the water velocity in this pot too much, both to prevent settling of particles and also to reduce the amount of fouling. Good results are obtained by using an eccentric outlet from the pot; here the outlet pipe from the pot is not in the centre but to one side, and a swirl is therefore created inside the pot. Another possibility is to avoid the use of the outlet pot

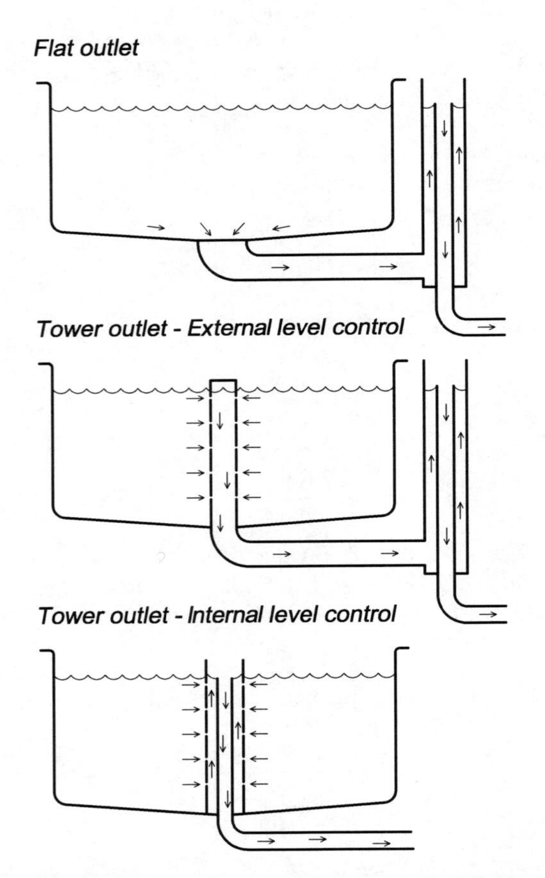

Figure 13.11 Flat or tower outlets are typical designs.

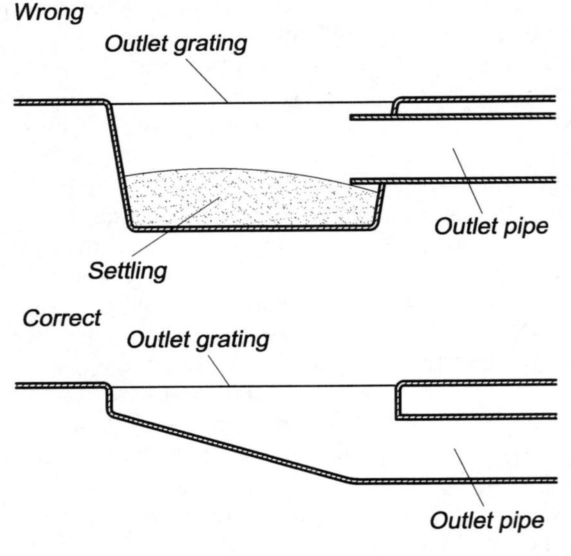

Figure 13.12 Correct design of the outlet pot is important.

totally by having the outlet pipe directly into the tank bottom. This requires the use of another type of outlet screen inside the tank (not a horizontal one) to ensure enough hole or slot area. Outlet screens may, for instance, be shaped like a cone, pyramid or a small tower, not reaching the water surface.

There are two types of tower outlet, with either internal or external regulation of the water level. In both cases there will be a tower inside the tank going from the bottom to the water surface, consisting of a screen with holes or slots that prevent the fish leaving the tank. The disadvantage of this outlet is that the presence of the tower in the middle of the tank may make it difficult to handle the fish in the tank. With regard to self-cleaning, no differences have been observed between flat and tower outlet systems.

Instead of taking the outlet from the bottom of the tank, a siphon can be used. Here the outlet from the tank goes over the top edge of the tank; however, this will require a fixed water level. The advantage with the system is that it is not necessary to have any pipes going out of the bottom of the tank, which is therefore only a basin. This reduces the cost and increases the second hand value. The disadvantage is that it is not possible to reduce the water level because it is fixed; a pump must be used to drain the tank completely.

Both flat and tower outlets, include a level control to keep the water inside the tank at a given height. Level controls can be adjustable or fixed. Adjustable level controls can be incremental or continuous. Fixed level controls have no possibilities for adjustment and therefore there must be a bypass for complete drainage of the tanks. For large volume tanks (>100 m^3) adjustable level controls are quite expensive and seldom used. Usually fixed or perhaps incremental controls are installed. A continuous level control may be built by having a sliding pipe inside another pipe. The same principle may be used for a stepwise control, but here the level is controlled by addition and removal of pipe parts.

13.10 Dual drain

It is also possible to use a double outlet, i.e. a dual drain (Fig. 13.13).[14,17–19] Such an outlet allows part of the water flow (up to 5% of the total flow) to be taken through a separate centre drain located in the bottom of the tank, while the main water flow is taken out from the water column above the tank bottom through some type of tower outlet. This can be a traditional tower outlet or a small tower reaching 5–10 cm up from the tank bottom. In this way the tank is used as the first effluent treatment system. A particle purification step is achieved because the tank functions as a swirl separator using the 'tea-cup' principal. The circular flow pattern in the tank will cause the solids to be colleted in the centre of the bottom where the particle outlet is located. Normally over 90% of the solids be taken out via the bottom outlet. Hence the particle concentration will be significantly reduced in the main outlet that is taken out slightly higher up the water column.

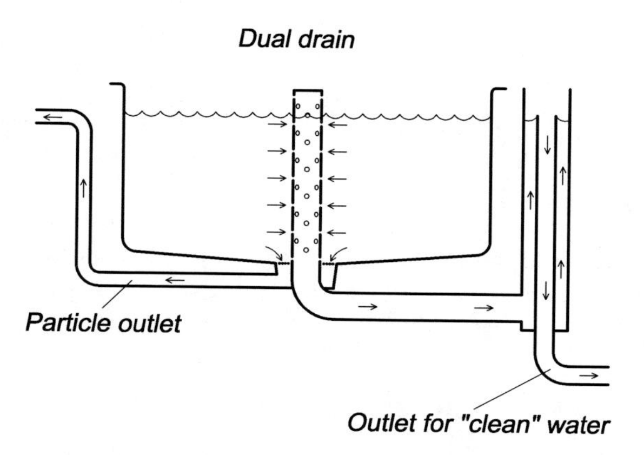

Figure 13.13 The double or dual drain outlet will separate particles from the water.

The main water flow may also be taken out via a sidewall.[4] By taking out up to 20% through the bottom and the rest through the outlet in the upper part of a tank wall, good separations have been achieved. A specially designed square tank with cut corners, with the main inlet and main outlet in the corners and the particle outlet in the centre, has also been used with good results.[20] Here less than 1% of the total water flow was taken out via the particle outlet. In both the above two systems the flow pattern inside the tank gave satisfactory self-cleaning. In the last system the place on the tank bottom where the main effluent collected was easily regulated by changing the direction of the nozzles in the inlet pipe; these were adjusted to force the effluent towards the middle of the tank where the particle outlet was located.

By using dual drain systems the cost of particle purification will be reduced because only a minor part of the total water flow has to be purified; the size of the particle filter will be reduced dramatically. If using a water re-use system (Chapter 10), the water in the main outlet may be re-used without going through a particle separation step as the concentration of particles is already low enough.

13.11 Other installations

There may also be other installations that can be integrated into the tank. By using a specially designed outlet screen that can be opened from the surface, it is possible to knock dead fish lying on the outlet screen out of the tank and through the outlet system; it is then necessary to have a collector for the dead fish in the outlet piping system. This avoids dragging dead fish up through the production water.

Handling facilities can also be an integral part of the tank construction. Additional outlets used only for guiding the fish to a common centre (see Chapter 17), or hatches in the tank walls for transport of fish through channel systems, can be integrated.[21] The tank may also be constructed to include guides for adapting tank internal grading grids (see Chapter 17).

A feed detection unit may also be integrated into the tank outlet system.[22]

References

1. Wheaton, F.W. (1977) *Aquacultural enginering*. R. Krieger.
2. Cripps, S.J., Poxton, M.G. (1992) A review of the design and performance of tanks relevant to flatfish culture. *Aquacultural Engineering*, 11: 71–91.
3. Lawsons, T.B. (1995) *Fundamentals of aquacultural engineering*. Kluwer Academic Publishers.
4. Timmons, M.B., Riley, J., Brune, D., Lekang, O.I. (1999) Facilities design. In: *CIGR Handbook of Agricultural Engineering, Part II Aquaculture Engineering* (ed. F. Wheaton), pp. 245–280. American Society of Agricultural Engineers.
5. Øiestad. V. (1995) Shallow raceways as the basis for industrial production centers of seafood. In: *Quality in aquaculture*. European Aquaculture Society special publication no. 23.
6. Solaas, F., Rudi, H., Berg, A., Tvinnereim, K. (1993) Floating fish farms with bag pens. In: *Fish farming technology, Proceedings of the first international conference on fish farming technology*. (eds H. Reinertsen, L.A. Dahle, L. Jørgensen, K. Tvinnereim). A.A. Balkema.
7. Tvinnereim, K. (1994) Hydraulisk utforming av oppdrettskar. Brukerrapport. *SINTEF report STF60 A94046* (in Norwegian).
8. Chen, S., Coffin, D.E., Malone, R.F. (1993) Production, characteristics, and modeling of aquacultural sludge from a recirculating aquacultural system using a granular media filter. In: *Techniques for Modern Aquaculture, Proceedings of aquacultural engineering conference, Spokane, Washington*, (ed. J-K. Wang), pp. 16–25. American Society of Agricultural Engineers.
9. Wong, K.B., Piedrahita, R.H. (2000) Settling velocity characterization of aquacultural solids. *Aquacultural Engineering*, 21: 233–246.
10. Skybakmoen, S. (1991) *Kar og karmiljø, temahefte*. Aga as (in Norwegian).
11. Jobling, M, Jørgensen, E.H., Christiansen, J.S. (1993) Growth performance of salmonids exposed to different flow regimes. In: *Fish farming technology. Proceedings of the first international conference on fish farming technology* (eds H. Reinertsen, L.A. Dahle, L. Jørgensen, K. Tvinnereim). Balkema.
12. Jobling, M., Jørgensen, E.H., Arnesen, A.M., Ringø, E. (1993) Feeding, growth and environmental requirements of Arctic charr, a review of aquaculture potential. *Aquaculture International*, 1: 20–46.
13. Davidson, W. (1997) The effects of exercise training on teleost fish, a review of recent literature. *Comparative Biochemistry and Physiology*, 117a: 67–75.
14. Timmons, M.B., Summerfelt, S.T., Vinci, B.J. (1998) Review of circular tank technology and management. *Aquacultural Engineering*, 18: 51–69.
15. Lekang, O.I., Andreassen, I., Nergård, R. (2003) *Design of start feeding tanks for wolf fish.* ITF conference report 194. Norwegian University of Life Science.
16. Piper, R.G., McElwain, I.B., Orme, L.E., McCraren, J.P., Fowler, L.G. Leonard, J.R. (1982) *Fish hatchery management.* US Fish and Wildlife Service.
17. Lunde, T., Skubakmoen, S. (1993) Particle separation integrated in enclosed rearing units. In: *Fish farming technology. Proceedings of the first international conference on fish farming technology* (eds H. Reinertsen, L.A. Dahle, L. Jørgensen, K. Tvinnereim) A.A. Balkema.
18. Twarowska, J.G., Westerman, P.M., Losordo, T.M. (1997) Water treatment and waste characterization evaluation of an intensive recircualting fish production system. *Aquacultural Engineering*, 16: 133–147.
19. Davidson, J., Summerfelt, S. (2004) Solids flushing, mixing, and water velocity profiles within large (10

and 150 m^3) circulation 'Cornell-type' dual drain tanks. *Aquacultural Engineering*, 32: 245–271.
20. Lekang, O.I., Bergheim, A. & Dalen, H. (2000) An integrated waste treatment system for land-based fish-farming. *Aquacultural Engineering*, 22: 199–211.
21. Lekang, O.I., Fjæra, S.O., Thommassen, J.M. (1996) Voluntary fish transport in land based fish farms. *Aquacultural Engineering*, 15: 13–25.
22. Summerfelt, S.T., Holland, K.H., Hankin, J.A., Durant, M.D. (1995) Hydroacoustic waste feed controller for tank systems. *Water Science and Technology*, 31: 123–129.

14
Ponds

14.1 Introduction

Earth ponds are the most used unit for fish production worldwide, and more than 40% of world aquaculture production is performed in ponds.[1] Ponds are used both for fish and shellfish and at different life stages. Important species grown in ponds, include different types of carp, catfish, shrimps and prawns.[2,3] Ponds are normally used in extensive production[4,5] and to some extent in more intensive production; here, however, the construction is less optimal (see Chapter 13).

An earth pond for aquaculture farming is usually defined as a pond where a natural ecosystem is created inside. This is the major difference between earth ponds and other closed production units, and the reason why they are described separately. The water exchange in the pond is normally very small and it will also function as a settling pond, so faeces and particles will settle on the bottom. There will be none or very little self-cleaning in the pond. When establishing the pond this must be taken into consideration, so that there is some spare capacity.

Inside the pond there can be a monoculture or polyculture. If using a polyculture, the natural food created in the pond (phytoplankton, zooplankton, aquatic insects, benthic organisms and the vegetation) can be utilized optimally by different species.

14.2 The ecosystem

An established ecosystem in the pond includes full algal photosynthesis. During the day the algae produce oxygen by photosynthesis, while during the night they consume oxygen. Thus there will be a daily fluctuation in the oxygen level in the pond, and special care must be taken during the night when it may be necessary to supply additional oxygen. Similarly, the pH may fluctuate because photosynthesis fixes carbon dioxide and therefore the pH will increase during the daytime, while at night the algae release carbon dioxide and the pH will drop.

The ecosystem created will affect all nutrients since a nitrogen cycle will occur in the pond: nitrification will transform TAN to NO_2^- and further to NO_3^-. If there are areas in the pond lacking oxygen, dentrification of NO_3^- to N_2 will also occur.

A major benefit of a pond is therefore that it is possible to utilize this biological production, which includes prey that is food for the fish. In addition, there will also be decomposition of waste. However, to achieve this state the water exchange rate must not be too high (see Section 14.3). The major disadvantages with production in ponds is the low production per unit volume, and the difficulties of maintaining control over the water quality and the actual fish biomass.

14.3 Different production ponds

Ponds can be separated into those for fry production and those for on-growing production (Fig. 14.1); the difference is normally the size of the ponds. However, full production ponds are also possible. In such ponds, spawning, fry production and on-growing all occur, although harvesting can be quite difficult. Full production ponds may, for instance, be used in crayfish production (Fig. 14.2).

Figure 14.1 Ponds for on-growing production during (A) the summer (catfish) and (B) the winter (rainbow trout). Warm groundwater is utilized in the winter season to keep the pond free from ice. (C) Empty ponds reveal the construction more clearly.

In a pond for fry production, it is especially important to have a well functioning ecosystem, including photosynthesis. Eggs or newly hatched fry are released into the pond where the on-going ecosystem will produce natural prey for the fry. As the fry grow they will gradually feed on other prey that are also available in the pond. Depending on the desired production, additional feeding of the fry may not be necessary.

To stimulate and increase the development of the natural ecosystem, it is possible to fertilize the pond.[6–8] This increases production of algae and hence production of higher organisms that function as natural prey for the growing fry. It is, however, easy to lose control of the ecosystem, and total breakdown may occur. If fertilizing, it is therefore of major importance regularly to monitor and control changes in the water quality, for instance by monitoring the oxygen content in the pond water.

In on-growing ponds, there is often some type of additional feeding, but this depends on the species. Some species will utilize the plants growing in the pond and the organism created by the ecosystem, but this is normally not enough if high production is wanted; an example here is grass carp. Other species may only use supplied artificial feed, such as catfish and rainbow trout. In on-growing ponds it is easy to overload the system when adding formulated feed and cause problems in the ecosystem which will be put out of balance so that the pond functions in an uneconomic way.

The water flow through a pond having a natural ecosystem must not be too high, otherwise algae and natural prey may flow out with the outlet water, and an imbalance in the ecosystem will occur. Many earth ponds are, however, used in this way. If the fish densities are high the water requirements increase with the fish density and the ponds function as raceways, for which the pond construction is sub-optimal (see Chapter 13). The results are large variations in pond water quality and accumulation of faeces and feed loss.

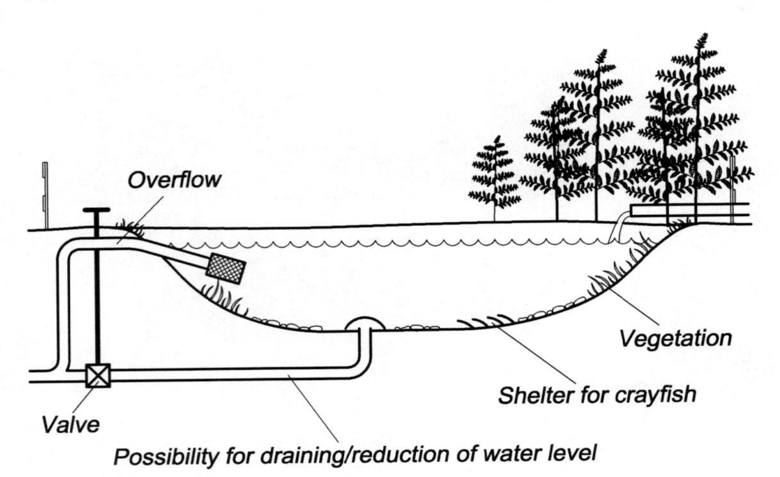

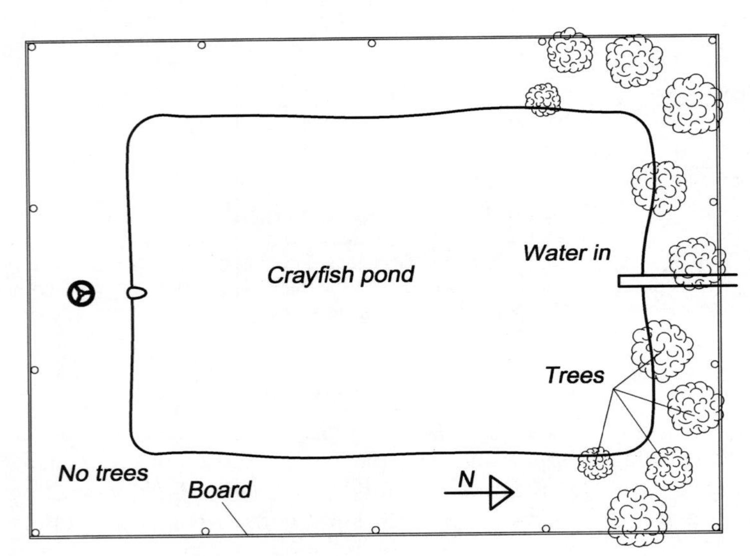

Figure 14.2 A crayfish pond in cross-section and from above.

14.4 Pond types

There are several ways to classify ponds: one is based on construction and another on whether it is possible to drain the pond or not.

14.4.1 Construction principles

Based on their construction three types of ponds can be identified: watershed, excavated and embankment or levee ponds[7] (Fig. 14.3). Watershed ponds utilize the terrain features; for instance, a ravine can be dammed so the construction is quite simple. However, there are few sites that satisfy the requirements for a watershed pond, so this is not a very common pond type. An excavated pond is simply a hole in the ground which is filled with water. Part can be below the water table and in this way water infiltrates into the pond, but this construction is little used.

The main type of construction is the embankment or levee pond. There are several ways to establish such ponds: they can be at ground level, or the levee can be above and the bottom below ground level. For the first type it is necessary to supply material; for the second type the excavated material can be used to construct the leveess which will reduce the cost of establishing the pond. Levee

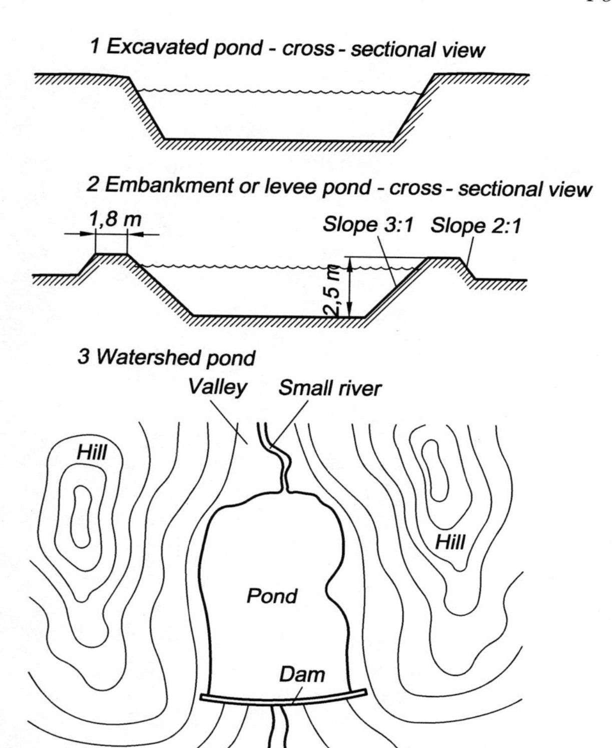

Figure 14.3 Ponds can be divided into: 1, excavated; 2, embankment or levee; 3, watershed ponds.

ponds can be constructed in a flat landscape, and large areas can be used for pond production. It is important that the levee is sufficiently wide to carry traffic, for instance for feeding, maintenance or harvesting.

When constructing the pond it is an advantage to ensure that it is possible to tap the water to a lower level to make drainage possible. Eventually a canal drainage system can be excavated in conjunction with the pond area.

14.4.2 Drainable or non-drainable

Depending on the construction of the outlet system, traditional earth ponds can be divided into drainable and non-drainable ponds (Fig. 14.4); non-drainable ponds are normally larger, up to several hectares. Small natural lakes used for aquaculture production function like a non-drainable pond. The same is the case if a ravine is dammed, watershed ponds are used or if a hole is excavated in the ground. Low establishment cost is the main advantage with non-drainable ponds. Natural lakes can also be used and create a low cost farming volume.

The advantage with drainable ponds is the possibility for a more effective harvesting process and to have control over the water level, because the water can be drained out of the pond. It is simpler to fertilize/feed, and also to supply additional air. While non-drainable ponds are normally run extensively, drainable ponds can be run more intensively, depending on the amount and growth rate of the fish. In intensive drifted ponds the fish can be fed (Fig. 14.5), and additional air supplied periodically.

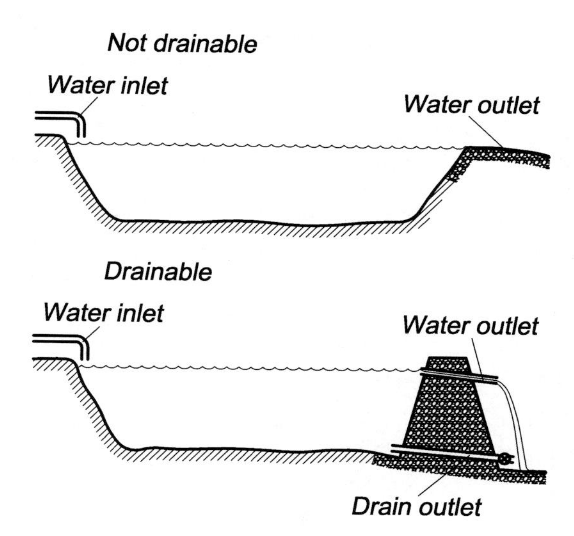

Figure 14.4 Drainable and non-drainable ponds are constructed differently.

Figure 14.5 Feeding from a tractor-trailer with a feed blower to increase production in the pond.

This may also be done without damaging the ecosystem in the pond, if extra care is taken.

In ponds the production is given in kg per hectare of pond surface area. This varies with water temperature, environmental conditions, pond type and the fish species, so it is difficult to give a general value.[9,10] A rough estimate is 1000 kg/ha, but this can vary by a factor of 10 in both directions: over 15 000 kg/ha can be achieved for channel catfish by use of additional feeding and continuous aeration.[2,7]

14.5 Size and construction

The size of the pond varies with species, fish size and site conditions, from fractions of a hectare to several hectares. As the ponds become larger, control becomes more difficult. The same is the case with harvesting. To carry out the harvesting the pond must normally be emptied and/or drained several times, otherwise the fish density will be too high when lowering the water level and harvesting. A seine net may also be used for harvesting. Commonly, relatively small ponds are used for brood stock, fry and juvenile production, while larger ponds are used for on-growing.

Pond depth is usually between 0.5 and 2.4 m, depending on what the pond is used for. For on-growing fish it is normal to choose a depth sufficient to prevent any light reaching the bottom of the pond. In this way growth of vegetation at the bottom is prevented and harvesting is easier. Ponds for fry are normally shallower because the bottom vegetation may function as shelter. However, the depth must not be so great that temperature layers occur; deep water will also increase the pressure on the sides and bottom of the pond together with possibilities for increased seepage. This increases demands for compaction of the material used for the bottom and sides when constructing the pond.

It is important to have a slope towards the outlet on the bottom to make drainage possible and harvesting easier: this can be in the range 1/1000 to 1/100, with the largest slope in the smallest ponds.[7,10] The slope of the walls inside the pond is between 2.5:1 and 4:1; 3:1 is quite common, but this varies with the composition and angle of repose of the material used for construction. Exterior walls are normally slightly steeper than 2:1, but of course this also depends on the angle of repose for the material. The recommended width of the pond crest varies with depth; for 3 m deep ponds a crest width of 1.8 m is recommended;[11] for shallow ponds the crest width can be reduced depending on the material used for construction.

The length–width ratio of ponds is normally about 2:1, but of course is adapted to the site conditions. The shape of watershed ponds depends on the terrain. Harvesting with a seine net is easier if the pond is rectangular. If ponds are too wide, harvesting will be more difficult.

14.6 Site selection

Ponds should be as close to the water source as possible, to avoid long inlet pipes or channels. In addition, there must be enough clay in the earth to prevent leakage. A rule of thumb is that the material must consist of at least 20% clay particles with diameter below 0.002 mm in a 1.5 m deep core taken where the pond is to be established.[12] If the material contains too much sand, it will be porous and water will drain out much faster. The seepage loss in sand is reported to be between 25 and 250 mm/day, in loam 8–20 mm/day and in clay 1.25–10 mm/day.[13] Furthermore, the earth must be free of toxic substances, for instance copper.

There are several methods to prevent leakage from ponds. If the leakage is only slight, a solution is to break down the earth structure, reduce the aggregate size and puddle the bottom. Breaking up the lumps in the surface layer achieves this and is quite commonly done on rice fields. Addition of chemicals may also reduce the aggregate size.[14] Compression of the surface may also be used to reduce the water loss, for example by using a road roller. Several thin layers of compressed earth are better than one thick layer. If the natural soil is unsuitable, a membrane of clay or plastic may be used. A clay layer transported to the site must be about 30 cm thick for a 3 m deep pond.[11] This, however, represents increased costs for establishing the pond. To avoid the layer of clay crumbling as a result of drying or freezing, a covering layer of sand or gravel may be used; this can be from 30 to 45 cm thick on clay and 15–20 cm thick on plastic.[14] The plastic membrane should also be covered to avoid breakage from plants growing through it. Material used to construct the pond may also be sprayed with plant poison before laying the membrane.

After ponds have been in use for a time, water leakage will normally be reduced because settled materials block the cracks in the earth.

14.7 Water supply

The volume of water supply depends on the amount of fish in the pond and the intensity of farming in relation to evaporation.[15,16] In addition, the acceptable time for filling the pond with water is important. Normally this is the value specified for the water supply, because the water flow into a pond under normal farming conditions is very low.

The value used to calculate the necessary water supply should be sufficient to fill the total pond volume within 2–3 days. The whole farm, including all the ponds, should be filled during a 20 day period.[12] The amount of water needed is, of course, affected by evaporation. The total evaporation depends on temperature, cloud conditions, wind conditions and pond construction; Normally it is in the region of 0.25 to 1 cm per day in temperate areas.[9]

Example
Calculate the necessary water supply to cover the evaporation loss to a 100 m² pond where the evaporation is 0.5 cm per day. The water supply is equal to the evaporation volume.

$$\text{Evaporation volume} = 100\,\text{m}^2 \times 0.005\,\text{m} = 0.5\,\text{m}^3/\text{day} = 0.35\,\text{l/min}$$

If the water supply and exchange are too high, the production of algae will not be adequate. The algae may flow out through the outlet and the ecosystem will not flourish. The ponds are now functioning as raceways, for which the construction is sub-optimal. The ecosystem inside the pond becomes unstable when the daily water exchange is too high, i.e. between 25 and 33% of the total volume, which equates to retention times of 3–4 days.[9]

In addition to supplying water, it may be necessary to supply extra air or oxygen[17] (Fig. 14.6). Control of concentrations CO_2 and ammonia in critical situations may also be necessary; additional inlet water may be required in such circumstances.

14.8 The inlet

The water can either be supplied by a pump or under gravity; the latter is the best solution. Seawater ponds may be filled at high tide.

Channels or pipes can be used to distribute the water from the source to the ponds. If channels are used, gravity flow is necessary. If pumps are used they must eventually lift the water from the source into the distribution channel.

Figure 14.6 A surface aerator can be used to increase the supply of oxygen to the pond. The photograph shows a paddle aerator on shore.

Since the amount of inlet water is normally quite low, the actual design of the inlet pipe is of less interest than for other closed production units. Therefore where and how the inlet water is taken into the pond is less important than for other closed units. If using channels, the inlet from the distribution channel into the pond is just a hatch in a small channel branching from the distribution channel. Water is usually supplied to the ponds continuously, but batch filling is also possible.

14.9 The outlet – drainage

How the outlet is constructed depends on whether or not there is a collection basin for the fish, and if this is inside or outside the pond. The design of the level control will also influence the construction of the outlet system. If open channels are used, there are separate ones for the outlet water and level control.

Normally a standpipe inside the pond functions as a level control. The water has to pass through this standpipe, which may be variable or fixed, to flow out of the pond. A swivel can be used to control the level of the standpipe; this will again control the water level in the pond. If using a double standpipe, it is possible to take the outlet water from some depth in the pond. The standpipe and level control may also be placed outside the pond.

Special material must be used at the end of the outlet pipe to prevent erosion when tapping down the pond (shock tapping); concrete is usually used here. Concrete can also be used to construct the collecting basin for fish so that fry can be harvested or collected. When the outlet pipe is laid through the pond levee it is important to use mooring blocks, for instance of concrete, which are clamped on the pipe to prevent it being dragged out by the frictional forces of the water on the pipe. When both the flow rate and velocity are high, this is especially critical.

If channels are used for drainage, a specially design outlet is commonly placed in a small channel where the water leaves the pond; this is known as a weir gate or monk[15,18] (Fig. 14.7). It is normally constructed with two plates vertically installed in the channel, like hatches. The water has to pass below the first and above the second. The level of the second controls the level in the pond. From the pond the outlet water continues through a common drainage channel out of the farm. For large water flows open channels represent a simple cost effective solution.

The methods chosen for handling the fish may influence the construction of the pond and hence the outlet system. To collect the fish from the pond it is normal to either reduce the water level with the help of the outlet or to use a seine net, or eventually a combination of both methods. When using water reduction as a collection system, a collection basin, either inside or outside the pond, can be used to collect the fish. If the basin is inside the pond it

Figure 14.7 Outlet and inlet construction of a pond.

is normally small and at the bottom of the pond; it is commonly made of concrete or wooden planks. The outlet pipe from the pond is taken out from this basin. When the fish have been collected they can be removed from the basin using a net, pump or screw. The other possibility is to have the collection basin outside the pond. In this case the fish are tapped directly out of the pond together with the water and collected in the external collection basin. Both these solutions improve capacity and reduce the cost of fish handling.

A seine net may also be used for collection. Typically it is dragged through the pond and the fish are collected. The net can be hauled from the pond levees. If the pond is too wide, this will, however, be quite difficult, so widths of more than 20 m are not recommended.[19] The seine net may also be dragged mechanically, for instance by a tractor.

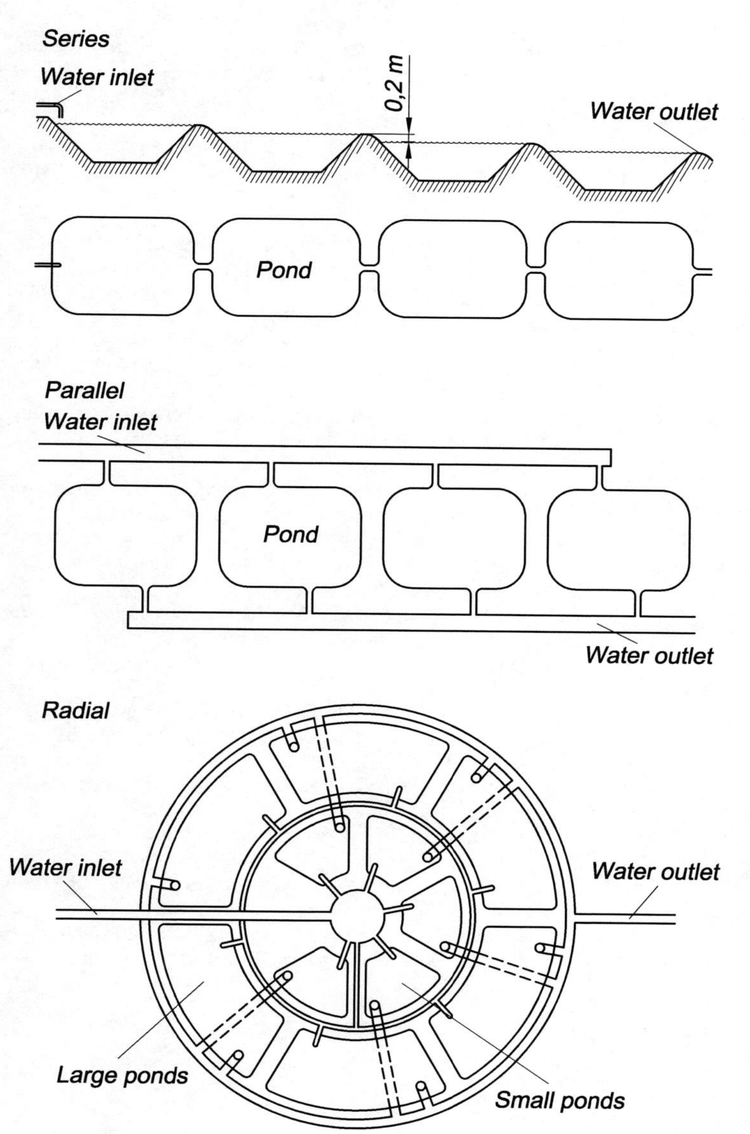

Figure 14.8 Several pond layouts are used: series, parallel and radial. The parallel layout is the most common.

14.10 Pond layout

A farm normally comprises several ponds. The arrangement of the ponds is important for optimal utilization of the area, and to ensure efficient water transport, fish handling and (eventually) feed handling. If watershed ponds are used, they must be adapted to the ground conditions and the layout is normally predetermined. If using levee or embankment ponds the layout is more important. Rectangular ponds are usually best regarding utilization of the area. Four main layouts may be used (partly from ref. 19) (Fig. 14.8):

- *Series* ponds are constructed so that the water flows from one pond into the next. The advantage is that gravity can be used to ensure the water flows though the entire farm. The main disadvantage is that the effluent water from one pond is the inlet water to the next pond and water quality decreases from pond to pond. Eventually disease pathogens will also follow the water and spread disease from pond to pond as isolation of a single pond is impossible. The water may be aerated when flowing from one pond to the next.
- *Parallel* ponds are set out beside each other, with a common water supply canal and a common effluent water canal. This is the most usual layout for a pond farm. The advantages are that the water quality is the same in each pond and it is also possible to increase and reduce the water flow to the separate ponds.
- *Radial* ponds are in a circle with the smallest close to the centre and the larger ponds outside this. In this way the size of the ponds increases with the radius of the circle. The great advantage with this system is that fish handling is very easy. If the fry are in the inner ponds they can gradually be moved to larger ponds when they grow and need a greater volume of water. The empty ponds can then be restocked with new fish.
- *Inset* ponds are small ponds placed inside a larger pond. This method can provide a nursery pond inside a grow-out pond.

References

1. Nash, C.E. (1988) A global overview of aquaculture production. *Journal of World Aquaculture Society*, 19: 51–58.
2. Stickney, R.R. (2000) Pond culture. In: *Encyclopedia of Aquaculture* (ed. R.R. Stickney). John Wiley & Sons.
3. Stickney, R.R. (1994) *Principles of aquaculture*. John Wiley & Sons.
4. Coche, A.G., Muir, J.F. (1992) *Pond construction for freshwater fish culture, pond farming structures and layouts*. FAO Training Series, FAO.
5. Coche, A.G., Muir, J.F., Laughlin, T. (1995) *Pond construction for freshwater fish culture, building earthen ponds*. FAO training series, FAO.
6. Kwei Lin, C., Teichert-Coddington, D.R., Green, B.W., Veverica, K.L. (1997) In: *Dynamics of pond aquaculture* (eds H.S. Egna, C.E. Boyd). CRC Press.
7. Boyd, C.E., Tucker, C.S. (1998) *Pond aquaculture water quality management*. Kluwer Academic Publishers.
8. Brunson, M.V., Haregreaves, J., Stone, N. (2000) Fertilization of fish ponds. In: *Encyclopaedia of aquaculture* (ed. R.R. Stickney). John Wiley & Sons.
9. Timmons, M.B., Riley, J., Brune, D., Lekang, O.I. (1999) Facilities design. In: *CIGR handbook of agricultural engineering, part II aquaculture engineering* (ed. F. Wheaton). American Society of Agricultural Engineers.
10. Hargreaves, J.A., Tucker, C.S. (2003) Defining loading limits of static ponds for catfish aquaculture. *Aquacultural Engineering*, 28: 47–63.
11. Lawsons, T.B. (1995) *Fundamentals of aquacultural engineering*. Kluwer Academic Publishers.
12. Summerfelt, R.C. (ed) (1996) *Walley culture manual*. North Central Regional Aquaculture Centre, Iowa State University.
13. Coche, A.G., Van der Wal, H. (1981) *Water, for freshwater fish culture*. FAO training series. FAO.
14. Wheaton, F.W. (1977) *Aquacultural engineering*. R. Krieger.
15. Yoo, K.H., Boyd, C.E. (1994) *Hydrology and water supply for pond aquaculture*. Chapman & Hall.
16. Kelly, A.M., Kohler, C.C. (1997) In: *Dynamics of pond aquaculture* (eds H.S. Egna, C.E. Boyd). CRC Press.
17. Boyd, C.E. (1998) *Pond water aeration systems*. Aquacultural Engineering, 18: 9–40.
18. Landau, M. (1992) *Introduction to aquaculture*. John Wiley & Sons.
19. Lucas, J.S., Southgate, P.C. (2003) *Aquaculture, farming aquatic animals and plants*. Fishing News Books, Blackwell.

15
Sea Cages

15.1 Introduction

A cage represents a delineated volume in the body of water where the aquatic organisms can be farmed. Cage aquaculture may date back to as early as the 1200s in some areas of Asia,[1] and is currently a major form of aquaculture in countries including Canada, Chile, Japan, Norway and Scotland, where it has been successfully used, mainly for salmonid farming. However, a large variety of species are grown in cages today and include seawater, freshwater and diadromous species. Therefore today cages are used worldwide in the sea, in lakes and large rivers.[2] The main differences are in the size and construction for withstanding waves and currents. Trends today are that new more weather-exposed sites are taken into use to ensure continuous growth in the cage farming industry. The number of good sites in less exposed locations is limited.

There are a number of approaches to designing a cage and also classifying the various cage systems.[1,3,4] One classification is based on where in the water column the cage floats. Three categories can be used: floating, submerged, or submersible. The last two types consist of a frame that can float on the surface and that maintains its shape when lowered below the water surface.

Another classification is according to the type of net used in cages: rigid or flexible. Rigid nets may be created by using a flexible net attached to a stiff framework to distend it. Instead of using a flexible net a rigid metal net may be used. A rigid net cage will maintain its original shape regardless of the waves.

Instead of using a floating construction, a fixed construction may be used. This can, for example, be pilings driven into the seabed to which the net is then fixed to fence in an area.

Another classification of sea cages divides them into two categories depending on the nature of the bag that makes up the cage; it may be an open bag of net, or a closed bag of plastic, for instance. A closed bag will normally require water to be pumped into it, and there is an outlet pipe from the bag. Actually, a closed production unit has been created.

Open offshore cages can be classified as follows:[4]

- *Class 1* Gravity cages that rely on buoyancy and weight to hold their shape and volume against environmental forces (the focus of this chapter)
- *Class 2* Anchor tension cages that rely on the anchor tension to keep their shape and volume
- *Class 3* Self-supporting cages that rely on a combination of compression in rigid elements and tension in flexible elements to keep the net in position so the shape and volume are maintained
- *Class 4* Rigid self-supporting cages that rely on rigid constructions such as beams and joints to keep their shape and volume.

In this chapter the focus is on open floating sea cages which are those most used for intensive aquaculture. A traditional open cage comprises the following main parts (Fig. 15.1):

- Net bag with weights in the bottom to spread the bag
- A jumping net above the surface fixed to the net bag to prevent fish escaping
- Cage collar for spreading out the net bag and give buoyancy to keep the bag in the correct position in the water column
- Mooring system.

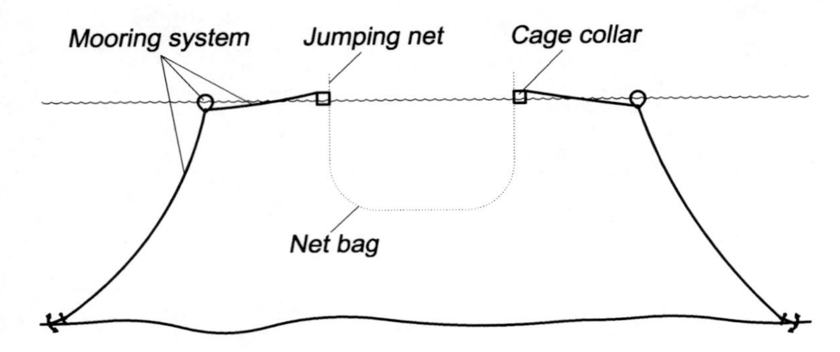

Figure 15.1 Major components in a traditional open sea cage farm.

When choosing technology and systems for farming in traditionally open sea cages, there are many conditions to be evaluated. This is also the case when the actual type of cage and mooring system is chosen and designed. The following list can be used to help when establishing a new sea cage farm:

(1) Choose a site that is suitable for farming
(2) Describe and calculate the environmental conditions on the site
(3) Choose farming systems, i.e. the cage and mooring system, adapted to site conditions
(4) Design the cages (normally done by the cage manufacture) and mooring system
(5) Set out the cages and mooring system
(6) Establish requirements for operational control of the system.

15.2 Site selection

Selecting a good site is of major importance for the future economic viability of the cage farm. A suitable site for cage farming must fulfill a number of requirements.[1,3] It is normally difficult to fulfill all of these, and they will depend also on the cage technology used; for example, the extent of wave tolerance. There are a number of ways to classify the factors that must be evaluated when selecting a site.[1,3] Several of the factors also affect each other directly.

The main factor is, of course, the water quality. This must be satisfactory for the cultured species, including temperature, salinity and oxygen content. A continuous supply of oxygen requires a current to exchange the water. This is also required to remove metabolic products from the cage area. A good water exchange will occur with a water velocity above 0.1 m/s. This is normally sufficient to supply enough oxygen and to remove fish excrement. However, the water currents ought to be below 1 m/s because velocities above this result in very large forces on the cage structures and mooring system; in these situations specially designed systems must be used. Fjords with a sill are not recommended because the water current and water exchange are normally too low to transport the faeces and eventual feed loss away so this will collect on the bottom below the cages and decomposition under anaerobic conditions may occur. Hydrogen sulphide (H_2S) gas which is toxic for the fish may then be released from the bottom sediments (Fig. 15.2). Areas where the water can be polluted with toxic substances must also be avoided; this can, for instance, be near various industries. Some areas are also more exposed to algal blooms and some sites are particularly exposed to fouling; this must be taken into consideration when selecting a site.

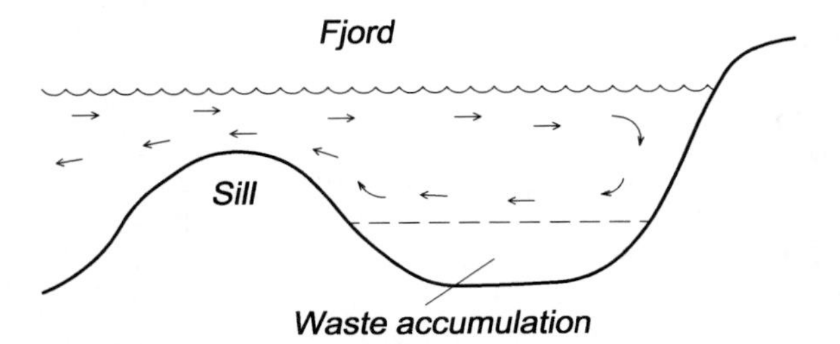

Figure 15.2 A sill fjord is not recommended for cage farming.

Shelter from the weather is also important. Wave height is normally the most critical parameter. It is usual to avoid areas with high waves, even if it is theoretically possible to build farms and mooring systems that can tolerate very large waves. However, these farms are difficult and expensive to operate when the waves are large and operational access is reduced. In addition, large expensive boats have to be used to operate such farms. If the wave height is below 2m the cage is easy to operate, and many available cages are constructed to tolerate such wave heights. Several suppliers deliver cages that may tolerate 4–5m wave height. Ocean cages can tolerate up to 7–8m, but special routines for operation of such farms must be taken into consideration before selecting such sites.

Another factor included in the geographical conditions on site involves water depth; a distance of more than 5m from the bottom of the net to the sea bottom is recommended, but this depends on the current conditions. Depths above 100m will greatly increase the costs of the mooring system because long mooring lines will be needed. Areas with frequently shipping traffic should be avoided because of disturbance to the fish and creation of waves. When selecting a site, good infrastructure, for example, proximity to roads, available electricity, will also be of benefit.

The legal requirements for fish farming in an area must be satisfied. There may be areas publicly designated for other purposes, or where cage farming is unwanted from an environmental point of view. For example, it will be difficult to establish a cage farm for salmonid production just outside an important salmon river because of the risk of escape. The legal requirements for access to land for an on-shore base and on-shore mooring are also included here.

Before choosing a site, the environmental conditions must be clearly known. This information may be obtained from government maritime departments or by the use of special oceanographic buoys that automatically monitor environmental conditions on the sites. Talking with people living in the area and local fishermen could also give valuable supplementary information.

15.3 Environmental factors affecting a floating construction

15.3.1 Waves

Waves are normally the limiting factor for site selection for cage aquaculture. If the wave height is too great it is very probable that this will affect the cage structure. Knowledge of the wave climate on the site will also be an important tool in choosing the correct cage technology and mooring system to avoid later breakages in cages and moorings.[5,6] The trend towards using an increasing number of wave-exposed sites for marine cage farming is proving the importance of this.

Several terms are used to describe a wave (Fig. 15.3):

- *Crest*: the high point of the wave
- *Trough*: the low point of the wave
- *Wave height*: vertical distance between trough and crest – H (m)
- *Maximum wave height*: highest measured wave height – H_{max}
- *Significant wave height*: average wave height of the highest one-third of the waves recorded in a period. During a recording period there will always be a variation in the wave height. The significant wave height corresponds quite well with what an observer will record as the approximate

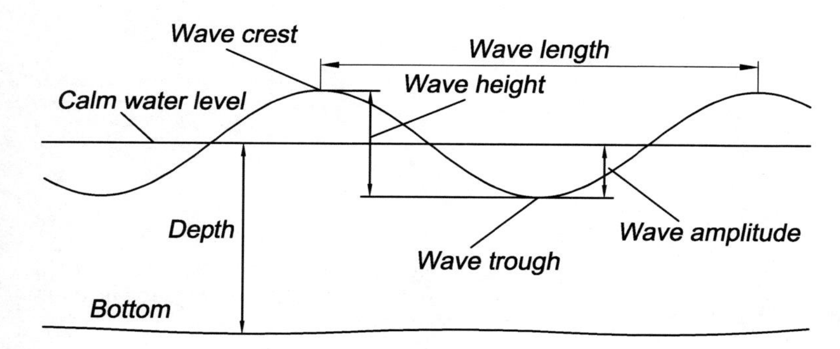

Figure 15.3 Important parameters describing a wave.

wave height when looking at the waves over a period – H_s

- *Wave amplitude*: distance from calm water level to crest or trough; wave height divided by two (H/2) – a (m)
- *Wavelength*: the horizontal distance between two following crests – L (m)
- *Wave period*: the time taken for a wave crest to travel a distance equal to one wavelength – T (s)
- *Wave frequency*: the inverse of the wave period; the same as the number of waves passing a given point per unit time – f (s^{-1}).

Wave calculations

The description and calculation of waves and wave forces are quite difficult, and in this chapter only a brief survey is given. Several textbooks are available and may be consulted for further information (see, for example, refs 7–12). To illustrate what a wave is and how it moves, the water volume may be represented by many single water 'particles' which will be transported both vertically and horizontally with the wave (Fig. 15.4); they will move up with the crest and down with the trough. There is no net transport of water particles as long as the wave is not breaking (see below); they stay in the same place, but rotate in an orbit depending on the wave height and wave description. It is normally the current that causes the net transport of the water particles in the sea.

To understand how a wave is created the following simplified explanation can be used. Imagine that a stone is dropped into the water. As it displaces the water, the water particles are forced down and away. The energy that the stone adds to the water will be used to force neighbouring particles up. In this wave a wave is created. The wave continues to disperse until the wave motion is damped by friction between the water particles and no energy is left.

If an object such as a sea cage is lying in the water, the energy from the wave will also be transferred into this, and it will follow the wave motion. If an object is lying in the sea, however, the wave motions will be reduced because energy is used to move the object (see section 15.4.4).

To describe the wave motion, actually how a single water particle moves, several theories are used. The linear wave theory is the easiest and is also quite easy to understand, but several simplifications are made compared to real waves.[11] In most cases, however, this theory will suffice. The waves are described as sine waves, and all the standard geometrical knowledge of sine waves can be used to describe them. The single water particle rotates in a circular orbit where both the acceleration and velocity vary depending on where the particle is in the orbit: on top of the crest or down in the trough. However, under practical conditions the waves will not behave as sine waves. The wave is the sum of several wave systems coming from different directions, with different wave periods, height and phases (Fig. 15.5). This can be described by an irregular wave spectrum. Such spectra will vary from sea area to sea area. Spectra that are fitted to describe the wave for the different sea areas can be made based on actual measurements and calculations. For calculating wave motions according to developed wave spectra computer programs are used. In the linear wave theory, which is a simplification using sine waves, a set of formulae have been developed

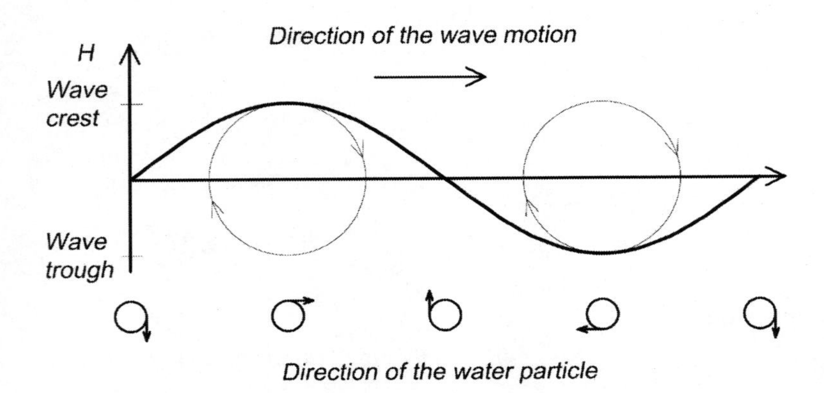

Figure 15.4 During a wave cycle the single water particles move in different directions.

to calculate the wave motion and the total energy in the waves.

Depending on the depth, the wave will 'travel' in different ways. In shallow water the waves will have an effect all the way to the bottom; at intermediate depths the wave motion will be reduced closer to the bottom. In deep water the wave motion will quickly decrease; half a wavelength down there is almost no wave motion left because so much energy is used to move the surrounding sea. If the cages are located in the sea, deep-water conditions are commonly chosen. Here also is the explanation for the advantages of using submerged cages in water with high waves; at a depth of one wavelength there are no effects of the wave in deep water conditions.

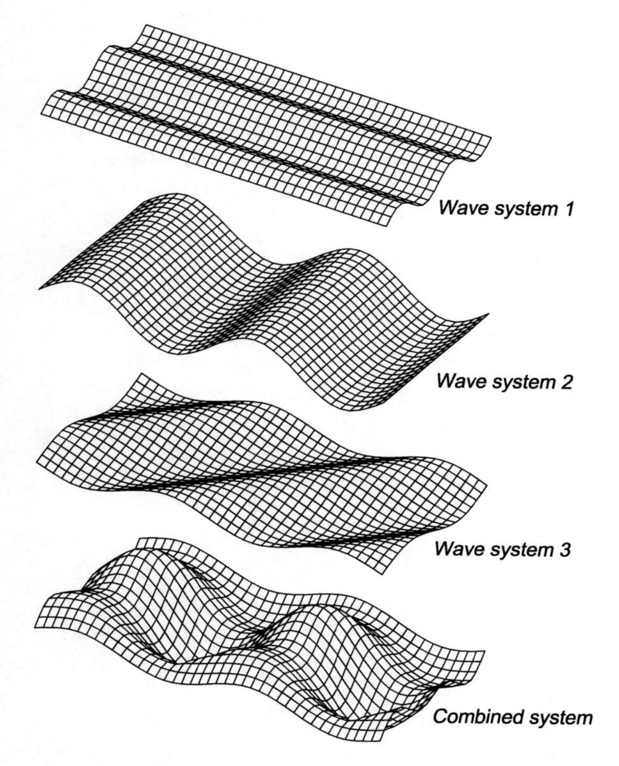

Figure 15.5 A wave comprises a number of single waves which create an irregular wave spectrum.

Breaking, reflecting and diffraction of waves

A wave 'breaks' when the height increases in relation to the wavelength (Fig. 15.6) and the wave gets steeper. White foam crests characterize a breaking wave visually. Much more energy is consumed when a wave starts to break. Under deep-water conditions wave breaking will occur under the following conditions:

$$H/L > 1/7$$

where:

H = wave height
L = wavelength.

When a wave breaks there is net transport of water in the direction that the wave breaks.

Example
The wavelength is 30 m and the wave height is 5 m. Does the wave break?

$$H/L = 5/30 = 1/6 > 1/7$$

Therefore the wave will break.

If the wave height decreases to 3 m will the wave break now?

$$3/30 = 1/10 < 1/7$$

Therefore the wave does not break.

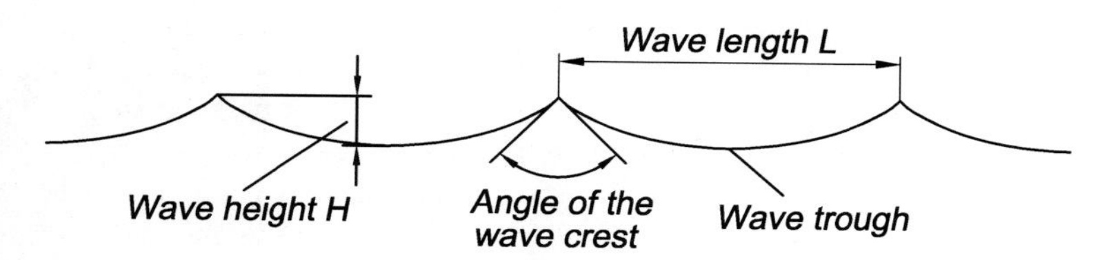

Figure 15.6 When the wave height increases relative to the wavelength the wave will start to break.

Reflection of a wave occurs when it hits an obstruction, such as the beach, a rock wall, a rock awash or a floating construction. If a vertical rock wall is hit by the wave an opposite wave motion is created and the energy in this wave will send the energy backwards. Under special conditions it can therefore be quite calm just outside such rocks, because the wave motions from the two different waves neutralize each other. If the wave hits the beach the depth will decrease and the wave will gradually start to break because it is forced up from the bottom and gets steeper. Breaking of waves can easily be seen when they hit the beach. Waves may not break before the beach, but when they hit the beach they start to break. The wave must get rid of all its stored energy when it reaches the land. How much is reflected is determined by the angle to the shore. Much energy is dispersed when the wave starts to break.

Diffraction occurs when a wave hits an obstruction; the angle at which the wave hits determines the direction of the diffracted wave. Diffraction may cause waves to be sent into areas that should be sheltered.

What creates waves?

Several factors may create waves but the most important are:

- Wind
- Human activity, such as shipping
- Special natural phenomena such as earthquakes, land slips and underwater volcanic eruptions create waves known as tsunamis
- Tide; waves with extremely long wavelengths are created.

Waves created by the wind are the most relevant to aquaculture facilities and are further described below. Shipping may also create waves that are unwanted on fish farms. To avoid such waves, sites close to heavily trafficked sea routes should be avoided. Waves created by exceptional natural phenomena (tsunamis) are difficult to avoid even if such phenomena occur more frequently in some areas than others. Tsunamis have a very long wavelength (>100 m) and a long period (around 1000 s). In such waves an enormous amount of energy is stored. They do not represent a great danger if a boat is on the sea, because the wavelength is so long. However, when they reach shallower water, and especially when they reach the shore and start to break, all the energy that is stored in them is released, and the consequences can be fatal. Such waves can be up to 30 m high when the shoreline has forced them to increase in height to dissipate their energy; they cause enormous destruction when they hit the shore. Waves created by the tide normally present no problems for cage farms. The wavelength here is so long that it is not interpreted as a wave, and the wave period is 12.5 h. Such waves can, however, create very strong tidal currents.

Wind created waves: The main ingredient in the formation of waves on the open ocean is wind. Wind created waves are normally also what inhibits site selection for cage farming. When winds blow across water, a drag is applied on the surface and pushes the water up, creating a wave. The height will increase as long as the wind is strong enough to add energy to the wave. After a period of time there will be equilibrium between the energy in the wind and the energy in the waves; the wave height will now be stable. Once a wave is generated, it will travel in the same direction until it meets land or is dampened by an opposing force such as winds blowing against it in the opposite direction, or by friction.

The height of wind created waves depends on the wind velocity (U_v), the duration of the wind (t_v), the fetch length (F) and the presence of other waves when the wind begins to blow. The fetch length is the distance where wave development can take place (Fig. 15.7).

The Beaufort wind scale gives the expected wind velocity for the different wind strengths (see below). Some scales also present the normal wave height with different wind strengths; however, this is on open sea with no protection from land or islands.

In protected water the fetch length where the wind can blow will limit the ability of the wind to create waves. The fetch length can be read from a chart and is the length of the free water surface. If the fetch where the wind is blowing is narrow, as in a fjord, the wind effect will be less because of the friction against land on both sides reducing the velocity. To calculate the effective fetch length a compensation factor called the fetch length factor

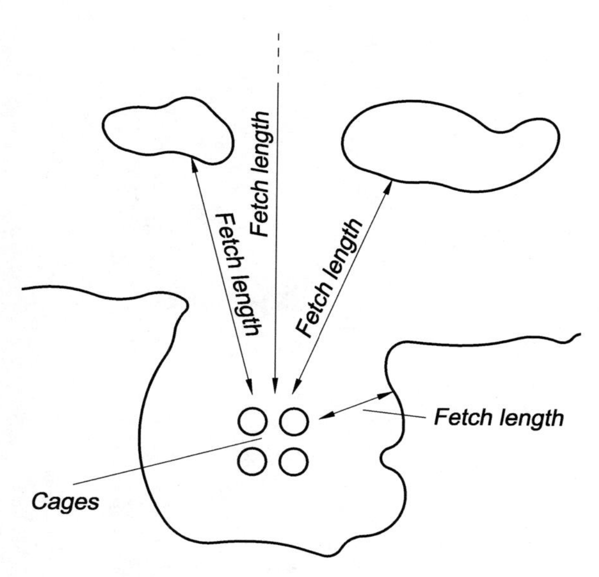

Figure 15.7 The fetch length is the length of the free water surface where the wind blows and can create waves.

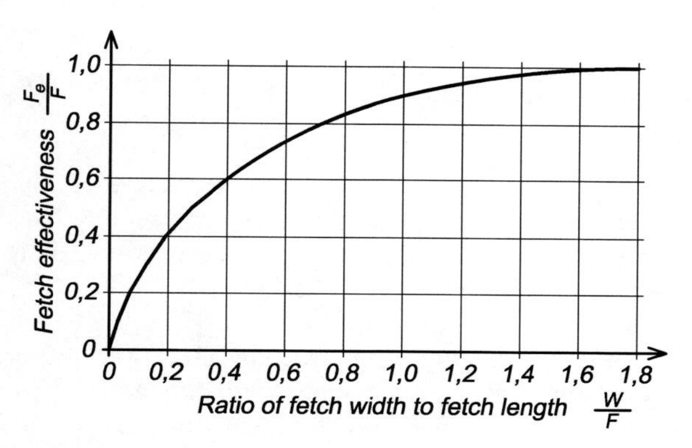

Figure 15.8 If the fetch is narrow the wind velocity will be reduced because of friction against land. The effective fetch length is calculated by the use of a compensation factor, the fetch length factor. (Adapted from ref. 13: Saville, 1954.)

is used[13] (Fig. 15.8). Calculations of effective fetch length in narrow fjords are given in the following two examples.

Example
The length where the wind blows is 10 km and the width of the fjord is 2 km.

Effective fetch length = fetch length × fetch length factor

Fetch length/Fetch width = 2/10 = 0.2

From Fig. 15.8 the fetch length factor is found to be 0.4.

Therefore the effective fetch length is

$$0.4 \times 10 = 4\,\text{km}$$

Example
Length where the wind blows is 10 km and the width of the fjord is 10 km.

Fetch length/Fetch width = 10/10 = 1

From Fig. 15.8 the fetch length factor is found to be 0.9.

Therefore effective fetch length is

$$0.9 \times 10 = 9\,\text{km}$$

In shallow water the Sverdrup–Munk–Bretsneider (SMB) method may be used to estimate wave height. Formulae and diagrams have been developed to find wave heights based on wind velocity (U_v), wind duration (t_v) and effective fetch length (F_e) (Fig. 15.9). It must be remembered that some diagrams use the traditional sea units of foot (ft), knot (kn) and nautical mile (nm) (1 ft = 0.3048 m; 1 kn = 0.5144 m/s; 1 nm = 1852 m).

To use the diagram in Fig. 15.9, knowledge of the three factors, wind velocity (U_v), wind duration (t_v) and effective fetch length (F_e) is required. First the wave height is found based on U_v and t_v, and afterwards based on U_v and F_e. Of the two different values found, the lower will be the wave height on the site under the specified conditions because either wind duration or fetch length will limit the maximum wave height. For instance, if the wind duration is short a maximum wave height will not be attained; if the fetch length is also short it will also inhibit maximum development of waves, even if the wind duration indicates higher waves.

Example
Use SMB to estimate the wave height and wave period for a site if a fresh breeze of 20 kn blows for 2 h and the fetch length is 10 nm.

First calculate wind velocity and fetch length in SI units:

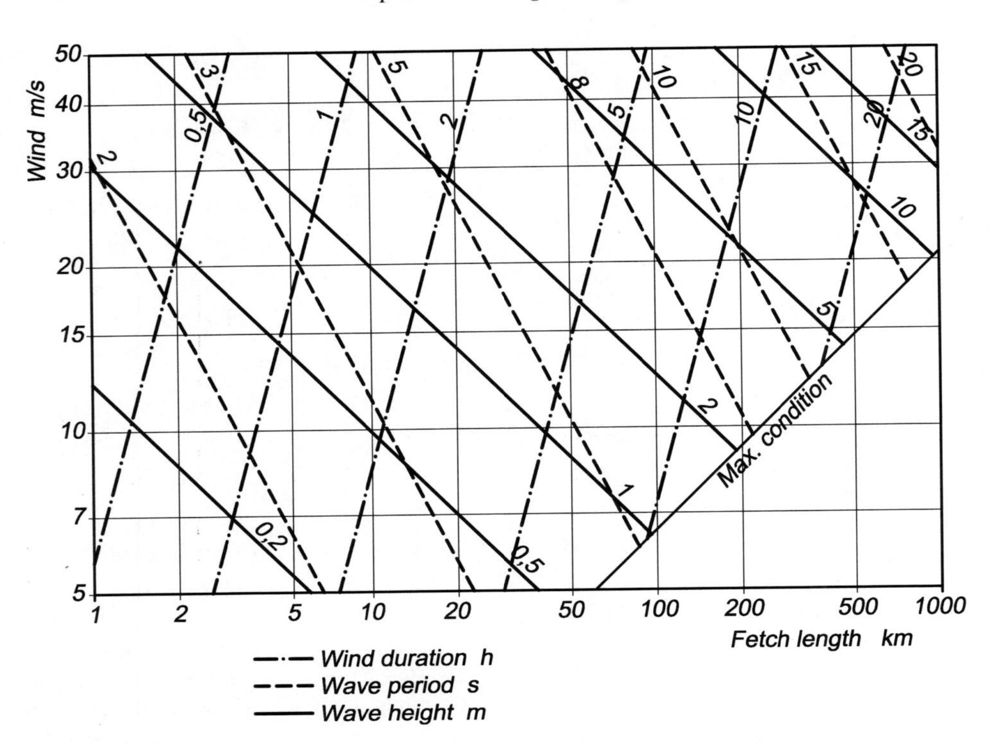

Figure 15.9 Diagram for calculation of wave height in relation to wind velocity, wind duration and effective fetch length (adapted from ref. 11 – see here for more details).

$$20\,\text{kn} = 20 \times 0.5144\,\text{m/s} = 10.29\,\text{m/s}$$

$$10\,\text{nm} = 10 \times 1.852\,\text{km} = 18.5\,\text{km}$$

Using wind velocity and duration criteria and Fig. 15.9

Significant wave height = ca. 0.6 m

Significant wave period = ca. 3.3 s

Using wind velocity and fetch length criteria and Fig. 15.9

Significant wave height = ca. 0.8 m

Significant wave period = ca. 3.9 s

This means that wind duration is the limiting factor for development of waves; the wind does not blow long enough to create maximum wave height in proportion to the fetch length. Critical values for the site will therefore be:

Significant wave height = 0.6 m

Significant wave period = 3.3 s

The SMB method with the values and diagram given above may be used for depths greater than 15 m, which is normal in sea cage aquaculture. For intermediate depths and shallow water other formulae and diagrams apply.[11]

In open sea conditions with no limitation of the wind duration, the wave height will only depend on the effective fetch length and the wind velocity. The wave height created is the maximum possible with the given fetch length. A simplified method can then be used to calculate wave height in shallow water[11] (Table 15.1):

$$H_s = 5.112 \times 10^{-4} \times U_A F^{1/2}\ (\text{m})$$

$$T_s = 6.238 \times 10^{-2}\ (U_A F)^{1/3}\ (\text{s})$$

$$U_A = 0.71 U^{1.23}\ (\text{m/s})$$

where:

U = wind velocity (10 min average value 10 m above sea level) (m/s)
U_A = adjusted wind velocity (m/s)
F = fetch length (m)
H_s = significant wave height (m)
t_s = significant wave period (s).

Table 15.1 Examples of wave heights with fully developed sea with different wind velocities and effective fetch lengths (F). Wind duration is without limitation.

Wind velocity (m/s)	Wave height (m)			
	$F = 1$ km	$F = 3$ km	$F = 5$ km	$F = 10$ km
5	0.08	0.14	0.19	0.26
10	0.20	0.33	0.44	0.62
15	0.32	0.56	0.72	1.01
20	0.46	0.79	1.02	1.45
30	0.75	1.30	1.68	2.38

The following may be used to estimate the maximum wave height from the significant wave height:[14]

$$H_{max} = 1.9H_s$$

Example
Use the simplified method to calculate the significant wave height and wave period for a near gale with wind velocity of 15 m/s and fetch length of 3 km.

$$U_A = 0.71 \times 27.96$$
$$= 19.9 \text{ m/s}$$

$$H_s = 5.112 \times 10^{-4} \times 19.9 \times 3000^{1/2}$$
$$= 0.56 \text{ m}$$

$$T_s = 6.328 \times 10^{-2} \times (19.9 \times 3000)^{1/3}$$
$$= 2.44 \text{ s}$$

Swell: Swell comprises wind generated waves created far away, which can also be called ocean waves; these may also affect cage farms when they come in from the sea. This is another reason for sitting cage farms in sheltered positions behind holms and breakwaters. Swells are characterized by quite large wave heights and long wavelengths. A swell can be recognized by its higher wave period compared to a local wind generated wave: typical swell periods are in the range 9–20 s, compared with 2–11 s for wind generated waves.

15.3.2 Wind

Wind is normally not directly harmful to sea cage farms. The area of the farm above the water surface, where the wind blows, is small. On large operational platforms with buildings, however, the wind will have some effect.

Wind can be separated into two components: one normal and the other fluctuating (gusts). The amount of gusting depends on the local topography. It is normally gusts of wind that cause damage, for instance to houses. The wind velocity increases with height above the ground. For meteorological purposes, wind velocity is measured 10 m above the ground (V_{10}). Because it varies continuously it is given as an average over a period, normally 10 min. Many meteorological stations measure the wind velocity; however, a large number are located at airports or lighthouses, where the landscape is quite flat with few mountains to create gusts. When transferring these wind data to other sites, this must be taken into consideration. An easy method to present the wind conditions on a site is by using wind roses; these show where the major wind is coming from and may also show the average strength of the wind from the various directions over a given period (Fig. 15.10).

15.3.3 Current

Water current is normally the dominant environmental force on a sea cage farm. Several factors may create a current in the water, including:

- Wind
- Tide
- Local water flows, such as rivers
- Large global oceanic currents or coastal streams such as the Gulf Stream.

Currents create both horizontal and vertical movements in the water. In fish farming the focus is normally on the horizontal currents.

There are large variations in the current from site to site, and the overall current picture comprises all the different single currents. A proper description of the current conditions on site must therefore be included in the site measurements (see below).

Wind generated current

Current is created in the water when the wind blows over the surface, because there will be a drag from the wind on the water surface. The velocity of the created water current depends on the strength of the wind. Because the drag is on the water surface, wind created current will be highest near the surface and decrease with depth. The following

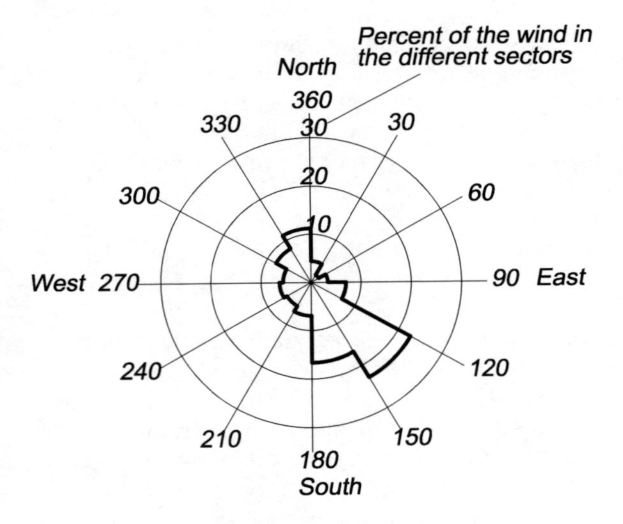

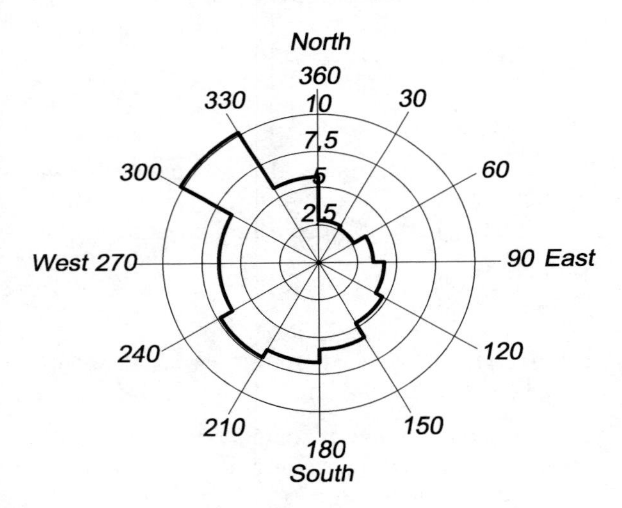

Figure 15.10 A wind rose may be used to show where the wind is coming from and how strong it is.

equation may be used to calculate the current created by wind in open water:[12]

$$U_{sv}(z) = 0.02U_{10}\left(\frac{50-z}{50}\right)$$

where:

U_{10} = wind velocity measured 10 m above the water surface

z = distance downwards from the surface to the depth at which the velocity of the wind generated current is to be found.

Example
Find the current generated by wind at depths of 5 m and 10 m. There is open water and the wind is near gale force.

First the near gale force wind velocity is found from the Beaufort wind scale to be between 13.9 and 17.1 m/s. Therefore use a velocity of 15 m/s.
The current velocity at 5 m depth is then

$$U_{sv}(5) = 0.02 \times 15 \times \left(\frac{50-5}{50}\right)$$
$$= 0.27\,\text{m/s}$$

At 10 m depth the wind creates a current of:

$$U_{sv}(10) = 0.02 \times 15 \times (40/50)$$
$$= 0.24\,\text{m/s}$$

These calculations also show that the velocity is reduced by the depth.

As can be seen from the first equation, in this subsection, the surface current velocity is 2% of the wind velocity (U_{10}) calculated for open sea. In shallow water the wind generated current velocity will normally be somewhat higher, partly due to stratification and reduced thickness of the water layers being dragged by the wind, and under practical conditions may be up to 5% of the wind velocity.

Tidal current

Tidal current is created by tidal range. The tidal range varies from site to site around the world. In Canada, a tidal range of up to 14 m occurs. In narrow fjords and narrow necks the tidal current is especially high, because the tide forces the water in and out. In North Norway a tidal current of up to 16 kn (8.23 m/s) has been measured in Saltstraumen in the inlet to Skjerstadfjorden, which is the world's strongest tidal race.

In open waters the current caused by the tide is not as high, and depends on the size of the tidal range in the area. If Norway is taken as an example, the tidal current varies from 0.2 to 0.8 m/s along the coast.[14] Tidal current is approximately equal through the total water column and does not vary with depth. The strongest tidal current occurs in the middle between high and low tide.

Oceani currents

Solar heating, variation in water density, wind, gravity and the Coriolis force have influences over the large ocean currents. The Gulf Stream is one of the strongest currents known. It starts in the Gulf of Mexico, passes along the east coast of the USA and crosses the Atlantic Ocean past Ireland and Great Britain and continues up past the west coast of Norway. Where its velocity is 0.4–0.5 m/s. Coastal currents present the largest velocity close to land and decrease with depth.

Measuring current

When evaluating a site, the water current must be measured. Specially designed instruments are used to measure the direction and the velocity of the current. They are an integral part of a floating buoy, and also include a recorder to store the results. Some also have a transmitter for downloading the monitored results. It is recommended that the measuring buoy stays out for quite a long period, preferably for a whole year or at least in the periods when the strongest and weakest currents occur. The results of the measurements can be shown in a current rose in the same way that the wind data can be presented.

15.3.4 Ice

In northern and southern regions near polar areas, ice in the water may be a problem for the development of cage aquaculture, especially in fresh and brackish water, but also in the sea. The problems are of three types:

- Surface ice
- Drift ice
- Icing up.

Surface ice is mainly a problem where there is a supply of fresh water which reduces the salinity of the top layer of the water. This causes the freezing point to change to close to 0°C, and the water surface might freeze. This will, of course, also be the condition on a freshwater site. From time to time there may be a thin layer of ice on the surface that is so sharp that it is able to cut the nets on the cages.

Drift ice does not normally present any problem for traditional aquaculture sites, but may occur near the polar areas from where it is released.

Icing up of parts of the farm above the surface might be a problem. Icing up occurs when sea spray or supercooled rain hits a construction cooled to below freezing point. When the water hits the construction, it will immediately freeze and coat the construction with ice. This can also happen with aircraft under certain weather conditions and is the reason they are de-iced before take off. The same phenomenon can be observed on fishing vessels working in polar areas; under unfavourable conditions the vessel may become totally covered with ice, the amount and weight which can be so large that the vessel will sink. When sea cages ice up the same thing can happen; the loads may be so large that construction breakage occurs. The weight of the ice can exceed the buoyancy of the collar and the cage will sink below the surface. Here, however, the ice will melt after a time and the cage will surface again. Windy conditions and relatively low temperatures may cause icing by sea spray. If the temperature is very low the water may freeze in the air before it hits the construction. Ice may cause supercooling of the water and possibly also fish death.

15.4 Construction of sea cages

A typical sea cage comprises several parts: the cage collar or support system (framework), the flotation system, the net bag, a jumping net, and weights to stretch out the net bag at the bottom and to stabilize it in the water column.

Three different methods may be used to construct the framework/collar for a sea cage (Fig. 15.11).

(1) *Stiff framework* The framework does not follow the wave movements. An example of a stiff construction is a boat. Some specially designed steel cages use a stiff framework. The construction is characterized by large forces transferred to the framwork.
(2) *Framework with movable joints* The framework will to some extent follow the wave movements. An example is a traditional steel cage

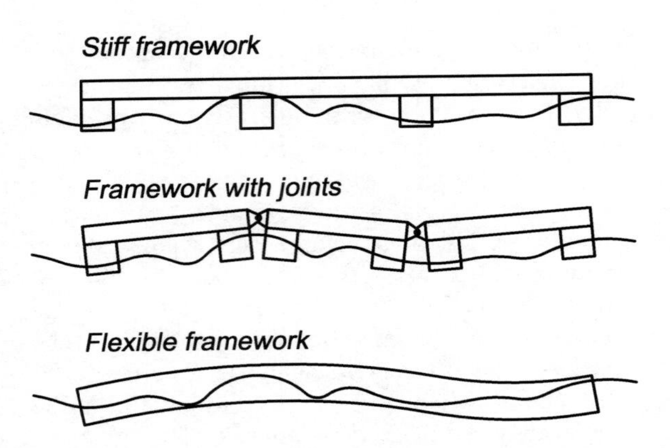

Figure 15.11 Different methods of construction for frameworks of traditional surface sea cages.

system, where joints are used to connect the single elements in the framework.

(3) *Flexible framework* The framework is quite flexible and will follow the wave movements well. These include frames made of plastic (for example, polyethylene, PE) which are flexible to some degree and those made of rubber (for example, ocean cages).

15.4.1 Cage collar or framework

The collar or framework may have several functions. It helps to support the cage safely in the water column, it helps to maintain the shape of the net bag, it may help with buoyancy and it may serve as a work platform.

The framework construction for stretching out the net bag can be combined with the buoyancy, as seen in plastic floating ring cages (PE or polypropylene (PP) pipes). Alternatively, the buoyancy can be independent of the cage collar as can be seen when using wood or steel for support systems with blocks of expanded polystyrene (PS), such as Styrofoam™, as buoyancy.

The buoyancy is necessary to keep the cage bag in the correct position in the water column. It must have a smooth surface to inhibit the accumulation of fouling. Fouling increases the weight of the collar, which results in increased requirements for buoyancy; furthermore, fouling will increase the friction between the flowing water and the sea cages which again increases the forces on the mooring system. Expanded polystyrene is commonly used as buoyancy; if not covered with PE, exposure to sunlight causes it to age. It turns yellow and becomes brittle. Uncovered polystyrene will also be very prone to fouling, because the surface becomes so rough. The use of uncovered polystyrene in sunlight is not recommended; to increase its durability it is quite common to put it into PE cylinders or rhombs.

If too much buoyancy is added the cage collar will float high up in the water column and fully follow the wave motions, floating on top of the water column throughout. Unnecessarily large forces on the cage bag and mooring system from the induced vertical motion result if there is much wave action in the area. The cost of the buoyancy will also be unnecessarily high. Buoyancy elements ought to have an aerodynamic shape to reduce the forces transferred from the water current. The current forces on the collar are, however, much smaller than the forces on the net bag.

The framework or collar can be of circular, polygonal or square construction. It is best to use a round framework because the forces are equal all around the circumference; polygonal or square frameworks will have large forces in the corners and eventually breakages in the construction will occur here (see, for example, ref. 15). For this reason, good connections at these points are important. Wooden frameworks are sometimes used to construct sea cages; only bolts, nails or ropes are used to connect the planks at the corners. If these cages are used in exposed sites with fast currents and high waves, the framework will break at the weak points in the corners.

Polygonal collars are better than square collars because there are more corners to share the total forces, and the force in each corner is therefore reduced.

Different materials may be used in the framework (Fig. 15.12), ranging from steel, aluminum, wood and concrete which are rigid, to more flexible materials such as PE and rubber. The modulus of elasticity for the material is a measure of its rigidity and is given by its E value, the load in relation to non-permanent deformation. Steel has a high E value, while wood has a rather lower value;[1] that for PE is even lower. Bamboo is also used in cage collars, but only on low exposed sites.

The risk of corrosion of the framework in seawater when using steel or aluminum must be taken

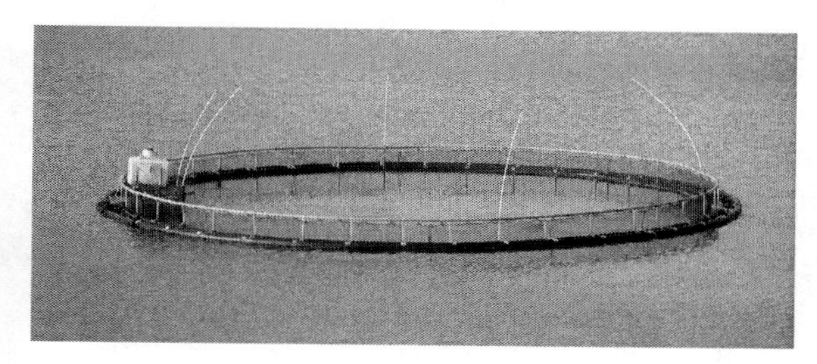

Figure 15.12 Different designs of framework.

into consideration. If steel is used it must always be covered, for example with paint or zinc, to avoid corrosion.

15.4.2 Weighting and stretching

Weights on the bottom of the net bag are used to keep the net bag down, and to maintain as much effective volume as possible for the fish. Lead rope may be used for the bottom line in the net bag together with lump weights. The lump weights are normally added at the corners and in the centre. For example, on a 15 m × 15 m square cage, the total amount of weights can be 150–200 kg, divided into lump weights of 25 kg in each corner and in the centre, and the rest evenly spread along the bottom line as lead rope.

Rapid currents will decrease the effective volume and adding more lump weights can inhibit this; however, care must be taken because this will increase the current forces on the net bag. The need for buoyancy will also increase, and the same will be the case for the size of the mooring system. Use of weights will also increase the dynamic forces on the net bag caused by the waves (stretch and slack).

Instead of using weights, stays may be used to stretch out the net bag; this gives a rigid construction. The environmental forces will be greatly increased by doing this and the bag will require a much larger mooring system.

15.4.3 Net bags

Net bags can be constructed in different ways and with different materials.[16–18] In the past materials such as cotton and flax were used. These materials get heavy in water and their strength is rapidly reduced; in addition they are not very durable. Today synthetic plastic materials, such as polyamide (PA; nylon) predominate. This material is cheap, strong and not too stiff to work with. PE is also used to some extent because it is more resistant to fouling as the surface is smoother; it is however, stiffer to work with. Polyester (PES) has also been tried.

Nylon used for nets is made as a multifilament consisting of several thin threads spun together to make a thicker one. The advantage with multifilament is that the thread is easy to bend, easy to work with, tolerates more loads and is more resistant to chafing. In contrast, monofilament is a single thread as used in a fishing line. It can be made of PE; it is stiffer and more vulnerable to chafing than a multifilament.

Nets are either knotted or knotless, in which case they are sewn together. In the past there were problems with knotless nets because they came unstitched, but today this problem has been overcome and both knotted and knotless nets are in common use.

The normal mesh shape is square; hexagonal meshes are also used, but to a lesser extent. Hexagonal meshes are more commonly used for trawling bags on fishing vessels.

A number of dimensions are used to describe the mesh. Bar length is the distance between two knots while mesh size is the distance between the knots on a stretched mesh. Mesh size may, however, also refer to bar length, which makes this expression rather confusing. In this chapter mesh size is given as stretched mesh. In a hexagonal mesh, the mesh size is given as the distance between the two longest parallel bars (Fig. 15.13).

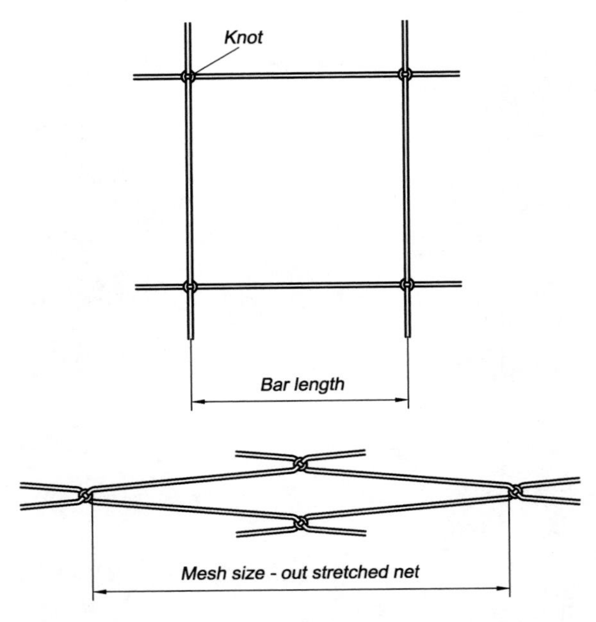

Figure 15.13 Important dimensions used to describe a mesh.

Another expression that indicates how the net panel is standing in the sea is how the net is stretched in the x and y directions. This can be called the hanging ratio of the net (E). This is the ratio between the length of the stretched net panel (L_y) and the length of the line where the net is fixed (top line) (L_x):

$$E = L_X/L_y$$

Normally E for net bags for fish farming is in the range 0.6–0.9, while for a fishing net, for instance, E is between 0.4 and 0.6, meaning that fishing nets have meshes that are more stretched out (Fig. 15.14).

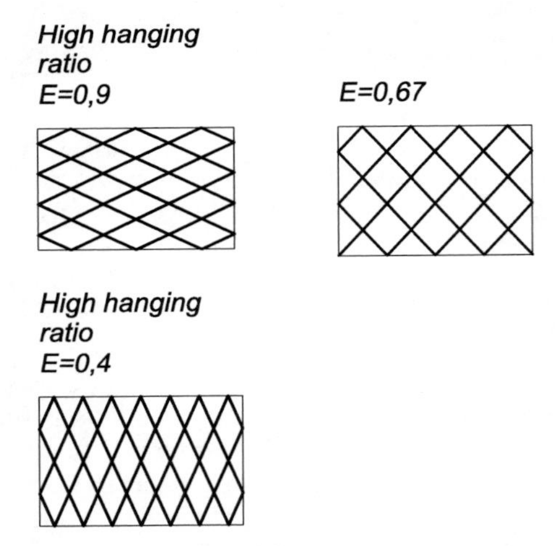

Figure 15.14 How the net is stretched in the x and y directions will result in different shapes of the mesh, and the net panel will appear different in the sea.

Solidity is used to describe how 'tight' a net is and is the ratio between the total area that the net covers, compared to the area covered with threads including knots. This relation is important when the resistance against water flow through the net is to be calculated. Fouling on the net will increase the solidity, because the covered area is increased.

The material strength of net panels exposed to sunlight (UV), wind, rain, acid rain, etc. is reduced. This process is called weathering. Polyvinyl chloride (PVC) is the material that is most resistant to ageing, followed by PE and PA; PP has the shortest lifetime.[16] The lifetime can be increased by adding a coloured (black) antioxidant, so the development of weathering is reduced. Today, however, white untreated material is commonly used for net bags. As it is usual to add some type of antifouling agent to the nets, this will also cover the multifilament.

The normal lifetime of a net bag will vary with the site conditions; in Norway, for example, the lifetime of a net bag is usually set as 5 years.[19] Another way of controlling the duration of a net bag is to carry out a strength test. In Norway, the breaking strength of the net bag below the surface must not fall below 65% of the initial strength.

When the bag is exposed to water currents the volume is reduced by deflection. Because of this, the net bag must be correctly constructed. Narrow deep nets are not recommended on sites exposed to currents. Recommended net depths are 0.8–1.25 times the diameter of the bag.[20]

Because wave motions decrease significantly with increasing depth, it is an advantage to place the cage bags at some depth on exposed sites; 15–25 m deep is normal for large cages with a circumference above 60 m. The recommended depth of cage bags in some exposed sites is more than eight times the significant wave height.

The merits of vertical or sloping sides in the net bag are as follows. Volume reduction is limited by using sloping sides, but if the amount of lump weights has to be increased to maintain bag volume, the forces are reduced if the bag has vertical sides.

15.4.4 Breakwaters

On sites exposed to waves, breakwaters may be used to reduce wave height and impact. Hence the environmental loads on the cages lying behind the breakwater will also be reduced. Breakwaters may be constructed in different ways.[1] One method is to use concrete blocks or a steel construction fixed to the bottom; however, these are expensive to install and little used for protection of sea cages in deep water, although they may be used in shallow water. Pneumatic barriers with air bubbles may also occasionally be used. Most usually a breakwater made of rubber tyres is used. Old tyres from trucks or cars are tied up with wire to form a fleet. The width of the breakwater is important for its effect on the breaking waves. It is normal to use several breakwaters and their total width must be at least as great as the width of the farm to be protected. The width of a breakwater is recommended to be at least 1.5 times its length.[21] This is because the most damaging wavelengths are 0.5–1.25 times the length of the structure. A simple pipe will to some extent function as a breakwater, but it is much less effective than specially designed breakwaters; eventually several pipes can be placed adjacent to each other.

Breakwaters can be moored like sea cages. Some distance is necessary between the breakwater and the farm: up to four wavelengths is recommended.[1] The breakwaters will then create a shadow where the cages are placed.

The breakwater decreases the wave height by reducing the energy in the wave. This is because there is:

- Reflection and waves travelling in the opposite direction are created
- Distribution in the breakwater
- Transfer of energy to the breakwater
- Deflection of waves hitting the corners of the breakwater.

15.4.5 Examples of cage constructions

Below, examples of designs used for sea cages are shown to give some idea of the dimensions.

Plastic cages

Plastic collar cages made either of PE (actually high density polyethylene (HDPE)) or PP are often circular, but may also be made quadrangular and be used as a system farm in less exposed sea areas.

In circular cages it is normal to have two pipes of diameter 200–315 mm. Both may be filled with PS, or one filled with PS while the other is air filled. A wide range of circumferences are available, commonly between 30 and 120 m. Between the two pipes some type of fitting is used, either of plastic or steel, to ensure strength and make a base for the walkway. Some also use a chain all round the circumference to improve the strength. Examples of required buoyancy are from 40 to 120 kg/m depending on the cage dimensions.

Steel cages

Steel cages are constructed with pontoons to ensure buoyancy, while the steel framework gives strength and stretches out the net bag. The steel construction is normally galvanized but can also be painted. Typically there is a 2–3 m wide walkway that eventually can be used by a small forklift truck. Normally the buoyancy is in the range 800–4000 kg/m^2, the highest value being for walkways for driving. The walkways around the cages are smaller (up to 1 m wide) and have lower buoyancy of around 500 kg/m^2. Between the centre gangway and side-ways there are special movable hinges. The pontoons are normally made of PE infilled with expanded PS.

Ocean cages

One type of cage suitable for large waves is made of rubber pipes with a typical exterior size larger than 400 mm. The cages are made as a quadrangle, hexagon or octagon. Steel pipes are used in the corners and to connect the parts made of rubber pipes. The rubber pipes will follow the wave motion very well. The cages are reported to tolerate very rough weather conditions, such as wave heights of up to 8 m.[22]

Ocean spar technology is another technology available for ocean cages, and these have no typical cage collar.[23] In one system vertical cylinders (spars) are placed in each corner of a quadrangular cage bag to stretch it out; the horizontal areas affected by the waves are thus reduced. Another type includes a central spar and rim held together with tension stays. It forms a cube-like construction that is only partly above the water surface which will be dragged below the surface when there is much wave activity or strong currents.

15.5 Mooring systems

The function of the mooring system is to keep the farm in a fixed position and to avoid transfer of excessive forces to the cages, especially vertical forces. Different methods are used for mooring depending on the type of cage, how exposed the sites are to the weather, and the requirement for position exactness. Two major systems are used for mooring: pre-stressed and slack (Fig. 15.15). Slack mooring is used to moor ships which drift around one anchorage point. Such mooring systems are well adapted for stiff constructions such as ships. Few cage farming systems are stiff and therefore pre-stressed mooring systems are most often used, but slack mooring has also been tried.[24] Pre-stressed systems are well adapted for use in flexible constructions, and in correctly designed systems the forces will be equally spread over the entire farm. Pre-stressing of the mooring system is performed at high tide and forces can be up to several tens of kilonewtons.

A special type of mooring is needed for tension leg cages[25] in which the forces are taken up in the tension legs which, regardless of the weather conditions, are always under tension. In this way the dynamic loads resulting from the weather, that affect traditionally moored cages and create slack and tension in the lines, are avoided. The challenge when constructing tension leg moorings is to find technical solutions where the legs will always be tensioned.

A pre-stressed mooring system contains three major parts (Fig. 15.15):

(1) Mooring lines which include the point of attachment to the cages
(2) Buoys
(3) Anchors.

Later in this chapter the design and construction of different mooring systems are described with primary emphasis on mooring of seawater cages exposed to some waves (>1 m) and current (up to 1 m/s). In well protected seawater sites and freshwater sites the environmental loads transferred to the cages are reduced and a smaller mooring system can be used. However, the same basic principles can be used for design and construction.

15.5.1 Design of the mooring system

The design used for the mooring system depends on the type of cages to be moored: these may be as follows (Fig. 15.16):

- Single cages
- System for mooring several single cages
- Single cages with walkway
- Single cages with walkway and landing

Prestressed - normal condition - no load

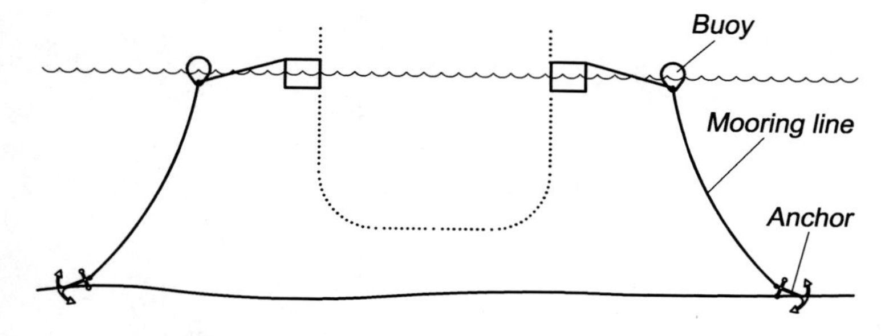

Prestressed - loaded condition

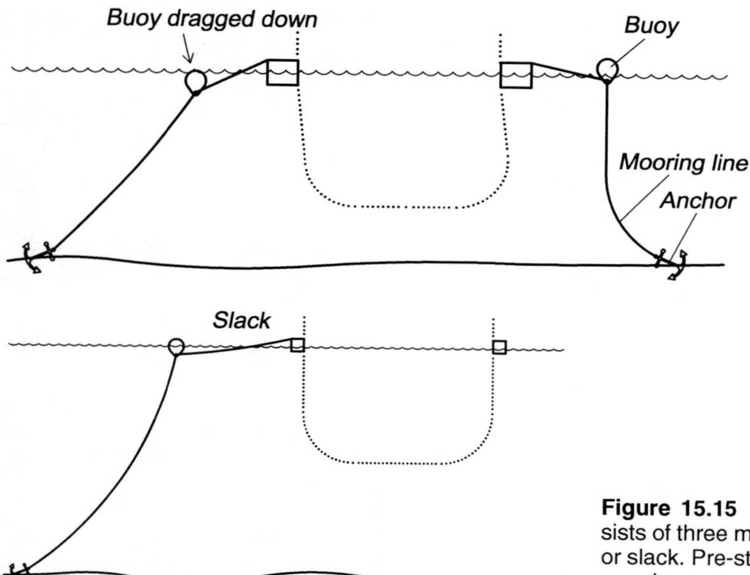

Figure 15.15 The mooring system of sea cages consists of three major parts; the system can be pre-stressed or slack. Pre-stressed systems are commonly used when mooring sea cages.

- Group of cages – platform cages
- Ocean cages
- Cages lying on sway.

A single cage may be moored by between four and six single buoys attached to anchors and to the cage by mooring lines. It is recommended that the single mooring line be divided into two before the fixing point to the cage; this is known as the hen foot mooring; it reduces the forces at the points of attachment to the cage framework, because the number of fixing points is doubled. The points of attachment are critical in the mooring system, because all forces are transferred through these points. The cost of mooring single cages is high; the system is most viable for circular and polygonal cages.

Today, systems for mooring several cages are more usually used. Two or three longitudinal mooring lines are attached to each other with transverse mooring lines; in this way a frame where the single cages can be fixed is built. By lowering the frame 1–2 m below the water surface, access to the cages by boat is improved. Here again, the single cages are attached to the frame with a hen foot mooring. One single cage may be removed, taken out from the system and transferred to another site.

Single cages may also be moored to walkways. If the walkway has a landing, the requirement to keep the cages in the same position is greater (high

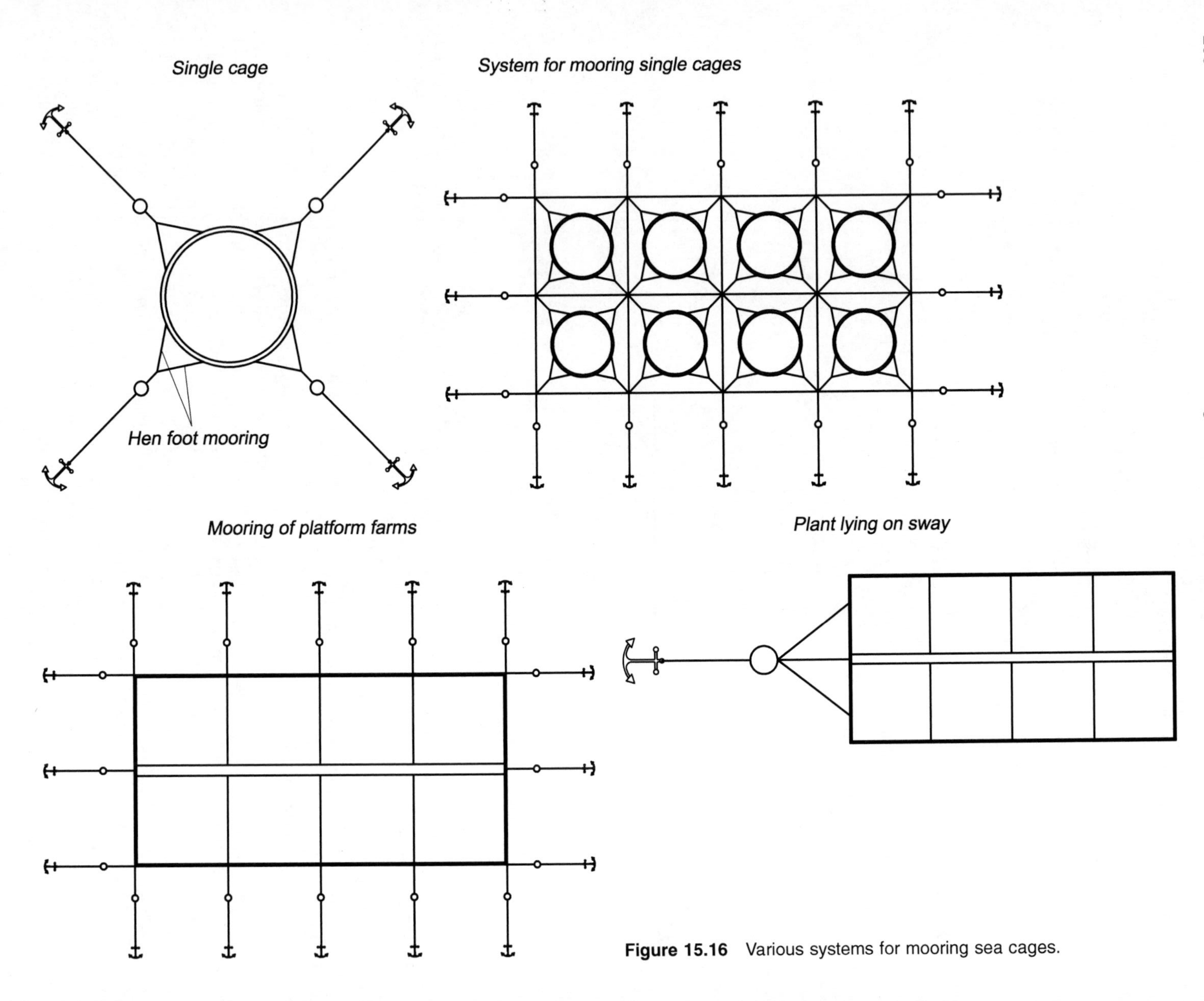

Figure 15.16 Various systems for mooring sea cages.

position exactness); the pre-stress in the mooring lines must also be higher. With such moorings the problem is that the wave is transmitted through the cage and the walkway with different velocities, resulting in large forces at the connection points. The connection point may therefore be subject to material failure. To reduce the forces it is advantageous to use several connection points. Alternatively, the connection point must be flexible. Fixed connections to walkways are not recommended in exposed water.

In ocean cages the requirements for the moorings system are large because the forces are so great. Ocean cages are moored individually and the producer will normally have their own mooring systems, which could be an integral part of the cage construction.

In a platform cage farm the mooring line is typically withdrawn from each corner and from the middle of the farm. Such farms are normally sited in protected water because they tolerate lower waves, due to their construction with linkages.

If the sway principle is used, only one mooring point is necessary; it is actually a slack mooring. This mooring point is exposed to large forces and a big anchor and buoy are necessary.[24] Such farms need a large area because the unit will drift around the mooring point. Ageing of the sites is, however, delayed when using this arrangement. However, the method is seldom used, mainly because such a large area is necessary.

15.5.2 Description of the single components in a pre-stressed mooring system

Fixing point

The mooring line is fixed to the collar by a shackle. If a rope is used this should be spliced and a thimble used to reduce the bending of the rope. All bending will weaken ropes to some extent. A bending diameter of three times the rope thickness is necessary to avoid significant weakening. A knot can reduce the strength of the rope by 50%.[19] Rings may also be used in the connections because they tolerate chafing better. It is an advantage to over-dimension the connection, for instance by doubling the size of the maximum transferred forces.[26]

The force through the connection point can be divided into horizontal and vertical components. To avoid breakage of the cage collar by the tide, the vertical component should be transferred so that it is as low as possible, almost negligible. By using several connection points to the collar, the forces transferred at each point will be reduced, although this will increase the mooring costs. Normally at least four points are used. As said previously, it is advantageous to split the single mooring line before attaching it to the collar, so more connection points are achieved; such an arrangement is known as a hen foot mooring from its design.

To secure the fixing point against breakage, a secondary fixing may be used (Fig. 15.17). An extra rope or wire transfers forces directly from the collar to the mooring lines if the main fixing point breaks.

Mooring lines

Different materials and designs can be used for the mooring lines which are often made of synthetic rope. When choosing rope, the breaking strength is the most important factor, but price and duration are also major determinants. The elasticity, given by the E modulus, must also be taken into account. How much a line stretches lengthwise when loaded can be given in the mooring line characteristics.

Lines of synthetic rope are often stretched permanently after the first load so are beyond their elastic range and do not return to their original length. It may be advantageous to pre-stretch the rope before it is used; alternatively, the rope must be tightened after exposure to the environmental loads.

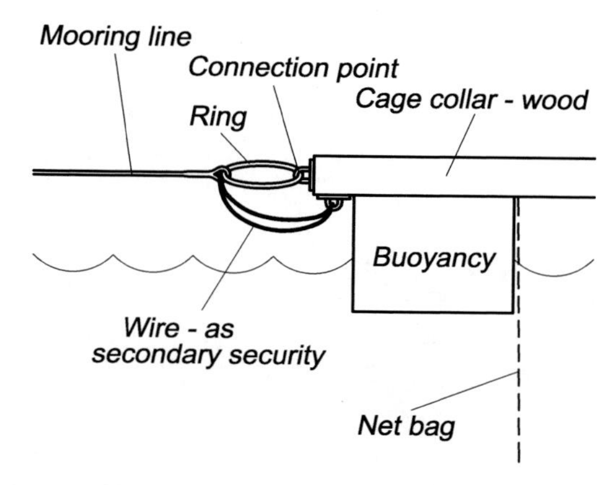

Figure 15.17 Double fixing is recommended where the mooring lines are connected to the collar.

Different materials are used for synthetic ropes, such as PA (nylon), PE, PES and PP, all of which have advantages and disadvantages. PA tolerates the highest forces with a given diameter, while PP ropes have the lowest weight. Metal or Kevlar threads may also be integrated into the rope to increase its strength, but this also increases the cost.

Ropes may be exposed to chafing and this must be avoided. The rope may be covered, for instance with a PE pipe, to reduce chafing. Variable loads, such as occur in mooring lines, will decrease the strength of the rope more than constant loads.[19] The lifetime of synthetic rope in a mooring system can be set as 4 years as a starting point, but will of course depend on the rope and the loads on it.

If a chain is used in the mooring system, it will be exposed to corrosion. Good quality therefore confers an advantage, but this will be a question of price compared to duration. Chain is heavier to handle than rope and therefore more handling time is necessary. Chain tolerates more mechanical influence (chafing) than rope. For this reason chain is often used near the bottom, close to the anchor, normally for the first 15–20 m of the mooring lines. Afterwards rope is used because it is so much easier to handle. To avoid the connection point being exposed to chafing against the bottom, a small buoy may be added here.

Wire or metal rope, may also be used for mooring, but even though it is strong, it is expensive and difficult to work with, so it is seldom used. It may, however, be a good alternative secondary fixing for cages.

Buoys

Buoys are used to hold the mooring lines up so that vertical forces on the collar are avoided (Fig. 15.18). They will also take up the weight of the mooring system so this is not transferred to the cage if placed some distance (normally 15–20 m) away from the collar. By limiting this distance, spreading of the mooring system will also be avoided. Furthermore, to ensure horizontal transference of forces, the buoys have a major roll in pre-stressing the farm by pre-stretching the mooring lines. Buoys will also function like extension springs and damp wave movements, for instance. When a load is added to the cage the buoy will be dragged down and not so much will be seen on the surface; if the load is removed it will float up again.

Typical buoy sizes are from 200 to 700 l. The buoys can be filled with air or foam. To avoid puncturing reducing the buoyancy, the use of foam filled buoys is highly recommended, and in some places is mandatory. Expanded PS or polyurethane (PU) is commonly used. If an air-filled buoy is dragged below the water surface the buoyancy will decrease rapidly because the buoy is compressed; reduced volume results in reduced buoyancy. Foam-filled buoys will also be compressed when beneath the water surface. PVC foam filling will tolerate more pressure, but is more expensive than PU. The buoyancy of the buoys is recommended to be somewhat higher than the calculated requirement, twice what is necessary, for instance. Round, cylindrical and polygonal buoys are used in mooring systems. Experiments in which single buoys were replaced with multiple floats covering a distance of the rope showed that these gave a much smoother dynamic response on the anchor lines.[24]

A lump weight can be added to the mooring line between the cage collar and the buoys. This makes it possible to go over the lines by boat, and will also function as an additional pre-stressor of the farm and reduce jerks in the lines. However, the wear

Figure 15.18 Buoys are used to reduce the vertical forces on the collar, and birdnets are used to avoid birds taking small fish.

on the mooring lines is increased by this method, which is therefore not recommend.

From the buoys the mooring lines go to the mooring points. Depth–length ratios are commonly set to 2.5–4. Long anchor lines reduce the vertical forces and reduce the required buoyancy in the buoys, but have the disadvantages that a greater area is needed and the cost of the mooring lines is increased.

The magnitude of the pre-stress in the mooring lines for a cage farm depends on the site and other factors. One method is to pre-stress at high tide so that 75% of the volume of the buoy is below the surface and 25% is above to take additional loads. Inspection of the buoys could then constitute a simple control of the mooring system. Since all mooring lines are equal and equally pre-stressed, all buoys will have the same volume above the surface. It can be an advantage to have distinct marks on the buoys to control more easily how much of the buoy volume is below the surface. This check must, however, be done when no environmental forces affect the cage. If, for instance, a current is coming in from one side, the buoys on this side will be dragged down and those on the opposite side will float higher in the water as a result of the unequal loads on the mooring lines.

Anchors

The simplest type of anchor is the dead weight or block anchor. Theoretically all heavy objects, such as old engines may be used, but they may be difficult to handle. In addition, this may result in chafing of the mooring lines. Concrete blocks are most commonly used as weight anchors and vary from some hundreds of kilograms to several tonnes. As the density of concrete is quite high (around 2.4 t/m^3) it will easily sink to the bottom, stay there and take up forces. The great disadvantage with block anchors is that they are heavy to handle. Large cranes are needed on the boats setting them out and precautions must be taken to avoid tilting the boat during this operation.

To prevent block anchors being displaced on the bottom, good friction between anchor and the bottom is necessary; this depends on the bottom conditions and is given by the friction coefficient. Sand and clay have high friction coefficients while that of rock is low (between 0.1 and 0.5), meaning that an anchor will slide easily on rock. Block anchors are not recommended for use on rocky ground; here either drag anchors or bolts should be used. Friction coefficients of 0.5 for sand and 0.3 for clay may be used as a basis, for anchor choice if no measurements are done.[26]

To increase the friction coefficient, the bottom of the anchors can be rough, even having iron bars emerging from the base. Another method of increasing the friction is to add an iron jacket around the bottom of the block. This will increase the forces holding the anchor on the bottom, because it is sucked to the bottom like a sucking disc, assuming suitable ground conditions such as clay. The angle at which the mooring line is joined to the block is, of course, also important in inhibiting horizontal movement (see section 15.7.2). To avoid tilting of block anchors, the width of the blocks ought to be large in relation to the height (>2:1), or the block anchor may tilt over the mooring lines and cause chafing.

A drag anchor or ebbing anchor is another type much used (Fig. 15.19). The aim of the design is for the anchor to be dragged down into the ground like a plough and become fixed. Various designs of ebbing anchors are available, and the different suppliers normally have their own designs. The old traditional one is the stock anchor as used on boats; today more effective designs for mooring of fish farms are available. How well fixed an ebbing anchor is to the bottom depends on the bottom conditions and the design of the anchor; the angle of the mooring line is also important. The optimum angle depends both on the bottom conditions and anchor design. The angle for ebbing anchors used in sand can be 30–35° and in clay 30–50°.[1]

An ebbing anchor will tolerate a large horizontal force, but tolerance of vertical forces is low. To improve this, a heavy chain may be used before the anchor. Another method is to use a small block weight on the mooring line before the anchor. An ebbing anchor is of much lower weight than a block anchor, because it is based on a totally different principle, but is more expensive. From the back edge of the ebbing anchor there should be a rope to the surface; to keep the rope on the surface where it can be reached, it is attached to a buoy. This must be done to ensure the anchor can be released when removal is necessary, otherwise this might be impossible as the anchor is normally

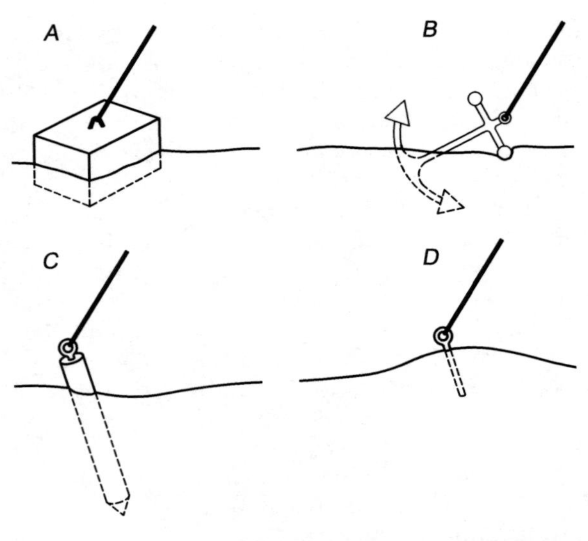

Figure 15.19 Different anchors are used for mooring sea cages: block (A), drag (B), pile (C), bolt (D). The photograph shows drag anchors.

completely buried in the bottom after the mooring system has been pre-stressed.

Bolts can be used to advantage where it is possible to fasten the mooring lines in rock, usually where the mooring lines lead to land, but bolts may also be use under the water surface. Galvanized bolts are set into drilled holes; either they can have an expanding construction or expanding slurry can be injected into the holes. The bolts are either of the eye or T-type.

Piles are an alternative in sand and clay bottoms, but the forces they can take up are normally quite low; otherwise they must be very large which increases the cost considerably.

There are several other special anchors that could be used, but they are normally more expensive. One type used in the offshore oil industry is the suck anchor. In principle the anchor is a large specially designed sucking cup that sucks down to the bottom. This can be made from a wire-spoked wheel with a plastic cover.

15.5.3 Examples of mooring systems in use

The mooring system will naturally be individual to each site, and depend on the environmental forces. Some examples are given below for some exposed sea sites. The system will also depend on the chosen quality of the rope, the chain and the type of drag anchor.

A system mooring is used to moor eight circular plastic cages with a diameter of 15 m. The system mooring is prefabricated and consists of a frame of chain (13 mm × 88 mm alloy). Within the frame a 36 mm synthetic rope is used. The three main longitudinal mooring lines are of 58 mm rope; this is also used in the ten side moorings. Anchors are 32 mm eye-bolts and drag anchors with a weight of 1.2–1.8 t In the last part of the mooring lines to the anchors/bolts chain is used to avoid chafing.

To moor a platform cage farm with a total of 16 cages with a diameter of 15 m × 15 m and a bag depth of 10–15 m, 32 mm synthetic ropes are used.[27] 16 mm chain with a weight of 3.8 kg/m is used on the last part to the anchors. A lump weight of 100 kg is used at the connection point between the chain and the rope. Eye-bolts or drag anchors of 800 kg are used as anchors and the buoys are 500 l.

Typical anchor sizes for mooring offshore rubber cages are 10–30 t for block anchors or 3–5 t for drag anchors. The mooring ropes are typically 50–70 mm diameter.[22]

15.6 Calculation of forces on a sea cage farm

To be able to design a cage with its mooring system, it is necessary to calculate the environmental forces affecting the farm. On ordinary sites it will be the current that causes the highest forces, while on more exposed sites the wave forces will also be

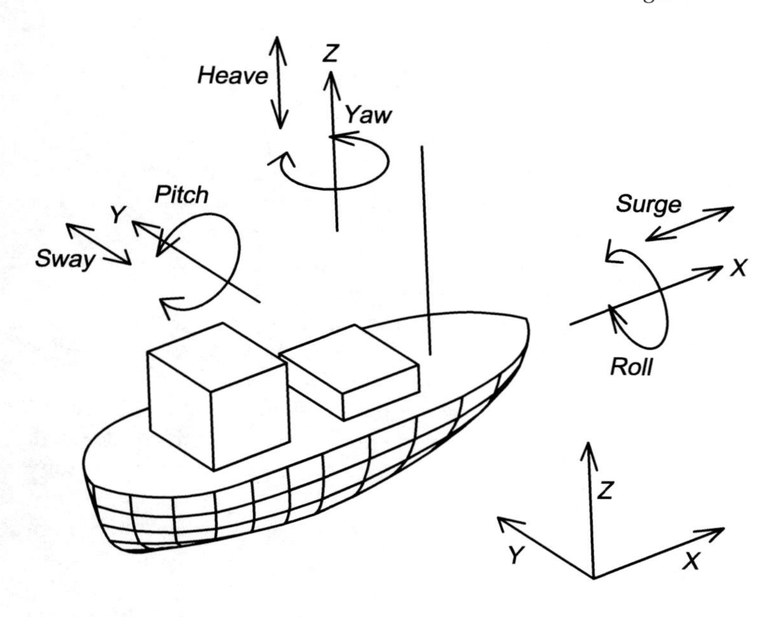

Figure 15.20 Like a boat, the top of a sea cage top will be subjected to forces as a result of current, waves and wind. (Adapted in part from ref. 1.)

considerable. The direct wind forces will be low due to the small area above the water. Both static and dynamic (from the waves) forces will affect the cages. The size of the forces must be known to find whether the cage construction will tolerate the forces on the site, or if the construction will break down. The supplier of the cage ought to have done these calculations and know whether the equipment will tolerate the forces at the site. However, the environmental forces will also determine the size and design of the mooring system, and these must be determined on site, because the environmental loads will vary with the site.

Calculation of the forces that affect the sea cages is quite difficult, especially for open ocean cages; normally computer programs utilizing numerical methods are employed (for the latest information see, for example, refs 28–30). In this section only a brief overview is presented, focusing on simplified solutions to increase the reader's knowledge of how environmental forces affect floating constructions.

15.6.1 Types of force

A construction floating in the water, for instance a boat or a sea cage, will be exposed to forces as a result of current, waves and wind. The forces can be resolved into three linear components in the x, y and z directions; the torque occurs around the same three axes (Fig 15.20):[1]

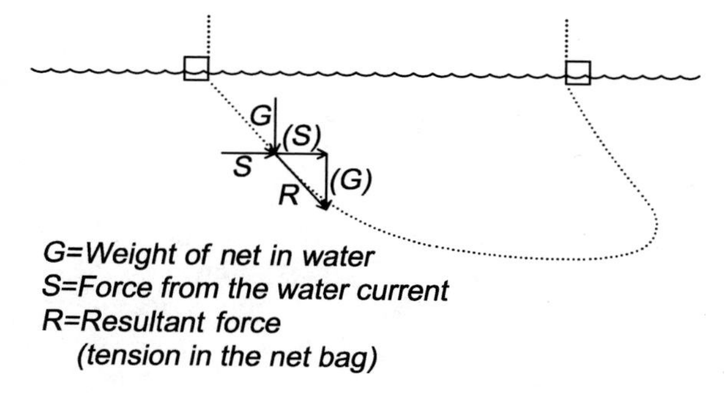

Figure 15.21 Resolution of the forces affecting the cage farm into horizontal and vertical components.

- Linear movements
  - heave: vertical motion
  - surge: horizontal motion along longitudinal axis
  - sway: horizontal motion along the transverse axis
- Rotating movements
  - yaw: rotation about the vertical axis
  - roll: rotation about the longitudinal axis
  - pitch: rotation about the transverse axis.

When doing calculations, the six forces may be resolved into two forces: a resultant horizontal force and a resultant vertical force. Even if the accuracy is less, the calculation is greatly simplified. Thus there are two forces affecting the farm (Fig. 15.21):

(1) Drag force parallel with the current direction, F_D
(2) Lift force normal to the current direction, F_L.

To give some idea of the environmental forces involved, the following minimum values have been recommended in Norway[26] when designing a mooring system for sheltered offshore farms:

- Significant wave height: 1.5 m
- Current velocity: 1 m/s
- Wind: 30 m/s over 10 min period
- Storm flood tide: 1 m
- Fouling: 30% of the area underwater is fouled completely.

These values will, however, vary from site to site around the world, depending of the location. During the past few years an increasing number of exposed sites have been used for farming purposes where environmental forces are greater than those given above.

15.6.2 Calculation of current forces

General methods

All parts of a cage farm below the water surface will be affected by the current. This includes the net bag, the cage collar (pontoons), the buoys and the mooring lines. The size of the forces on the elements depends on the area affected by the current. Thus the net bag is affected the most because it has the largest area.

To calculate the current forces affecting a submerged construction, the first part of Morrison's equation can be used. The same equation may also be used to calculate the wave forces, but here an acceleration term is also included. As the current velocity is constant, this term can be neglected when calculating the forces.

Morrison's equation without the acceleration term and with forces only in the x direction can be written as follows:

$$F_D = \frac{1}{2} \rho C_D U_c^2 A$$

where:

F_D = drag
ρ = water density
A = area affected by the current
U_c = current velocity
C_D = drag coefficient.

The drag coefficient gives a picture of the current resistance between the object and the water current and is determined by experiment. If the object is square the coefficient will be higher than for an aerodynamically shaped construction. The drag coefficient can be found from the bar diameter and bar length of the net used in the cage bag.[31] It can also be found from the solidity, Reynolds number and the angle of attack of the current.[32]

Fouling on the net bag will increase the current forces because the area of the net is increased, and by this the area that effects the current. Solidity increases with fouling because the bar diameter increases. To calculate the current forces on the net bag is, however, not easy. Not only does the water pass through the net bag, but the current will also cause the net bag to deflect (Fig. 15.22). By having several net bags one behind another, the current velocity will be reduced from net panel to net panel as a result of the resistance to flow through the net bags; therefore more and more water will gradually flow below and around the bags behind the first one.

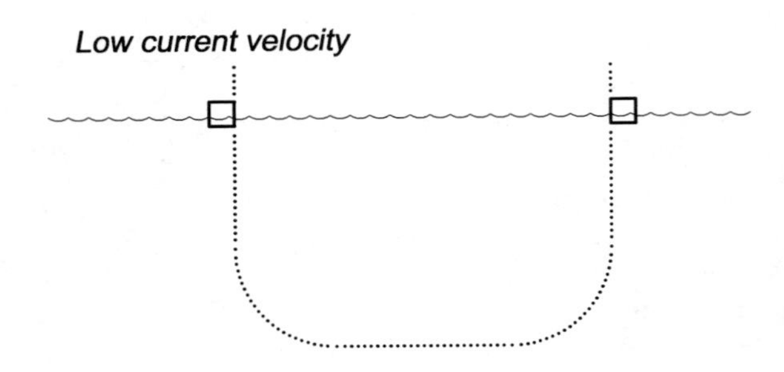

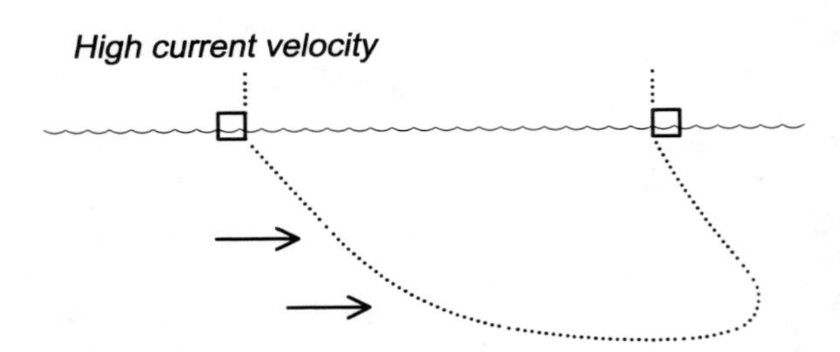

Figure 15.22 A normal net bag will be deflected by the current.

The net bag will be exposed to a lifting force due to the deflection because it is fixed on the surface by the mooring system. More accurate equations for the current forces affecting the net bags which are not rigid, and deflected by an angle α, will therefore be as follows:

$$F_D = \frac{1}{2}\rho C_D(\alpha)U_c^2 A$$

$$F_L = \frac{1}{2}\rho C_D(\alpha)U_c^2 A$$

where:

F_D = drag
F_L = lift
α = angle of deflection of the net
ρ = water density
A = area affected by the current
U_c = current velocity
C_D = drag coefficient
C_L = lift coefficient.

These equations will also only be partly correct because the net bag is not deflected by a fixed angle towards the surface, but in a curve (Fig. 15.23). By dividing the bag into several sections with different angles it is possible to compute the drag and the lift, but this requires a lot of calculations. The coefficients will depend mainly on the solidity of the net and the attack angle.

Because it is fairly complicated to calculate the forces on the entire net bag in one go, a method which divides the net bag into single net panels has been utilized.[33,34] For example, a square cage contains four side panels and one bottom panel. The force is then calculated for each panel and the total force represents the sum of the forces on the separate panels: a trial showed that the calculated forces were in the range 0.9–1.3 of the measured forces.

The amount by which the net is deflected depends on the current velocity, the weights in the bottom of the bag, and the bag design. Experiments have shown that the reduction in volume of the net bag can be over 90%, when the velocity is increased from 0 to 1 m/s.[35] To reduce the deformation, increased lump weights in the bottom could be used. However, this will increase the horizontal force on the net several times over and may not be advisable.[33] In addition, if there are waves extra weights in the bottom of the bag will increase the wave loads on the net bag.

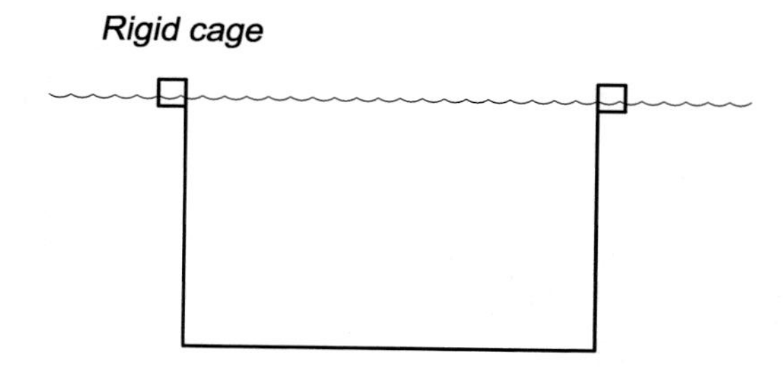

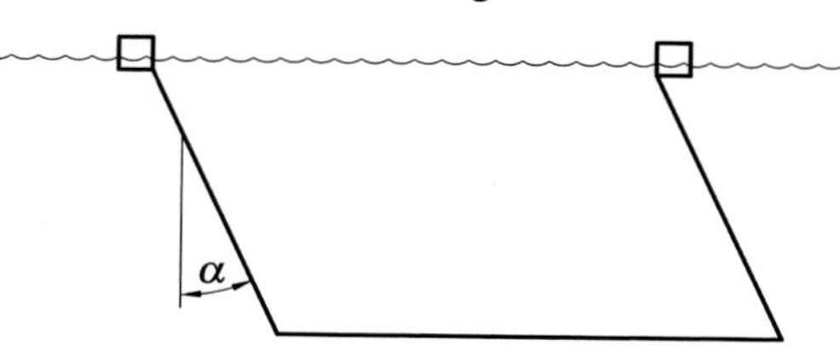

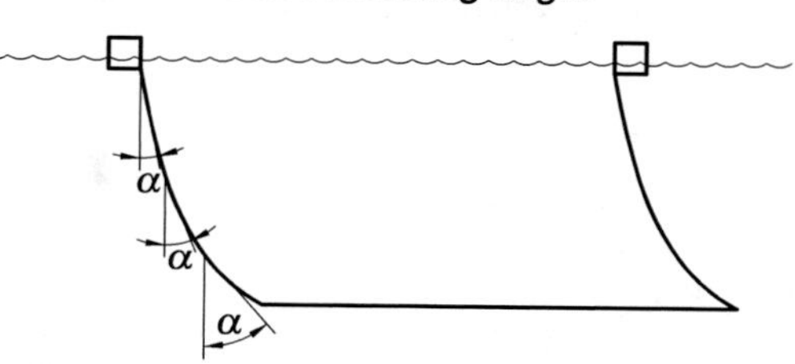

Figure 15.23 The cage is not rigid but will be deflected at an increasing angle by the current.

Reduction in velocity

When the water current passes one net panel the current velocity will be reduced by friction from the net; much of the water is then forced to go under and around the following nets. When the water current hits the next panel, the force on this panel will be reduced because the water velocity is reduced. To be able to calculate the forces on the following net panels it is necessary to know the size of this reduction, particularly since the velocity is squared in Morrison's force equation so gives a large contribution.

The following equation may be used to estimate the current velocity U_i after i net panels:[33]

$$U_i = U_c r^i$$

where:

i = number of net panels that the water has to pass
U_c = current velocity of the water when it hits the first panel
r = reduction factor (depending on the solidity, S_n, for the net).

The velocity reduction factor may also be expressed by the drag coefficient C_D, which again is affected by the solidity, S_n:

$$r = 1 - 0.46C_D$$

If S_n is between 0.1 and 0.35 C_D may be calculated as follows:

$$C_D = 0.04 + (-0.04 + S_n - 1.24S_n^2 + 13.7S_n^3)\cos\alpha$$

where:

α = angle of deflection of the net.

If using rigid cage bags the equation will be as follows:

$$C_D = S_n - 1.24S_n^2 + 13.7S_n^3$$

As seen there is a dramatic reduction in velocity for every net panel the water has to pass.

In addition, to reduce the forces on the net bags that lie behind the first one will result in a reduction in supply of new oxygen-rich water. When having several cages, one after another in the direction of the current, lack of oxygen might occur in the last cages that the water passes. As the water exchange is reduced there will also be a reduction in removal of metabolic waste substances, which also shows the importance of correct individual placing of the cages in relation to the current direction.

By increasing amount of fouling the current velocity will be reduced even more, as a result of increased solidity, so it is important to minimize fouling on the nets.

Example
Three cages are lying behind each other. The velocity of the current that hits the first net panel is 0.7 m/s; C_D is 0.32. Calculate the current velocity in each of the three cages assuming rigid nets.

First the velocity reduction factor (r) is calculated from C_D

$$\begin{aligned} r &= 1 - 0.46(0.32) \\ &= 1 - 0.15 \\ &= 0.85 \end{aligned}$$

Then the velocity inside cage 1 (after 1 net panel), cage 2 (after 3 net panels) and cage 3 (after 5 net panels)is calculated:

$$\begin{aligned} U_1 &= 0.7 \times 0.85^1 \\ &= 0.6\,\text{m/s} \end{aligned}$$

$$\begin{aligned} U_2 &= 0.7 \times 0.85^3 \\ &= 0.43\,\text{m/s} \end{aligned}$$

$$\begin{aligned} U_3 &= 0.7 \times 0.85^5 \\ &= 0.31\,\text{m/s} \end{aligned}$$

Simple method for calculating the current forces with rigid nets

As shown, there is a lot to consider when calculating the current forces on sea cages, and usually specially designed computer programs are used for this purpose. For rigid nets, however, it is possible to calculate the forces employing quite simple methods. Use of rigid nets will, however, result in overestimation of the forces compared to the real situation, but will show the principles. Using the values obtained to calculate the size of mooring lines and anchors will also ensure that the mooring is large enough, although the mooring system will be more expensive than necessary because it is overspecified.

A simple set of equations and diagrams have been developed for calculation of the current forces on rigid sea cages.[33] The method assumes rigid mesh/bags with no deflection, so therefore no lifting forces. In addition, only forces normal and parallel to the current on the farm are taken into consideration.

First the drag on the net panel normal to the current direction, F_{DN}, must be calculated; then forces parallel to the current direction F_{DP} are calculated; these are then added to give the total forces F_{Dtot}:

$$F_{DN} = \frac{1}{2}\rho C_{DN} U_c^2 (BD) m \left(\frac{1-r^{4n}}{1-r^2}\right)$$

$$F_{DP} = \frac{1}{2}\rho C_{DP} U_c^2 (B+2D) lm \left(\frac{1-r^{4n}}{1-r^4}\right) r^2$$

where:

$F_{Dtot} = F_{DN} + F_{DP}$
ρ = density of the liquid
C_{DN} = drag normal to the current direction
$= S_n - 1.24S_n^2 + 13.7S_n^3$

$$S_n = solidity$$
$$= 2 \times \frac{bar\ diameter}{bar\ length}$$

C_{DP} = drag parallel to the current direction (set to 0.04 based on the shape of the bag)
U_c = current velocity
B = width of the cage bag
D = depth of the cage bag
l = length of the cage bag
n = number of cages parallel the current direction
m = number of cages normal to the current direction
r = reduction factor
$= 1 - 0.46C_{DN}$.

The above equation is used to find the force when the farm is parallel to the current direction. To use it when the farm is normal to the current direction requires only that the values of n and m in the equations be exchanged.

Example
A rectangular system farm includes two cage bags, one after another and eight side by side in relation to main current direction, a total of 16 cages. The size of the single cage is: width 10 m, length 15 m, depth 12 m. The design water flow is 0.8 m/s. Calculate the current forces that affect the farm, presuming rigid nets.

The drag coefficient C_{DN} must be calculated first:

$$C_{DN} = S_n - 1.24S_n^2 + 13.7S_n^3$$
$$= 0.3 - 1.24 \times 0.3^2 + 13.7 \times 0.3^3$$
$$= 0.56$$

Calculate r

$$r = 1 - 0.46C_{DN}$$
$$= 0.74$$

Then the forces on the net panels normal (F_{DN}) on parallel (F_{DP}) with the current direction can be calculated.

Drag normal to flow direction:

$$F_{DN} = \frac{1}{2}\rho C_{DT} U_c^2 (BD) m \left(\frac{1-r^{4n}}{1-r^2}\right)$$
$$= \frac{1}{2} \times 1025 \times 0.56 \times 0.8^2 \times 10 \times 12 \times 8$$
$$\times \left(\frac{1-0.74^{4\times 2}}{1-0.74^2}\right) = 354\,724\ \text{N}$$

Drag parallel to flow direction:

$$F_{DP} = \frac{1}{2}\rho C_{DP} U_c^2 (B+2D) lm \left(\frac{1-r^{4n}}{1-r^4}\right) r^2$$
$$= \frac{1}{2} \times 1025 \times 0.04 \times 0.8^2 \times (10 + 2 \times 12) \times 15 \times 8$$
$$\times \left(\frac{1-0.74^{4\times 2}}{1-0.74^4}\right) \times 0.74^2 = 38103\ \text{N}$$

Total drag:

$$F_{Dtot} = F_{DN} + F_{DP} = 354\,724 + 38\,103$$
$$= 392\,827\ \text{N}$$
$$= 392.8\ \text{kN}$$

Another very simple method is to use Morrison's equation on the total length of the rope used to create the net panel. By calculating the current resistance from the total length of the rope the total forces affecting the bag can be found.[18]

$$N = \frac{L}{I_m f}$$

$$T = \frac{D}{I_m f}$$

where:

N = number of meshes in the length of the panel
I_m = bar length
f = coefficient of decrease (normally around 0.7)
L = length of the panel
T = number of meshes in the depth of the panel
D = depth of the panel.

Next the length of the filament (rope) per side (L_T) is calculated:

$$L_T = 2NTI_m k_n$$

where:

k_n = knot factor.

Therefore the total drag force on the net panel can be calculated by Morrisons equation:

$$F_D = \frac{1}{2}\rho C_D v^2 L_T d$$

where:

d = rope diameter (bar diameter)
C_D = drag coefficient for a cylinder is used since this is the same as that of the rope, and this is 1.2.

15.6.3 Calculation of wave forces

Waves are important when designing sea cages. The wave forces will influence design of both the cage collars and the net bag. If the wave forces are too high the collar may break. The wave forces will, however, also affect the mooring systems and must be taken into consideration when calculating the size of the mooring system, even if they are smaller than the current forces.

Calculation of wave forces normally involves the use of computer programs and numeric solutions. Methods for calculating the wave forces include, for instance, Morrison's equation, diffraction theory and Froude–Krylov forces.[7,12] Calculations for traditional fixed offshore constructions will overestimate the forces for a pre-stressed cage farm floating on the surface and following the sea.

Compared to currents waves apply dynamic forces to the construction. Use of Morrison's equation can illustrate this. Morrison's equation has two additional terms, one for the velocity of the water particle in the wave and one for the acceleration of the water particle:[27]

$$F_i = (C_M + 1)\rho V a_i + \frac{1}{2}\rho C_D u_i |u_i| A$$

where:

F_i = forces on the object in the x, y or z direction
ρ = density of the liquid
A = area of the object
C_M = mass coefficient
C_D = drag coefficient
a_i = water particle acceleration in direction i
u_i = water particle velocity in direction i
V = volume of the object.

Without going deeper into this equation, it shows that in addition to the drag term, as there was for current forces, there is a mass force term that is proportional to the acceleration. The total force is the sum of these two forces.

Wave forces are dynamic, which means that they come again and again. This will result in reduced loads being tolerated by structures before breakage occurs. Experiments have shown that a cage may only tolerate a dynamic load which is 10% of one single static load.[5] Waves impose additional forces on the mooring system. On normal, partly protected off-shore sites, the wave forces will be much lower than the current forces. Usually waves add up to 20–30% to the current forces, but this is of course site-dependent.

Due to the complexity of calculating wave forces and measuring water particle velocity and acceleration, this will not be further described here. Specialized literature is recommended, as mentioned earlier.

15.6.4 Calculation of wind forces

Since such a small part of a cage farm is above the water surface where the wind is blowing, the forces transferred to the cage and further to the mooring system will be very low compared to the forces from the current. To calculate the forces on the jumping net the following equation may be used:[32]

$$F_w = \frac{1}{2}\rho U^2 \mu A n$$

where:

ρ = density of air
U = wind velocity
μ = drag coefficient
A = effective area of the jumping net
n = protection factor (for instance, distance from other jumping nets).

The wind forces will be considerably higher if, for instance, a feed barge is moored together with the cages.

15.7 Calculation of the size of the mooring system

15.7.1 Mooring analysis

A calculation of the size of a mooring system must include a performance analysis under extreme conditions,[36] for instance for an intact mooring system, with a break in one of the mooring lines and with an increase in the water level compared to normal due to a storm tide of 1 m for example (requirements in the Norwegian standard for cages).[14] The analysis must show that the mooring system will withstand such situations without breakdown.

What is then happening when the environmental forces affect a pre-stressed moored sea cage farm? A total environmental force (F) will try to move the cages out of their original position. In all mooring lines on the side where the forces F is acting there will be additional tension, and a corresponding reduction in the lines on the opposite side. Depending of the degree of pre-stress and the elasticity of the lines, there will be a drift away from the equilibrium position caused by the acting forces, but the farm will not drift freely. The mooring lines will gradually create an opposing force towards F that prevents the cage farm drifting freely. A new equilibrium position will be established where forces from the mooring lines are opposite and equal to the forces created by the environmental factors. The tension in the mooring lines on the side where the environmental forces are acting is now much higher than when the cages were in their original positions; there might also be slack in lines on the opposite side.

If, in addition, there are waves additional dynamic forces will be imposed on the farm and it will oscillate around the equilibrium position as long as the mooring system does not break. The maximum load in the mooring lines will therefore be higher than from the static current force.

If there is a break in one of the mooring lines due to unforeseen circumstances the farm will drift into a new equilibrium position. This movement will be determined by the tension in the broken line, the weight of the farm and the resistance against movement from the net bag and cage collars. When the farm is drifting and needs to be stopped, it is important that the tension in the remaining lines does not exceed their breaking strength. If this happens other lines will break and most probably this will result in progressive breaking of all the remaining mooring lines so the cages will drift freely. If one mooring line to a cage breaks, there is also the possibility that the cage will crash into another cage or fixed construction such as the walkway whilst drifting towards its new equilibrium position. This may cause a material break in the cage collar, for instance. When doing mooring calculations a break in one of the mooring lines will normally be tolerated, but progressive breaking must not occur in such situations.

15.7.2 Calculation of sizes for mooring lines

There will always be some inaccuracy when describing and calculating the environmental loads affecting cage farms. For example, this can be that the current velocity or wave height varies and might be slightly higher than expected. A load factor is recommended to compensate for the possible inaccuracy. Normally the load factors are between 1 and 1.5 depending on the uncertainty in the calculations of environmental loads. A load factor of 1.5 means that the environmental loads can be up to 50% more than those calculated and the mooring will still hold. The total force that is used in further calculations is then:

$$F_t = \gamma_f F_E$$

where:

F_t = total force
F_E = calculated environmental forces
γ_f = load factor.

In the new Norwegian standard for mooring analysis, the standard load factor is set at 1.15 for unmanned farms and 1.3 for continuously manned farms when doing static analysis i.e. the safety factor is larger.[14]

The breaking strength of the different types of mooring line is also given with some accuracy by the suppliers and is based on a number of measurements. To take care of possible inaccuracy when testing the material, and minor variations in the materials, it is recommended that a material factor (γ_m) be used. However, will this vary with the material and the degree of testing of the material that has been performed. Normally it lies between 1.1 and 5. To find size of the mooring lines necessary this must be taken into consideration, which gives the following equation:

$$F_R = F_T \gamma_m$$

where:

F_R = mooring forces (the force that the mooring lines shall tolerate)

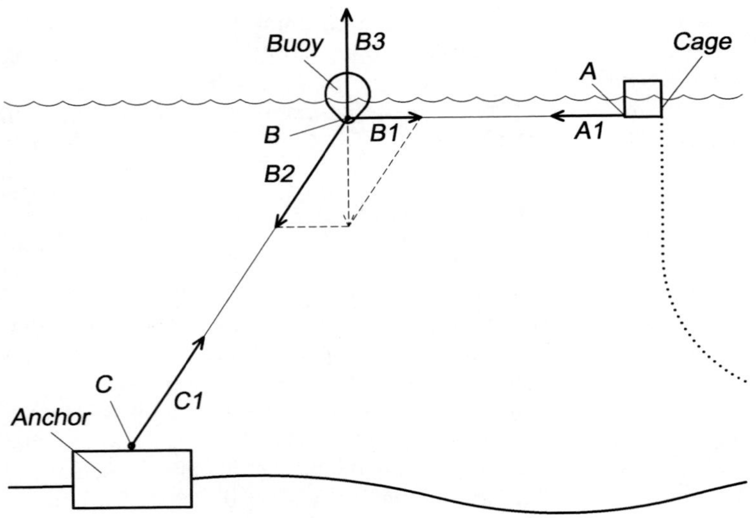

Figure 15.24 Environmental forces will affect the forces in the mooring system.

F_T = calculated total forces including load factor
γ_m = material factor.

The following material factors are used in the Norwegian standard for cage farms:[14] chain, 1.5; synthetic rope with knot, 5.0; synthetic rope, 3.0; synthetic rope specially resistant to ageing, wave and water absorption 1.5.

Example
To show how a mooring analysis can be performed an example is shown where the mooring line, buoy and anchor are to be dimensioned (Fig. 15.24). On a cage farm the calculated environmental force that an anchoring line must take up is 1 t or approximately 10 kN. The length and type of mooring line, buoy type and size, and anchor type and size should be found to keep the farm in position. The depth of the site is 30 m and the bottom is sand. Current velocity is set to 0.2 m/s.

First the design of the mooring system must be found. Choose to have the buoys 15 m away from the cage collar. The mooring line is set to 3× the depth, and becomes 90 m. The total length of the mooring line is therefore 90 + 15 = 105 m. The forces on the buoys can then be found. First the angle α that the mooring line has from the bottom and to the surface must be found:

$$\sin\alpha = \frac{30}{90}$$
$$\alpha = 19.47°$$

Then calculate the force in the direction x on the mooring line:

$$\cos\alpha = \frac{10}{x}$$
$$x = \frac{10}{\cos\alpha}$$
$$x = 10.61\ \text{KN}$$

The mooring line must therefore tolerate a force of 10.61 kN.

Calculate the force in the y direction:

$$\sin\alpha = \frac{y}{x}$$
$$y = \sin\alpha x$$
$$= 3.54\ \text{kN}$$

The buoy will be dragged down with a force of 3.54 kN.

Now the buoy can be described. The requirement for buoyancy is set to twice the force F in the mooring line which is 7.08 kN. Archimedes law is used to calculate the buoyancy; the density of seawater, $\rho_w = 1025\ kg/m^3$.

$$F = \rho_w g V$$
$$V = \frac{F}{\rho_w g}$$
$$= \frac{7080}{1025 \times 9.81}$$
$$= 0.70$$

where:

F = *buoyancy*
ρ_w = *density of the displaced liquid*
g = *acceleration due to gravity*
V = *displaced volume.*

This means that the buoy needs a volume of 700 l or more to stay in the correct position on the surface. In addition the buoy must have buoyancy that covers its own weight; this depends on buoy type and is given by the supplier.

The next step is to calculate the size of the anchor. A block anchor is chosen. First the size to withstand the vertical lifting force (y) is calculated:

$$y = 3.54\,\text{kN}$$

Choose a concrete block anchor with density f_c of 2500 kg/m³. The weight of the anchor (G) is given by:

$$\begin{aligned} G &= mg \\ &= \rho_c V g \end{aligned}$$

where:

m = *mass of anchor*
V = *volume of anchor*
g = *acceleration due to gravity*
ρ_c = *density of cancrete.*

The buoyancy (F_0) that will lift the anchor is, from Archimedes law:

$$F_0 = \rho_w V g$$

where:

ρ_w = *density of seawater.*

The following equation may be used to find the necessary volume of the block anchor (y):

$$\begin{aligned} y &= G - F_o \\ &= \rho_c g - \rho_W V g \\ &= Vg(\rho_c - \rho_W) \end{aligned}$$

$$\begin{aligned} V &= \frac{Y}{g(\rho_C - \rho_W)} \\ &= \frac{3540}{9.81(2500 - 1025)} \\ &= 0.245\,\text{m}^3 \\ &= 245\,\text{l} \end{aligned}$$

The horizontal force is calculated using a friction coefficient of 0.5 for the sand bottom. The horizontal force (F) that will try to move the anchor is 10 kN, while the weight G will keep the anchor in place. In addition the buoyancy F_0 of the anchor will have effect, because it will reduce the weight compared to when it is on shore. The following equation can be set up:

$$\begin{aligned} F &= f(G - F_0) \\ &= fVg(\rho_c - \rho_w) \end{aligned}$$

$$\begin{aligned} V &= \frac{F}{fg(\rho_C - \rho_W)} \\ &= \frac{10000}{0.5 \times 9.81(2500 - 1025)} \\ &= 1.382\,\text{m}^3 \end{aligned}$$

where:

f = *friction coefficient for the block anchor.*

Therefore the volume of the block anchor must exceed 1.382 m³ or the mass be above 3.46 t. In practice two or three anchors will be used.

15.8 Control of mooring systems

After setting out the mooring system, it is important to do necessary checks to avoid breakages. Insurance companies will normally require some type of checking of the mooring system. There may also be national standards to prevent breakage and possible escape of fish. For instance the following may be used:

- The supplier's specification must be used to set out the cage
- Parts in the mooring system above the surface must be checked daily
- The whole mooring system including under water installations must be checked every year.

Some time after the mooring system was first set out, a more comprehensive and systematic check should be carried out, for instance every 4 years. This includes load tests on important and heavily loaded parts. Components exposed to hard wear should be replaced, such as mooring ropes, chain, wire, fixing points and eventually links.

References

1. Beveridge, M. (1996) *Cage aquaculture*. Fishing News Books, Blackwell Scientific.

2. Liao, I.C., Lin, C.K. (eds) (2000) Cage aquaculture in Asia. In: *Proceedings of the first international symposium of cage aquaculture in Asia*. Asian Fisheries Society and World Aquaculture Society – Southeast Asian Chapter.
3. Huguenin, J.E. (1997) The design, operations and economics of cage culture systems. *Aquacultural Engineering*, 16: 167–203.
4. Loverich. G.F., Gace, L. (1997) The effect of currents and waves on several classes of offshore sea cages. Paper given at *International conference on open ocean aquaculture 97. Maui, Hawaii, USA*.
5. Cairns, J., Linfoot, B.T. (1990) Some considerations in the structural engineering of sea-cages for aquaculture. In: *Engineering for offshore fish farming*, pp. 63–77. Thomas Telford.
6. Pérez, O.M., Telfer, T.C., Ross, L.G. (2003) On the calculation of wave climate for offshore cage culture site selection: a case study in Tenerife (Canary Islands). *Aquacultural Engineering*, 29: 1–21.
7. Sarpkaya, T., Isaacson, M. (1981) *Mechanics of wave forces on offshore structures*. Van Nostrand Reinhold.
8. Sawaragi, T. (1995) *Coastal engineering – waves, beaches, wave–structure interactions*. Elsevier.
9. Sorensen, R.M. (1993) *Basic wave mechanics*. Wiley-Interscience.
10. Boccotti, P. (2000) *Wave mechanics for ocean engineering*. Elsevier.
11. US Army Corps of Engineers (1984) *Shore protection manual, vols I and II*. US Government Printing Office.
12. Faltinsen, O.M. (1990) *Sea load on ships and offshore structures*. Cambridge University Press.
13. Saville, T. (1954) *The effect of fetch width on wave generation*. Technical memo. 17. Beach Erosion Board, US Army Corps of Engineers.
14. NS 9415. *Marine fish farms. Requirements for design, dimensioning, production, installation and operation*. Norwegian Standardization Association.
15. Linfoot, B.T., Cairns, J., Poxton, M.G. (1990) Hydrodynamic and biological factors in the design of sea-cages for fish culture. In: *Engineering for offshore fish farming*, pp. 197–210. Thomas Telford.
16. Klust, G. (1982) *Netting materials for fishing gear*. Fishing News Books, Blackwell Science.
17. Klust, G. (1983) *Fibre ropes for fishing gear*. Fishing News Books, Blackwell Science.
18. Karlsen, L. (1989) *Redskapsteknologi i fiske*. Universitetsforaget (in Norwegian).
19. Tygut (1997) *Regelverk og veileder for dimensjonering og konstruksjon av flytende oppdrettsanlegg*. Fiskeridepartmenetet (in Norwegian).
20. Rudi, H., Oltedal, G. (1993) Metode for vurdering av lokalitet for matfiskoppdrett. Rapport marintek (in Norwegian).
21. Kerr, N.M., Gillespie, M.J., Hull, S.T., Kingwell, S. (1980) The design construction and location of marine floating cages. In: *Proceedings of the institute of fisheries management cage fish rearing symposium, University of Reading, 26–27 March 1980, pp. 23–49*. Janssen Services.
22. Gunnarson, J. (1993) Bridgestone Hi-Seas fish cage: design and documentation. In: *Fish farming technology. Proceedings of the first international conference of fish farming technology* (eds H. Reinertsen, L.A. Dahle, L. Jørgensen, K. Tvinnereim). A.A. Balkema.
23. Loverich, G., Forster, J. (2000) Advances in offshore cage design using spar buoys. *Marine Technology Society Journal*, 34: 18–28.
24. Lien, E. (2000) Offshore cage systems. In: *Cage Aquaculture in Asia. Proceedings of the first international symposium of cage aquaculture in Asia* (eds I.C. Liao, C.K. Lin). Asian Fisheries Society and World Aquaculture Society – Southeast Asian Chapter.
25. Lien, E. (1993) Tension leg cage, a new net pen cage for fish farming. In: *Fish farming technology* (eds H. Reinertsen, L.A. Dahle, L. Jørgensen, K. Tvinnereim), pp. 251–258. A.A. Balkema.
26. Det Norsk Veritas (1988) *Det Norske Veritas tentative regler for sertifisering av flytende fiskeoppdrettsanlegg*. Rapport Det Norske Veritas (in Norwegian).
27. Rudi, H., Lien, E., Slaatelid, O.H. (1994) *Håndbok for design og dokumentasjon av åpne merdanlegg*. Rapport Marintek (in Norwegian).
28. Tsukrov, I., Eroshkin, O., Fredriksson, D., Robinson Swift, M., Celikkol, B. (2003) Finite element modeling of net panels using a consistent net element. *Ocean Engineering*, 30: 251–270.
29. Fredriksson, D.F., De Cew, J., Robinson Swift, M., Tsukrov, I., Chambers, M.D., Celikkol, B. (2004) The design and analysis of a four-cage grid mooring for open ocean aquaculture. *Aquacultural Engineering*, 33: 77–94.
30. Suhey, J.D., Kim, N.H., Niezrecki, C. (2005) Numerical modeling and design of inflatable structures – application to open-ocean-aquaculture cages. *Aquacultural Engineering*, 33: 285–303.
31. Milne, P.H. (1972) *Fish and shellfish farming in coastal waters*. Fishing News Books, Blackwell Scientific.
32. Rudi, H., Aarsnes, J.V., Dahle, L.A. (1988) Environmental forces on floating cage systems: mooring considerations. In: *Aquaculture engineering, technologies for the future*. ICemE symposium series no. 111. Hemisphere.
33. Løland, G. (1993) Current forces on, and water flow through and around, floating fish farms. *Aquaculture International*, 6: 33–37.
34. Løland, G. (1993) Water flow through and around net pens. In: *Fish farming technology* (eds H. Reinertsen, L.A. Dahle, L. Jørgensen, K. Tvinnereim), pp. 177–183. A.A. Balkema.
35. Aarnes, J.V., Rudi, H., Loland, G. (1990) Current forces on cages, net deflection. In: *Engineering for offshore fish farming*. Thomas Telford.
36. Lien, E., Rudi, H., Slaatelid, O.H. (1996) *Håndbok for design og dokumentasjon av åpne merdanlegg*. Rapport Marintek (in Norwegian).

16
Feeding Systems

16.1 Introduction

16.1.1 Why use automatic feeding systems?

Feeding can be done by hand, or by automatic feeders or feeding systems. The time used for feeding can be considerable for large farms with intensive production, and can justify the investment in a system for automatic feeding. For instance, the daily requirement of feed for a rainbow trout farm with a standing biomass of 100 t of 100 g fish is at least 3500 kg per day with a water temperature of 16°C.

For intensive fry production, several species require an almost continuous supply of food, especially in the first feeding stage. This requires a tremendous amount of work, and is therefore normally done by automatic feeders. Feeding systems are of most interest for intensive aquaculture systems because of the importance of getting as much feed as possible into the fish.

16.1.2 What can be automated?

How easy it is to automate the feeding depends on the feed type used.[1–3] Dry, extruded or pelleted feed is quite easy to deal with: the particles are fixed and hard. Wet feed or moist feed is rather more difficult to feed automatically. To find good systems for distribution of dense particles is also difficult. Wet feed may be fed through pump systems, but here it is difficult to obviate the possibilities for over feeding; possible environmental impacts are also much higher with this type of feed.

The size and shape of the dry particles will also influence the feasibility of feeding automatically: small feed particles, for example for marine or freshwater fry, might be a problem. If the feed is like meal, it might be difficult to get it through the feed dispenser; it might clog inside the hopper, and the sliding angle is very high.

16.1.3 Selection of feeding system

Today is it normal to use automatic feeding systems in all types of intensive fish farming. Which type of automatic feeding system to choose, however, depends on a number of factors of which the most important are: feed type, production species, production type, production size and access to electricity.

A feeding system could range all the way from simple dispenser with no need for electricity to an advanced computerised feeding system which controls the feeding on the basis of the appetite of the fish.

16.1.4 Feeding system requirements

The requirements for the feeding system depend on the chosen type. In the following, some general claims especially adapted to dry feed are presented:

- Simple operation
- Low maintenance
- Tolerate wind and sea (offshore farms)
- Tolerate high humidity
- Simple to fill with feed
- Simple calibration (to control the amount dispensed)
- High dispersion accuracy
- Cause few breakages

Of the more general engineering subjects that are of interest for increasing the basic knowledge of automatic feeding there are, for instance, solids handling, solid conveying and bulk solids handling (for example, refs 4–6).

16.2 Types of feeding equipment

Feeding equipment can be divided into groups based on its construction and function. One classification is as follows:

- Feed blowers
- Feed dispenser
- Demand feeders
- Automatic feeders, feed machines
- Feeding systems.

This is specifically for the dry feed that is most commonly used in intensive fish farming. For wet and moist feed other separations can be made, but only a few methods are used for this type of feed. Feeding equipment for wet and moist feed will not be dealt with here.

16.2.1 Feed blowers

A feed blower is only a tool to simplify hand feeding (Fig. 16.1). There are different blower types based on the 'carrier' used for the feed particles which is normally either air or water. The feed can either be sucked up from a tank or a bag by vacuum, or the feed can be filled into a hopper standing over a pipe with flow of air or water. The hopper can be fixed on a boat or be movable.

16.2.2 Feed dispensers

A feed dispenser is often confused with a feeding machine, but does not have the distribution unit. Actually it is therefore something between a feed machine and hand feeding. A weighed portion of feed is placed on the dispenser and the dispenser will empty it during a fixed period, normally from one to three days. It either goes continuously or stepwise controlled by a control unit. To get the wanted feed ration, the actual amount of feed must be put in the dispenser. This is normally weighed out.

A great advantage with the feed dispenser is its simple and robust construction. It is also easy to monitor visually whenever it is functioning and the amount of feed that has been dispensed. The construction is favourable to use in research operations because if you weigh out the feed exactly you can be sure that the dispenser will supply the exact amount to the fish. The great disadvantage, compared to a feeding machine, is that it takes quite a long time to measure and/or weigh the feed that should be placed in the dispenser.

Several designs of feed dispenser are used (Fig. 16.2). In a disc feeder a scraper rotates on a horizontally fixed circular plate, and the feed falls

Figure 16.1 A feed blower being used to simplify hand feeding.

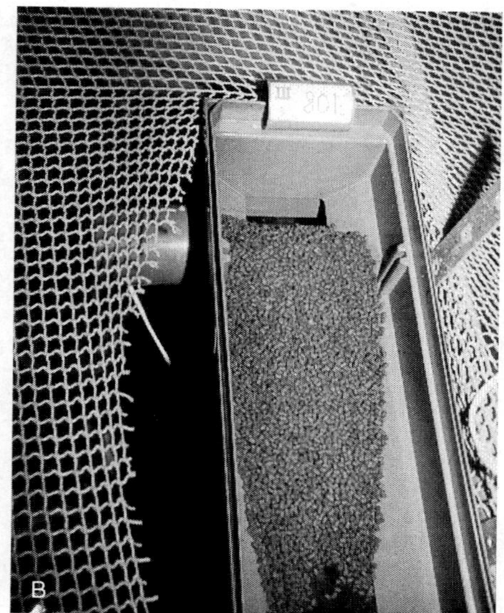

Figure 16.2 Typical feed dispensers: (A) disc feeder and (B) conveyor belt feeder.

off the edge of the plate and into the fish tank. A disc dispenser needs electricity to run the motor, normally 24 V a.c. The feeder normally goes stepwise, controled by a unit that regulates the start and stop intervals. Another much used construction is a rubber conveyor belt that is dragged along on rollers. When starting, the belt is dragged backwards so it creates a surface where feed is supplied. The end of the belt is fixed to rollers; when these rotate the belt will be dragged up and the surface where the feed is lying will gradually be decreased so that the feed falls off and into the fish tank. This type is either powered by electricity or by clockwork. An advantage with this type of feed dispenser is the possibility of running it without electricity. Feed is either dispensed continuously or stepwise.

16.2.3 Demand feeders

A demand feeder is normally a mechanical construction. A stick is attached to a slightly bowed plate sitting under a feed hopper (Fig. 16.3). The stick goes from the feeder down into the water. When the fish touch the stick, feed will be dispensed from the hopper. At the end of the stick is a knob, or something similar, which the fish touch. A great advantage with using demand feeders is that there is no need for an electricity supply. Furthermore, the design is simple with few moveable parts.

Figure 16.3 The fish operate a demand feeder when they move the stick hanging down in the water.

The fish operate the demand feeder themselves, and can therefore theoretically be fed according to appetite (*ad libitum*). However, some feed loss has been registered. The fish may use the demand stick as a toy and feed may be lost; demand feeders are also sensitive to movements in the water, such as waves; wind may also affect the demand feeder so shielding may be included.

Demand feeders are used for almost all species, even if some species, such as Atlantic salmon, are slow to learn the system. The fish need a training period to learn how to operate the system.[1] Compared to hand feeding, both improved and less good growth results have been shown.[7]

In electronic demand feeders feeding is triggered by electric signals. The mechanical stick is replaced by an electric cable with a pressure sensor at the end. When the fish touch this sensor a signal is given to the feeder which starts. This system allows extra control over the demand feeding, for instance by setting fixed interval for the operation of the feeder or by setting a maximum limit of distributed feed per portion or per day. A more advanced control system is, however, required in this type of feeder.

Much literature is available on the use of demand feeders, including the possibility for controlling the appetite of the fish experimentally.[8–12]

16.2.4 Automatic feeders

A feeding machine or an automatic feeder consists of four major components (Fig. 16.4): a feed container (hopper), a mechanism for feed distribution, an electrical power supply for the distribution mechanism and a control unit for starting and stopping the distribution mechanism. The feed distribution mechanism is the main component in an automatic feeder and distinguishes it from a feed dispenser. The feeders are fixed in a rack on the tank or on the cage, but may also be included in a buoy (see, for example, ref. 13). When using an automatic feeder, the amount of feed that has to be distributed over a period of time is known and the distribution unit runs for the period that satisfies this requirement.

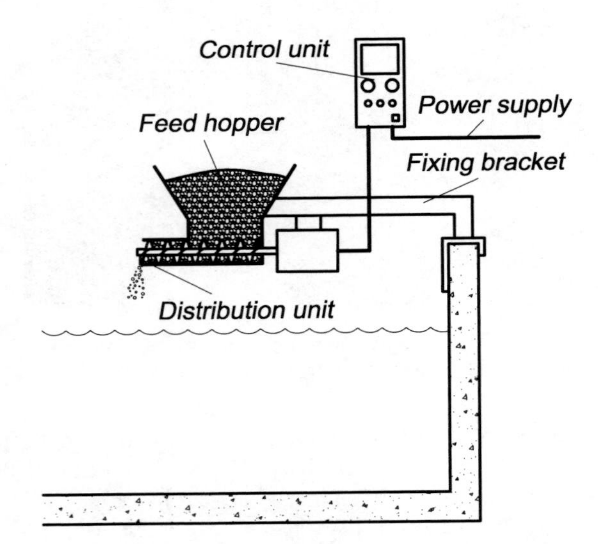

Figure 16.4 An automatic feeder consisting of a feed container (hopper), a mechanism for feed distribution, an electrical power supply and a control unit.

Feed is distributed by volume. In specially designed and more expensive feeders, the systems may also use feed mass. When using volume for distribution, the volume: mass ratio (litre/kg), i.e. the density, of the feed must be known. The density of the feed varies with formulation, from producer to producer, and also depends on the size of the feed particles. Because volume distribution feeders only distribute a certain volume of feed, and the in mass feed is of interest, calibration of the feeders is necessary. To calibrate the feeder, it is run for a known time period; then the exact amount of feed that has been dispensed is weighed so that the feed distributed per unit time can be calculated. This information is then used to find the necessary time that the feeder has to run to distribute a certain mass of feed.

Example
A fish tank requires 3 kg feed per day. For how long must the feeder be run to deliver this amount?
First the amount of feed delivered from the feeder per unit time must be found. The feeder is run continuously for 1 min and the amount of feed delivered is weighed and found to be 1 kg. The feeder is therefore delivering 1 kg/min; to deliver 3 kg to the fish tank, the feeder must be run for a total of 3 min per day.

If the feeder starts every 30 min throughout the day and night, it starts 48 times in total. Each time it must therefore run for: 3 min (= 180 s)/48 = 3.75 s.

Distribution mechanisms

Many mechanisms for feed distribution are available.[3,14,15] Some important types are described below.

Screw: A screw allows a specific batch of feed to be dispensed for every rotation (Fig. 16.5). The screw is installed under a hopper from which it is filled. The amount dispensed per unit time is related to screw diameter, design of the screw (rise of the screw thread), the speed of rotation, the degree of filling and the angle of the screw.

Vibrator: A plate that vibrates may be used to distribute a volume of feed. When the almost horizontally fixed plate starts to vibrate the feed on the plate will fall over the edge. One method of getting vibration is to attach a weight to one side of a vertically fixed shaft that rotates. When the shaft rotates there will be imbalance in the shaft and in the plate fixed to it. Another method is to use an electromagnet with an anchor fixed by a leaf or coil spring. When the electricity is turned on, the anchor is dragged towards the magnet by a varying magnetic field, making the whole vibrator shake; a slope on the feeding plate causes the feed to be shaken over the edge. Advantages of the electromagnetic vibrator are its simple construction and that it stops immediately the electricity is turned off. The amount of feed distributed is controlled by adjusting the voltage and hence the amplitude of the vibrator.

Cell wheel: A vertically installed rotating wheel with wings, cells or chambers sitting under a feed hopper may also be used to distribute feed (Fig. 16.5). When the wheel rotates it transports the feed in the cells; when the cells approach the lowest position, the feed is released. The amount of feed released depends on the number of chambers in the wheel and the speed of rotation. The leaf dispenser and drawer dispenser are similar in principle to the cell wheel.

Others: A number of other mechanisms might be used for feed distribution, of which a rotating disc with a scraper at the bottom of a cylindrical hopper

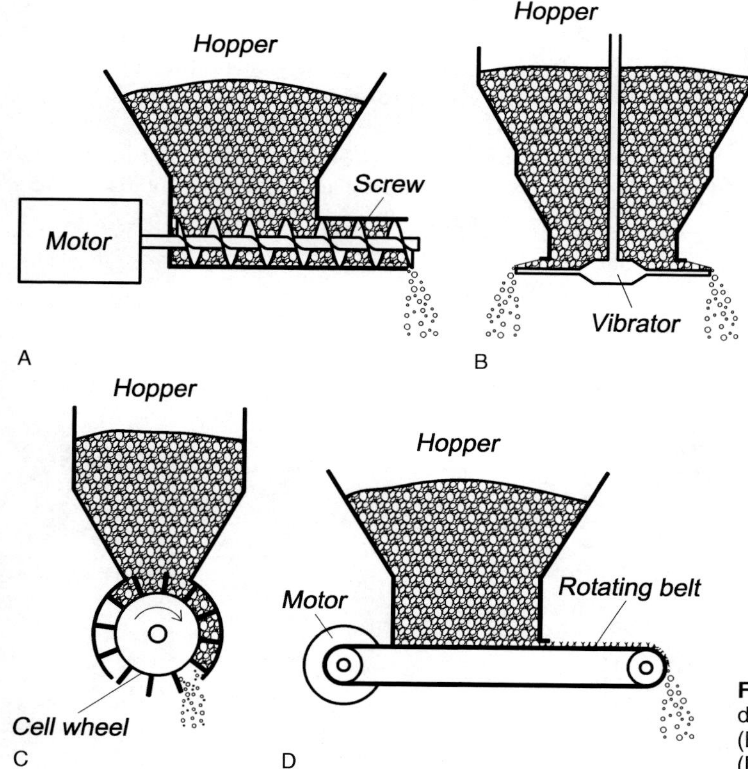

Figure 16.5 Different types of feed distribution mechanisms: (A) screw feeder; (B) vibrating feeder; (C) cell wheel feeder; (D) rotating belt feeder.

is one. When the disc rotates feed will be distributed with the help of the scraper fixed to the hopper. This system must not be confused with the disc dispenser which has no hopper. The distance between the rotating disc and the feed hopper regulates the amount of feed that is distributed. Use of a conveyor belt system is another method for feed distribution; the rotating belt is placed under the feed container and a distribution bar regulates the thickness of the feed layer on the belt and by this the amount of feed distributed.

For all methods the amount of distributed feed is, to various extents, dependent on the height of the feed in the hopper. When the hopper is full the pressure of feed is increased and more feed is distributed because the distribution mechanism is sited at the bottom of the hopper.

Another fairly new method used for distributing feed is a bowed screw with an open centre – actually a spring. The advantage with this arrangement is that the amount of feed dispensed does not depend so much on the level in the hopper because the screw is filled with feed from the side.

If the feed particles are very small (as in meal), the feed particles may clog around the inlet to the distribution unit. This is a particular problem for particles that have a high sliding angle. In fish farming today, the particles are larger and the sliding angle is quite low, so this is normally not a problem.

To achieve more exact dispensing and to avoid the chore of calibration, automatic weight control can be used as a supplement. However, this is quite expensive and has only been used in fish farming to a limited extent. This can, however, be a solution if requirements for dispensing are very precise. Electronic weight cells (tension and pressure) have been used, especially in feeding systems. The principle of weight cells is that they measure the temporary deformation of the material, which is related to the weight of feed/in them. When this system is used, a volume dispenser adds the feed, but is controlled by the weight cell, signals from which regulate the running time for the volume dispenser.

Feed hopper

Above the feeding mechanism is sited the feed container or hopper. Hopper size varies from some litres to several hundred litres, depending on the size of the fish to be fed. Hoppers are usually constructed of plastic or metal (aluminium) and must be designed in a way that gives easy access for refilling and ensures that all feed slides out easily.

Spreading of feed

On some feeders a unit for spreading of the feed is attached underneath the distribution mechanism. The purpose is to distribute the feed over a larger part of the pond, tank or cage. Three spreading patterns are used: point feeding (no spreading mechanism), sector feeding and circle feeding[16] (Fig. 16.6).

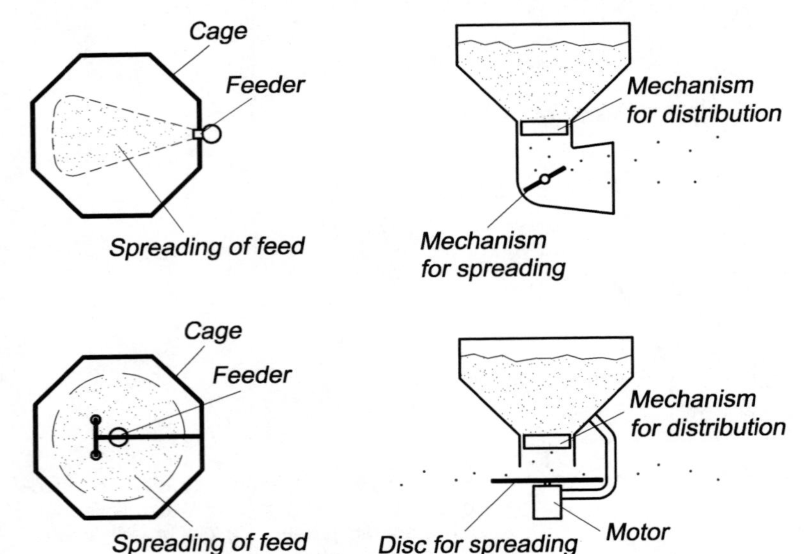

Figure 16.6 The feed can be spread in sectors of the cage, or in a circular pattern.

The sector feeding system normally consists of a vertical rotating plate or brush. When this rotates the feed is spread out in a sector in the cage or the pond. A circle feeding technique employs a centrifugal scattering system. The feed particles drop down from the distribution unit and hit a horizontally placed rotating disc. Because of the centrifugal forces the feed is spread out in a circle. The spreading unit may be integral part of the distribution system or it may be installed separately underneath the distribution unit. The velocity of the disc or plate determines the area of the sector or the circle where the feed is spread. However, a high velocity will increase the amount of breakage because the forces transferred from the spreading unit to the feed particles will increase.

Control units

The control unit manages the current to the motor on the dispensing system; this also controls the feeding. The simplest control unit sets the time interval between each meal and the running time, i.e. the length of each meal. Some control units are equipped with a photocell which only permits feeding during daylight. In more advanced units a daily increase in the running time may be added related to the expected growth rate of the fish in the production unit.

There may be an individual control unit for each feeder or the unit may control several feeders with the same feeding regime. There can also be one control unit with several channels, which means that it can control several feeders individually. The control unit can be a simple interval relay, where running time and time interval between each start are set (Fig. 16.7). Several relays may be set together in a multichannel control unit. The programmable logic controller (PLC) is a more advanced system carrying out the same tasks as a multichannel control unit. The input and output can be switched on or off, and it is quite easy to extend the system with more inputs and outputs. Each output channel can be programmed individually. In addition input signals can also be used to control the output signal. For instance, the output can only be started when the input signal from a light sensor registers that it is daylight; this means that feeding is only permitted during daylight.

A personal computer (PC) equipped with special 'cards' may also be used to control larger feeding systems. The PC can also collect data that can be used to control the feeding, such as water temperature and light intensity. PCs may also be used as a data logger to store, for instance, how much is fed every day.

Electric current

Both distribution and spreading mechanisms contain motors that normally need an external supply of power. This is normally electricity, either

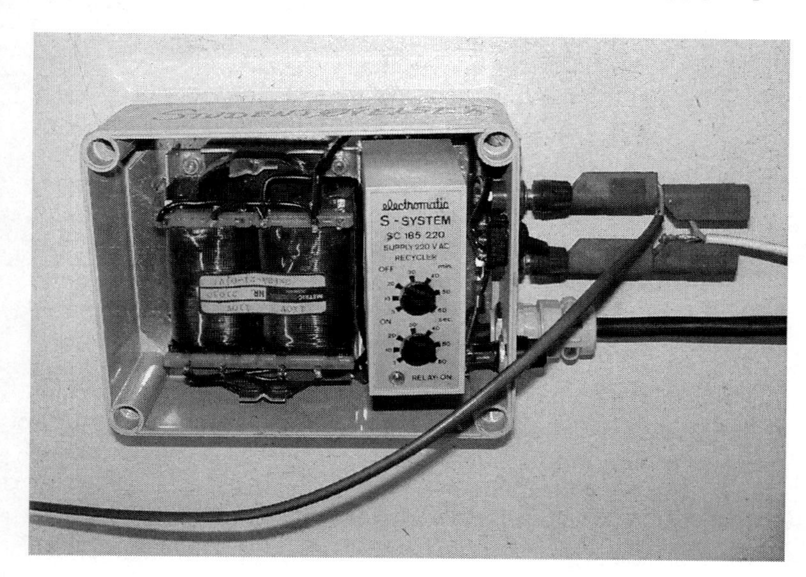

Figure 16.7 A feeder control unit with a transformer and relay.

from a central electric power station (alternating current) or from batteries (direct current). A clockwork mechanism may also be used on small feeders. Direct current motors normally require a supply of 12 V or 24 V. Alternating current motors for feeders are either low voltage (12 V or 24 V) or normal voltage (110 V and 220 V). The advantage with high voltage is that thinner cabling is required and current loss in the cables is reduced, particularly when using long cables. The disadvantage is that normal voltage, high humidity and free water surfaces can be a dangerous combination. When using normal voltage it is therefore important to be careful when laying electric cables and ensure that the feeding system is correctly insulated and earthed to avoid jump sparks. Qualified professionals must perform such work.

Normally the mains voltage is used as the electricity source. If low voltage feeding systems are to be connected, the use of a transformer is necessary. If direct current is to be used, a rectifier is also needed.

If there is no electricity supply, for instance on a sea cage, either batteries or a diesel-powered electrical generator must be used. Use of batteries requires direct current motors on the feeder. In these cases the battery must be taken out regularly for recharging; solar panels or windmills may also be used to charge batteries.

Direct current motors used on feeders have the advantage that the speed of rotation of the motor can easily be regulated. By adding a variable resistance, the size of the incoming current that is running the motor will be regulated. Regulation of the speed of rotation will control the amount of feed coming from the feeders.

The motor output should be matched to the need for forces to run the distribution unit. Motors that are too powerful can result in more breakage of feed and are not recommended. The main reason for adding a larger motor is to avoid wedging so that the system is more reliable. Care should, however, be taken regarding the possibility of breakage when feeders are equipped with large motors.

16.2.5 Feeding systems

The term feeding system refers to a complete system that takes the feed directly from the feed silo or hopper, transports it to the fish production unit, and at the end distributes it to the fish. A complete feeding system may comprise three parts: a storage unit, a transport unit and a feed distribution unit. Today feeding systems can be divided into two types:

(1) Systems where the feeder is centrally placed and feed is transported to the single fish production units (tanks, ponds or cages) through pipes, normally known as feeding systems.
(2) Systems where the feeder is installed on a rail system that covers several units, normally called feeding robots.

Central feeding system

A central feeding system consists of storage silos, a sluice valve, tubes with a flow of water or air for transporting of feed, a selector valve, and eventually a distribution unit (Fig. 16.8).

In this system the feed is delivered from the silo and into an auger that brings the feed particles into a hopper placed above a sluice valve. The sluice valve brings the feed particles from the hopper and into the pipes for further transport to the tanks or cages. To transport the feed particles, water or air is used as a medium and the sluice valve therefore also provides an air or water lock between the hopper and the transport medium; the sluice valve also represents the feed distribution unit.

Whether air or water is used as a transport medium, the velocity in the tubes is such that the feed particles will always stay in suspension. A blower or a pump ensures adequate velocity inside the tubes. During the past few years, air has become the major transport medium. After being transported for a short distance in the pipes (some metres) the feed enters the selector valve whitch determines the production unit to which the feed portion is sent. There are several designs of selector valve; normally a rotating or a horizontal moving selector is used.

After the selector valve the feed is transported to the production unit through the tubes. In sea cages the tubes may be up to several hundred metres in length. The silos and selector valve may be placed on-shore or on a barge. If the system is for large cages, a unit for spreading the feed may also be included.

Figure 16.8 A feeding system used on a fish farm: (A) feed silos with a rotating selector valve; (B) selector valve; (C) the end of the tube from which feed is spread into the tank.

Correct design and use of the feeding system is important to avoid feed breakage and dust production. Important factors are air temperature, pick-up velocity, material in the pipes, design and use of the selector valve and pipeline routings.[17,18]

A centrally placed computer controls this type of feeding system. The amount of feed to the different units can be set as fixed or be created automatically. Addition of the initial weight, water temperature, expected growth and mortality in to the computation ensures correct feeding. The computer also stores the inputs and is an important tool for production planning and production control.

Central feeding systems are also available for automatic feeding of moist feed.

Feeding robots

Put simply, a feeding robot is a feeder suspended from a rail system hanging above the fish tanks (Fig. 16.9). A motor to push the feeder along the rail

Figure 16.9 A feeding robot is simply a feeder suspended from a rail system hanging above the fish tanks.

system is included. The rail system is laid over the production units and under the feed silos. The robot may have its docking station under the silos where it enters for automatic refilling with feed when the hopper on the feeder is empty. When the robot is feeding it moves along the rail until it hits a chip attached to the rail over each tank. Based on information in this chip, the feeder (robot) recognizes the tank. On the robot there is a computer where the amount to be fed to the actual tank is programed. When the robot hits the chip it will therefore feed the programed amount of feed to the tank. After this it continues to the next tank, and so on. When the hopper on the robot is empty it automatically goes back to the silos for refilling. Several individual feeders can be attached to the same robot, so that it can deliver several feed sizes in the same operation. The electricity supply to the feeder and the motor may be an integrated part of the rail system, or it can be a battery which is recharged in the docking station. The great advantage with this system is that the same feeding mechanism can be used for feeding of several tanks. In this way more investment in this unit means it can be designed to feed more accurately. Accuracy can be improved by using double dispensers (multistage), as is in some robots.

16.3 Feed control

The appetite of the fish is affected by external factors, such as variation in water temperature, water quality, waves (in cages) and light conditions. With a normal feed control, the amount of feed to distribute during a given period is fixed. If there is variation in fish appetite there are no possibilities for controlling this. This requires one of two solutions: the use of restrictive feeding with no feed loss, (really underfeeding), or acceptance of a certain feed loss which is expensive and damaging to the environment.

Hand feeding is an old-fashioned system for regulating feed supply. The person who is feeding observes the appetite of the fish visually, and in this way regulates the amount of feed supplied according to fish appetite. One way to improve the feed control and utilization of feed is to use a feeder for basic feeding and hand feeding for topping up; this, however, requires manpower. Based on this, systems have been developed for automatic feed control which are of special interest in production units where large amounts of feed are used, such as large sea cages.

16.4 Feed control systems

Feed control systems can be divided into manual and automatic systems. In tanks, manual systems are used. The dual drain system with a particle trap represents such a system (see Chapter 13). If this system is correctly designed, it is easy to observe any feed loss in the screen or separation unit for the particle outlet. When screening the total outlet water from each tank, the feed loss may also be observed. This, however, requires a large screen on each tank, which is more expensive.

In sea cages a number of methods have been introduced. A manually operated method is to use a submersible video camera under the cage and watch randomly for feed loss when the feeder is running. Another manual method is to use a stocking (a small net bag in the shape of a tube) in the lower part and under the net bag. Here feed loss and dead fish are collected. An airlift pump brings the feed loss and dead fish to the surface and into a collecting bucket, where feed loss can be visually controlled and the daily feeding amount may be regulated (Fig. 16.10). Equipment to detect the gathering behaviour, such as infrared photoelectric sensors, has also been used to control feeding.[19]

Figure 16.10 By using an airlift pump the feed loss on the bottom of the cage is collected and brought to the surface and into a tray where it can be seen. The system can also be used for collecting dead fish.

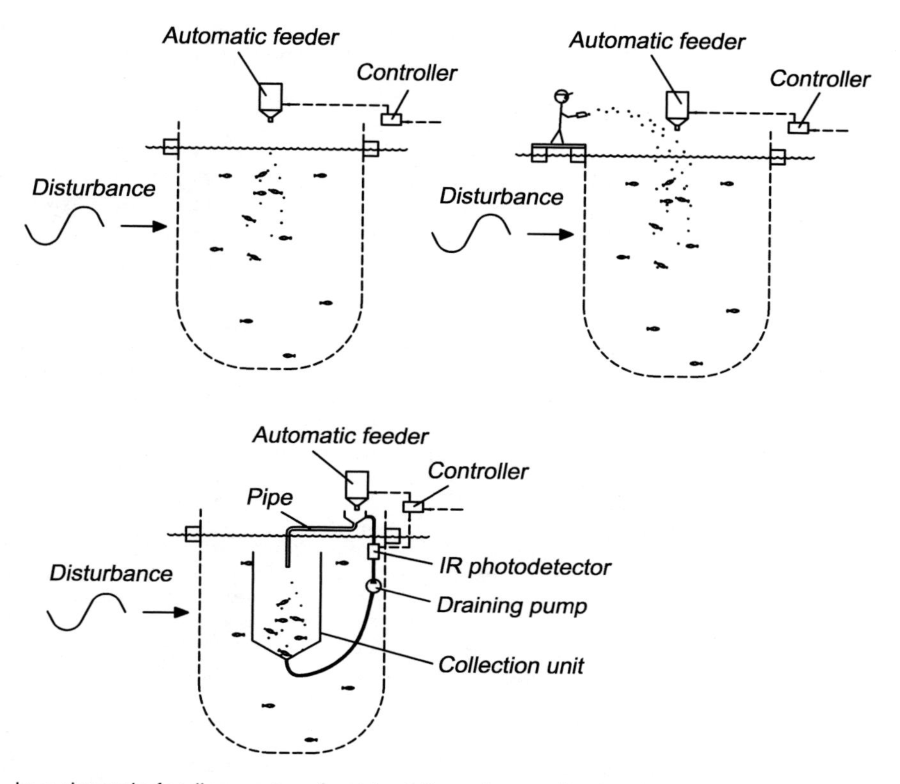

Figure 16.11 In a dynamic feeding system feed is delivered according to the appetite of the fish. (Adapted from ref. 27.)

Hydro-acoustic sensors, photocells, linear video and Doppler signals are all used for automatically measuring feed loss.[20–25] Either the sensor can be placed under the cage or inside it to measure the feed loss over a sample area. The sensor sends a signal to the feed controller to stop when feed loss exceeds a certain level. In tanks ultrasonic devices have been used to control the waste feed.[20] A system observing the gathering behaviour in relation to feeding has also been used for automatic feed control.[26]

16.5 Dynamic feeding systems

Dynamic feeding systems go even further and use the feed loss to control the amount of feed to be delivered. In one system a collector for feed loss is placed inside the cage.[27] When feeding starts, the pump takes the collected food lost from the previous feeding and pumps this through a pipe that delivers it to the top of the cage bag (Fig. 16.11). In this pipe circuit an infrared device detects eventually uneaten feed. If there are no feed particles left, the feeder starts to add a new portion of feed. The same procedure happens when starting the next feeding. The system follows the appetite of the fish in a dynamic way, with increasing appetite increasing the amount of feed and decreasing appetite decreasing the amount of feed. Such systems have no feed loss to the environment, and it is possible to get early warning of eventual abnormal behaviour of the fish.

References

1. Goodard, S. (1996) *Feed management in intensive aquaculture*. Chapman and Hall.
2. Swift, D. (1993) *Aquaculture training manual*. Fishing News Books. Blackwell Science.
3. Hochheimer, J. (1999) Equipments and controls. In: Wheaton, F. (ed.) *CIGR handbook of agricultural engineering, part II aquaculture engineering* (ed. F.

Wheaton), pp. 281–307. American Society of Agricultural Engineers.
4. Woodcock, C.R., Mason, J.S. (1998) *Bulk solids handling: an introduction to the practice and technology*. Springer Verlag.
5. Levy, A., Kalman, H. (2001) *Handbook of conveying and handling of particulate solids*. Elsevier Science.
6. Klinzing, G.E., Marcus, R., Rizk, F., Leung, L.S. (1997) *Pneumatic conveying of solids: a theoretical and practical approach*. Springer-Verlag.
7. Stickney, R.R. (1994) *Principles of aquaculture*. John Wiley & Sons.
8. Alanärä, A. (1996) The use of self-feeders in rainbow trout (*Oncorhynchus mykiss*) production. *Aquaculture*, 145: 1–20.
9. Covès, D., Gasset, E., Lemarié, G., Dutto, G. (1998) A simple way of avoiding feed wastage in European seabass, *Dicentrarchus labrax*, under self-feeding conditions. *Aquatic Living Resources*, 11: 395–401.
10. Gélineau, A., Corraze, G., Boujard, T. (1998) Effects of restricted ratio, time-restricted access and reward level on voluntary food intake, growth and growth heterogeneity of rainbow trout (*Oncorhynchus mykiss*) fed on demand self-feeders. *Aquaculture*, 167: 247–258.
11. Alänäre, A., Kadri, S., Paspatis, M. (2001) Feeding managment. In: *Food intake in fish* (eds D. Houlihan, T. Boujard, M. Jobling). Blackwell Science.
12. Rubio, V.C., Vivas, M., Sánchez-Mut, A., Sánchez-Vázquez, F.J., Covès, D., Butto, G., Madrid, J.A. (2004) Self feeding of European sea bass (*Dicentrarchus labrax*, L.) under laboratory and farming conditions using a string sensor. *Aquaculture*, 233: 393–403.
13. Fullerton, B., Robinson Swift, M., Boduch, S., Eroshkin, O., Rice, G. (2004) Design and analysis of an automated feed-buoy for submerged cages. *Aquacultural Engineering*, 32: 95–111.
14. Larsson, K. (1978) *Transport och portionering av kraftfôder vid mekanisk utfôdring*. Medelande nr 374, Jordbrukstekniske instituttet (in Swedish).
15. Sørlin, S. (1985) Teknik för mängdbästemning. Neddelande nr 407. Jordbrukstekniska Institutten (in Swedish).
16. Thomassen, J.M., Lekang, O.I. (1993) Optimal distribution of feed in sea cages. In: *Fish farming technology* (eds H. Reinertsen, L.A. Dahle, L. Jørgensen, K. Tvinnereim) pp. 439–442. A.A. Balkema.
17. Guajardo, M. (2004) *Relation between feed quality and handling in a feeding system*. Master Thesis. Norwegian University of Life Science.
18. Norambuena, F. (2005) *Aquaculture's feeding system, optimisation of pick up velocity based on feed rate and pipeline length*. Master Thesis. Norwegian University of Life Science.
19. Chang, C.M., Fang, W., Jao, R.C., Shyu, C.Z., Lia, I.C. (2005) Development of an intelligent feeding controller for indoor intensive culturing of eel. *Aquacultural Engineering*, 32: 343–353.
20. Blyth, P.J., Pursher, G.J., Russel, J.F. (1993) Detection of feeding rhythms in sea cage rearing of Atlantic salmon. In: *Fish farming technology* (eds H. Reinertsen, L.A. Dahle, L. Jørgensen, K. Tvinnereim), pp. 209–216. A.A. Balkema.
21. Dunn, M., Dallard, K. (1993) Observing behaviour and growth using the Simrad FCM 160 fish cage system. In: *Fish farming technology* (eds H. Reinertsen, L.A. Dahle, L. Jørgensen, K. Tvinnereim), pp. 269–274. A.A. Balkema.
22. Juell, J.E., Furevik, D.M., Bjordal, Å. (1993) Demand feeding in salmon farming by hydro acoustic food detection. *Aquacultural Engineering*, 12: 155–167.
23. Foster, M., Petrell, R., Ito, M.R., Ward, R. (1995) Detection and counting of uneaten food pellets in sea cage using image analysis. *Aquacultural Engineering*, 14: 251–269.
24. Kevin, D.P., Rayann, J.P. (2003) Accuracy of a machine-vision pellet detection system. *Aquacultural Engineering*, 29: 109–123.
25. Summerfelt, S.T., Holland, K.H., Hankin, J.A., Durant, M.D. (1995) Hydro acoustic waste feed controller for tank systems. *Water Science and Technology*, 31: 123–129.
26. Chang, C.M., Wang, W., Jao, R.C., Shyu, C.Z., Liao, I.C. (2005) Development of an intelligent feeding controller for indoor intensive culturing of eel. *Aquacultural Engineering*, 32: 343–353.
27. Skjervold, P.O. (1993) Fish feeding station. In: *Fish farming technology* (eds H. Reinertsen, L.A. Dahle, L. Jørgensen, K. Tvinnereim), pp. 443–445. A.A. Balkema.

17
Internal Transport and Size Grading

17.1 Introduction

Various forms of handling are necessary in all aquaculture activity. In extensive farming the fish are handled very few times, but frequency increases when the farming becomes more intensive. There are several reasons for handling fish and other aquatic organisms. They are transported within or between farms, or from farms to slaughterhouse; examples include the transport of fry to on-growing farms, and adult fish from on-growing farms to slaughterhouse (see Chapter 18). The need for this type of transport, of course, depends on the production strategy of the farm; however, most is performed inside the farm area. Internal transport is performed for various reasons in connection with other handling activities, such as division of fish groups, size grading, weight sampling and vaccination. In this chapter a description of different methods and equipment for handling of aquatic organisms inside the farm is given, mainly for fish, although some of the methods and equipment may also be used on shellfish and other aquatic organisms. A brief description of the advantages and disadvantages of fish handling is also included.

Independent of the handling procedures and equipment used, it is important that the operation is performed by trained personnel and in a way that minimizes the possibility for injury and stress to the fish.

When deciding on fish handling systems, it is important to consider the entire operation starting from when the fish are moved from the production unit and ending when the fish have been returned to the production unit or are in the unit that transports the fish out of the farm. Several handling operations may be necessary throughout these processes. One example of a handling line for internal transport of fish on a farm with tanks can be:

- Crowding in tank
- Dip net for lifting fish out of the tank
- Bucket for internal transport of the fish
- Dip net for lifting of the fish from the bucket an into the new tank.

It is important that the separate handling methods and equipment used in the handling line are compatible. The handling methods may also be an integral part of the farm construction,[1] and therefore be designed and the equipment selected before planning and building the farm. Use of 'alternative models', where alternative methods for performing the different handling operations in the line are included, can be an effective tool for selecting handling methods (see Chapter 22).

17.2 The importance of fish handling

17.2.1 Why move the fish?

The amount of fish or shellfish that can be produced on a farm depends on the fish density in the production units. Land-based fish farms equipped for intensive farming require high investment per unit farming volume. Continuous high fish density is therefore necessary to attain good production economy;[2–4] this will require frequent transport or reallocation of fish as they grow (Fig. 17.1).

How often the fish must be moved is mainly decided by their growth rate, but input density and maximum allowed density in the production units

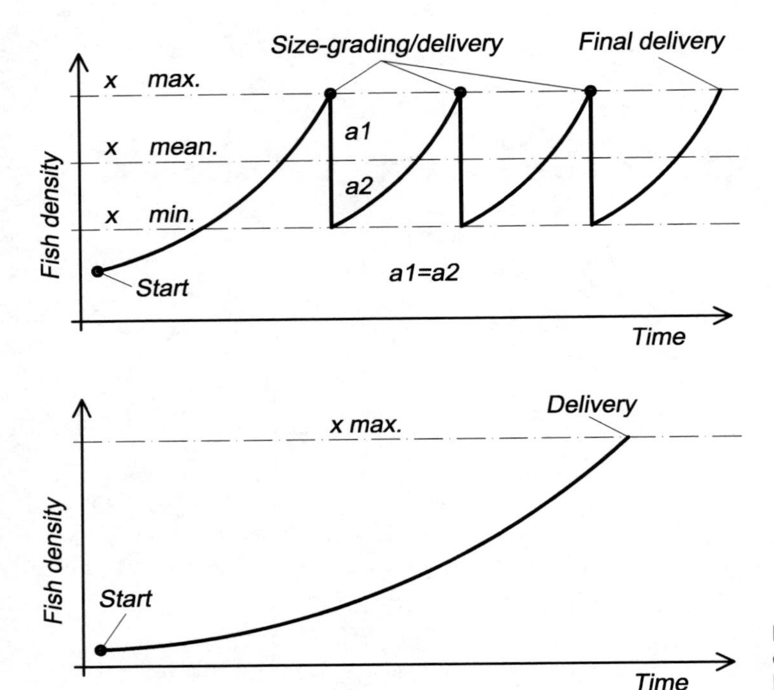

Figure 17.1 To keep high average fish density in the production unit, frequent handling is necessary.

are also important. An example can be used to illustrate this.

Example
On a farm for ongrowing fish, the fish will not be moved during production (from input to delivery). Average stocking density is set at 45 kg/m³, and the maximum density to avoid growth reduction is set at 100 kg/m³. Calculations show that the input density cannot therefore exceed 2 kg/m³ (determined by exponential growth in kg/m³). This shows that poor utilization of the production units will result if the fish are not moved. The length of production depends on input size, harvesting size and specific growth rate (SGR) in relation to fish size (daily growth rate expressed as percentage of body weight).

If higher average density is to be achieved, frequent moving of the fish is necessary, as illustrated by the following example. Maximum density must not exceed 100 kg/m³, and required average fish density in the production units on the farm is 70 kg/m³. The SGR is set at 0.9. The intervals between the movements for dividing/splitting of the fish group are calculated to be 3–4 months. The table illustrates the interval between handling (months) in relation to SGR, input density, average density and maximum density where the input weight of the fish is 100 g and the harvesting weight 4 kg. This clearly shows that increased growth rate increases the need for handling.

Fish density (kg/m³)			*Interval between handling (months) for different SGR*					
min	*avg*	*max*	*0.3*	*0.5*	*0.7*	*0.9*	*1.1*	*1.3*
40	*70*	*100*	*10–11*	*ca. 6*	*4–5*	*3–4*	*ca. 3*	*2–3*
40	*60*	*80*	*ca. 8*	*ca. 5*	*3–4*	*2–3*	*ca. 2*	*1–2*

17.2.2 Why size grade?

In a fish group there are several reason for size grading of fish.

Improved growth

In a large group of fish in tanks or cages, there is individual variation in the growth rate. Some individuals grow faster than others so differences in the individual weights for the fish in the group will develop over a period of time, even if all the fish in the group have exactly the same weight when the feeding period starts. This is a sub-optimal situation for several reasons.

An ordinary biological population will typically have a normal (Gaussian) distribution of weight resulting from genetic variation in growth rate. Under farming conditions with artificial feeding, this normal distribution may develop to form a hierarchy preventing smaller, less dominant individuals gaining access to feed. Because of this, the variation in size might become so large, the bigger fish will eat the smaller ones. How fast cannibalism develops in the population is species dependent. For instance, cannibalism will develop very rapidly if bass are not graded.[5] The distribution in the fish group may develop from normal distribution to a two group distribution, with typical winners and losers; this may reduce the total growth in a fish group[6,7] (Table 17.1).

The coefficient of variation (CV) can be used to describe the weight variation in the fish group. The CV is the standard deviation (δ) expressed as a percentage of the mean value

$$CV = (\delta/X_{mean}) \times 100.$$

If the mean weight of fish in group is 2 kg and the standard deviation is 0.5 kg, the CV will be (0.5/2) × 100 = 25%. In a fish group the CV varies with species, size, age and farming conditions. The CV in a fish group is related to the growth rate; faster development occurs with faster growth. For example, in salmon smolt production the CV can be up to 100%, while in on-growing production in seawater the CV is seldom above 30%.

When size grading a fish group into several weight groups, the individual weight variation in each of the new groups will be less than in the starting group. When dividing a fish group into two, fish smaller than the mean size are put in one group and those that are larger in the other. When grading into three or four groups, the size variation in each group is of coarse even less. On a typical salmon smolt farm between three and five gradings per year are normal.[8,9] For on-growing in cages, one grading is normally enough plus a possible grading in connection with harvesting.

Production control

To ensure good production control in an intensive drifted farm, it is necessary to maintain a small size variation, i.e. a low CV.[10] Production must be controlled to maintain a satisfactory growth rate on the farm in relation to budget. This requires regular weight sampling to ascertain average fish size in the group. In a tank or cage with tens of thousands of fish it is impossible to weigh all the fish individually, so only a sample of the fish is withdrawn for weight sampling; this sample must be representative of the whole group. However, this is difficult, especially if there is a large size variation in the group. Several methods, including the use of monitoring instruments (see Chapter 19), can be used for this purpose.

One manual method of weight sampling uses a dip net to take fish from a group, for instance, from a tank of juvenile fish.[11] Three samples of at least 50 fish each are taken out from the total fish group in the production unit using the dip net. A requirement of this method is that the CV of the three samples shall vary by less than 2%, otherwise more samples are needed. Below an example is used to illustrate how this functions in a population with some size variation.

Example
Three samples taken from a large fish group in a tank of juvenile fish give the following results:

Sample	*1*	*2*	*3*
No. of fish	*103*	*107*	*98*
Total weight (g)	*1102.3*	*1155.6*	*1009.4*
Av. weight of each fish (g)	*10.5*	*10.8*	*10.3*

The average fish weight based on the average of the three samples is 10.53 g. To simplify the calculations an equal number of fish between the samples is assumed. Between the three samples the standard deviation (δ) is 0.252. This gives a (coefficient of variation) (CV) of:

$$(0.252/10.53) \times 100 = 2.39\%$$

Table 17.1 Result from an experiment where the growth and coefficient of variation in a group of juvenile Atlantic salmon were studied for non-graded and groups graded into two or three weight classes.[7]

		20 Oct			21 Nov			19 Dec			18 Jan			16 Feb		
		X_{mean}	δ	CV	X_{mean}	δ	CV	X_{mean}	δ	CV	X_{mean}	δ	CV	X_{mean}	δ	CV
Not graded	**1A**	5.57	3.80	4.58	8.38	5.96	14.22	11.25	7.25	4.96	12.90	9.77	2.10	16.10	13.69	11.32
	1B	7.78	3.98	4.98	10.10	6.05	8.07	12.63	7.72	5.68	16.93	10.75	9.63	23.42	13.75	4.52
Divided into 2 weight classes	**2Aa**	4.13	1.93	5.96	5.59	3.31	3.69	7.04	4.82	10.95	9.02	7.15	13.92	11.81	9.58	7.75
	2Ba	4.17	2.01	4.16	5.59	3.71	8.25	8.25	5.60	2.68	10.86	6.95	0.96	12.42	10.06	15.02
	2Ab	11.21	2.59	2.51	5.93	3.60	2.56	21.20	6.81	1.15	28.98	5.74	4.26	34.00	9.09	3.19
	2Bb	10.87	2.66	2.83	16.71	3.52	2.33	22.39	5.68	2.56	28.20	6.16	2.77	34.50	9.97	0.44
Divided into 3 weight classes	**3Aa**	3.05	1.31	0.19	3.81	1.38	2.77	4.26	2.31	3.32	6.17	4.25	6.07	6.58	5.27	8.69
	3Ba	3.15	1.26	2.17	3.97	1.62	3.18	4.65	2.36	5.09	6.55	4.09	6.47	6.77	5.07	15.38
	3Ab	6.34	1.73	1.53	10.25	3.31	2.43	13.00	5.61	4.99	18.70	7.36	1.81	22.26	9.42	1.21
	3Bb	7.15	1.88	2.10	9.46	3.43	1.42	12.70	5.81	4.96	17.52	7.57	6.29	22.94	11.07	1.92
	3Ac	11.38	2.80	1.96	17.55	2.97	1.13	13.01	3.54	1.42	30.41	6.30	4.59	37.71	7.55	5.40
	3Bc	12.91	2.57	0.20	16.98	2.97	2.21	13.04	4.20	2.77	27.91	5.98	6.17	36.63	6.80	1.69

Key: X_{mean} = average value; δ = standard deviation; CV = coefficient of variation.

This does not fulfil our requirements for a CV below 2% and another sample needs to be taken.

Sample 4: 102 fish of total weight 1071.0 g, so average weight of each fish 10.50 g.

New average weight = 10.525 g new standard deviation (δ) = 0.206; new CV = 1.96%. This is satisfactory, below 2%.

This shows the difficulties of taking representative weight samples from a fish group. It is not normally possible to obtain an acceptable (CV) without size grading the fish.

Manual weight samples also include other sources for mistakes. Before taking out the fish for sampling it is important to mix them. Experience in tanks has shown that there is a tendency for the largest fish to be near the bottom; in cages the largest fish are always in the deepest layers when fish are collected for weight sampling (Fig. 17.2). Further, all the fish that have been withdrawn in the sample in the dip net must be weighed, not just the first individuals that are removed (see example below). If the fish are small it is quite easy to get far more than 50 fish in the dip net.

Example
Below the results from the author's experiment where the weight (g) of different fish withdrawn from a fish group collected in a deep net are shown. The results are given for the first 10 fish, the middle 10 fish and the last 10 fish withdrawn from a sample of 50 fish, for an ungraded and a graded fish group.

Ungraded fish group

Sample		*1*	*2*	*3*
Weight (g)	*First 10*	*16.1*	*15.7*	*13.6*
	Middle 10	*8.7*	*12.2*	*9.6*
	Last 10	*5.2*	*4.7*	*4.5*
	X_{mean}	*9.99*	*10.93*	*9.31*

Total $X_{mean} = 10.10\,g$, $\delta = 6.05$, $CV = 8.07\%$

Graded fish group

Sample		*1*	*2*	*3*
Weight (g)	*First 10*	*38.4*	*37.2*	*37.3*
	Middle 10	*38.1*	*37.0*	*40.2*
	Last 10	*32.1*	*33.5*	*33.7*
	X_{mean}	*36.92*	*35.93*	*37.07*

Total $X_{mean} = 36.63\,g$, $\delta = 6.8$, $CV = 1.69\%$

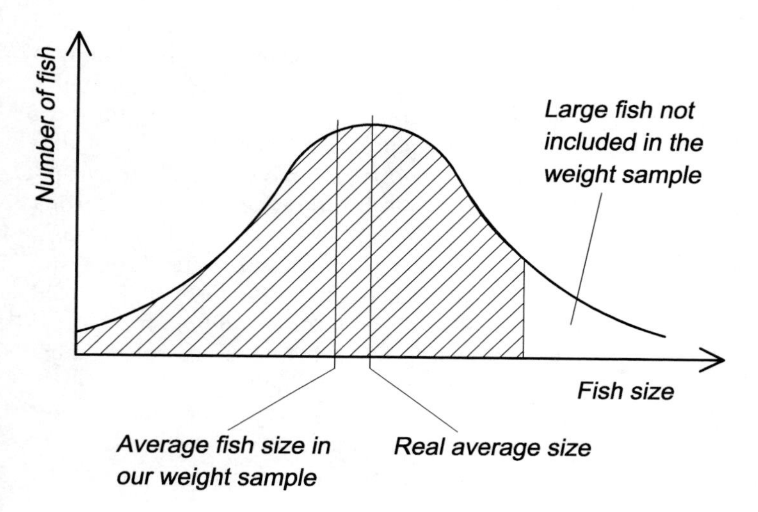

Figure 17.2 If the large fish are not included in a weight sample the wrong average weight of the fish group will be obtained.

These results show the largest fish are withdrawn first and the smallest at the end.

Harvesting of fish

When a pond, tank or cage is to be harvested, it is a great advantage if the fish in it have been size graded. If not, the slaughterhouse must deal with several sizes of fish which must go to the consumer in different weight classes. It is also more difficult to estimate precisely the amount of fish in each weight class and therefore to achieve a good price. An example is given below where the fish are ungraded (adapted from ref.10). Another solution is to size grade in connection with harvesting and send the fish that are too small back to the production unit.

Example
A fish population with 10000 fish (n) is to be harvested. The average fish size (x) is 4 kg and the standard deviation (δ) is measured as 1 kg giving a coefficient of variation (CV) of 25%. Find the number of the different weight classes represented, and the assumed number in each weight class of 1 kg if the fish group distribution is normal.

In this case a normal distribution table can be used, from which the value of δ is found to be 0.841. This means that 84.1% of the fish have a weight less than $x + \delta$ (5 kg). 15.9% of the fish in the group must therefore be larger than this, so of the 10000 fish, 1590 are over 5 kg.

Similarly the number of fish under $x - \delta = 4 - 1 = 3$ kg can be calculated and is 1590. Therefore between 3 and 5 kg there are 10000 – (1590 + 1590) = 6820 fish. These are equally divided between 3–4 kg and 4–5 kg weight classes, each of which comprise 3410 fish.

Next the numbers of fish larger than 6 kg and smaller than 2 kg are calculated; this represents $x + 2\delta$. Again the normal distribution table is used and $2\delta = 2$ gives the value 0.977, meaning that 97.7% of the fish are smaller than 6 kg and larger than 2 kg, so 2.3% are larger than 6 kg and 2.3% smaller than 2 kg. 2.3% of 10000 = 230 fish. Therefore there are 1360 fish between 2 and 3 kg and also between 5 and 6 kg. The numbers of fish below 1 kg and over 7 kg are very low (13 fish in each case) and can be ignored.

As this shows, there are six different weight classes that have to be sent to the slaughterhouse. This makes management quite difficult.

17.3 Negative effects of handling the fish

Even if handling is necessary, especially in intensive farming, it includes a number of possible adverse effects. Before selecting handling routines and equipment, this must be taken into consideration. Handling creates a stress response in the fish, which may affect the production results negatively. When the fish become stressed, the primary and secondary effects will not normally be discovered unless special measurements of heart rate, oxygen consumption or blood characteristics (for instance cortisol or glucose) are taken.[12–17] The farmer normally registers the secondary or tertiary effects of stress manifested by reduced growth and reduced immune defence,[18–21] which again may directly reduce productivity.

It is also important to consider the possible stress response involved in pre-harvest handling. This may increase the consumption of glycogen stored in the muscle (part of the stress response). The results of this may be an earlier occurrence and a shorter duration of rigor mortis after slaughtering, which again will reduce the fish quality.[22]

How much the fish is affected by handling is species dependent: some species are more tolerant of handling than others. Results also show that fish may adapt to handling procedures, and the stress responses will gradually be reduced. This can be seen, for example, when the fish tanks are washed.[14] The first time the tanks are washed it is possible to measure a high stress response, but this will gradually decrease as the fish begin to tolerate this procedure. Breeding programmes may also be used to adapt the fish to more and more of the normal handling operations in fish farming.[23–25] When starting to rear a new species, it is collected from natural wild stocks and put into farming conditions. The behaviour of such stocks differ from that of wild stocks that have been farmed and bred for generations as is clearly seen when looking over the edge of the tanks containing farmed and wild stock; difference in behaviour is also shown by the number of involuntary collisions between the fish and the tank walls.

Fish may also suffer physical damage if handled too roughly. Tolerance here, of course, also depends on species and life stage. The fish may be wounded, by rough handling leading to fungal attack. It is especially important to avoid physical damage in pre-slaughter handling because it may reduce the flesh quality and hence the price of the product.

All handling includes some kind of human work, which requires time and creates costs. The total economic cost and possible negative effects of handling must therefore be compared to the positive effects handling will have on the production. For this reason it is very important to use effective handling procedures and handling lines which affect the fish as little as possible.

17.4 Methods and equipment for internal transport

Two different principles may be used to move fish inside the farm:

(1) With a supply of energy
(2) By the use of signals or stimuli to get the fish to move voluntarily.

The first method is totally dominant, while the second is mainly the subject of research. The most common methods within each group are described below. For more information about integration of handling methods in farms see Chapter 21.

17.4.1 Moving fish with a supply of external energy

With supplied energy, the total internal transport process may again be divided into three phases:

(1) Crowding of the fish inside the production unit
(2) Vertical transport where the fish are lifted between the levels
(3) Horizontal transport of the fish between the units.

When moving the fish to a lower level no vertical transport is necessary; stored potential energy is used here as a source for the process. When moving fish between two equal levels, crowding can be used to force them to move.

Crowding

In almost all methods for vertical transport, crowding of the fish in a restricted volume is necessary. Too much crowding may, however, result in unwanted stress (Fig. 17.3). During the crowding process fish behaviour should be observed; if

Figure 17.3 Fish that are overcrowded.

odd behaviour occurs, further crowding must be avoided. Two methods are commonly used to crowd fish:

- Reduction of water level (tanks, ponds)
- Reduction of available volume (ponds, tanks and cages)

If reduction of water level is to be used to crowd the fish, the outlet of the tank or pond must be designed in a way that makes drainage possible (see Chapter 13). If there is no drain, a drainage pump can be put inside the tank or pond, but this makes the handling operation more difficult. Lack of oxygen in the water may occur during this operation and supplementary oxygen may have to be supplied; the oxygen level must therefore be monitored.

In cages, the net bag can be lifted to reduce the volume available to the fish, and through this crowd the fish. A seine net may also be used in ponds, large tanks or cages, but is less effective. In tanks, fixed or removable grids may be used to crowd the fish. A combination of collection grid and a decrease in water level can also be employed (Fig. 17.4).

Vertical transport

Dip net: The dip net is constructed with a round or rectangular frame with a net or tarpaulin bag inside (Fig. 17.5). Round frames are suitable for net bags to be used in cages, while rectangular frames are more suitable for use inside tanks. For a dry net, small mesh knotless netting should be used to avoid wounds. Plastic net have also been tried to reduce the possibilities for wounding.[26]

If a tarpaulin bag is used the fish will always be in water; this is also called a wet net. A hydraulic or mechanical crane must be used to operate a wet net because of the weight. A mechanism opens the bottom to empty the wet net of fish and water.

Use of nets is labour-intensive, especially for handling larger fish. Normally wet nets are between 100 and 500 l capacity. Fish densities above 50–70% fish compared to water are normally avoided to reduce the possibility of wounding the fish.

Pumps: A pump supplies energy to the water so it is either set under pressure or vacuum which causes the water and hence the fish to move. The fish are therefore in water throughout the entire handling operation. Several systems for pumping fish are available, including centrifugal, vacuum, ejector and airlift pumps; crucially these have an open construction that does not injure the fish.

Centrifugal pump: A centrifugal pump used for pumping fish utilizes the same principle as a centrifugal pump used for pumping of water (see Chapter 2). To avoid injuring the fish it does, however, have an open impeller with large channels and no narrow passages (Fig. 17.6). Because of its construction, it is not commonly used on fish larger than 1 kg. If such a pump were to be made for 4–5 kg fish the required dimensions of the impeller would be very large. It is normal to use submerged pumps or at least pumps with a supply pressure. Fish pumps can be made self-sucking with special adaptors. The impeller may be driven by hydraulic pressure. This makes it quite easy to move the pump around in the farm area, because only the impeller unit is moved. The pump may also be driven directly via a shaft from an electric motor. Centrifugal pumps are commonly used both in production farms and in well boats; they have been used for many years in traditional fishery well boats, among others, for pumping herring.

Vacuum–pressure pump: A vacuum–pressure pump consists of a tank to which inlet and outlet tubes are connected via valves (Fig. 17.7); a small pump is also attached. This pump can either pressurize the larger tank or withdraw the air from it, causing a partial vacuum. The function of the pump is first to evacuate the tank; then the valve to the inlet pipe is opened and water and fish are sucked into the tank; after this the inlet valve is closed and the tank is pressurized; lastly the outlet valve is opened and the fish are forced out through the outlet tube. The operation is repeated, and a new batch is pumped through.

The pump does not deliver water and fish continuously, because it operates in two phases: vacuum and pressure. However, two pumps can be used alternately to obtain more equal delivery of fish and water. A vacuum head of more than 5 m H_2O is normally avoided to prevent injuries to the fish; use of less than 40% water relative to fish should also be avoided for the same reason. Here the manufacturer's recommendations must be followed.

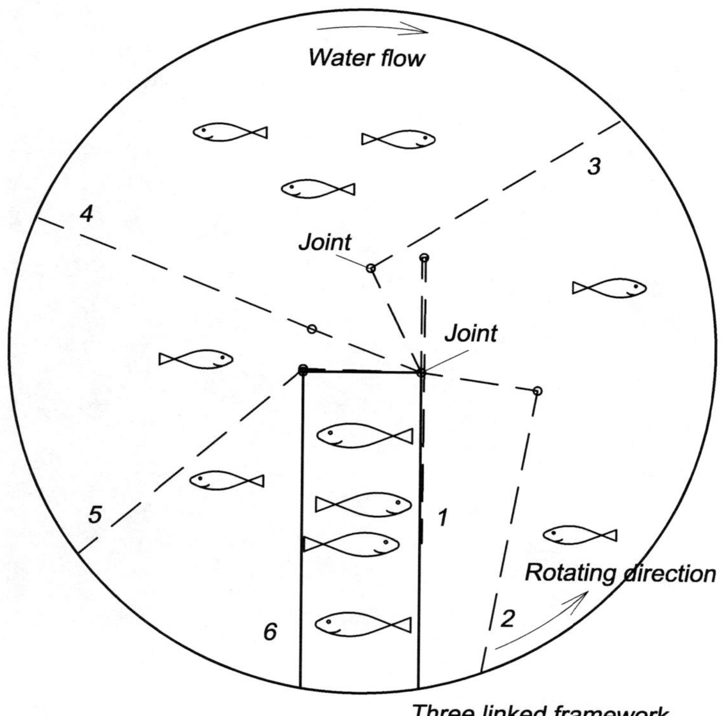

Figure 17.4 A grid can be used to crowd the fish.

Figure 17.5 Various types of dip net can be used to lift the fish.

Pumps of different sizes are required for handling small fish and harvesting large fish. The difference is the size of the tank, the pipes and the pump used to evacuate and pressurized the tank. Pumps are used on farms, in well boats and in slaughterhouses.

Ejector pump: In an ejector pump, a high velocity, high pressure part flow creates a region of low pressure (suck) in the larger main stream (Fig. 17.8). The fish travel with the water in the main stream. When the water flows past the ejector it will go from low to higher pressure. The pump can

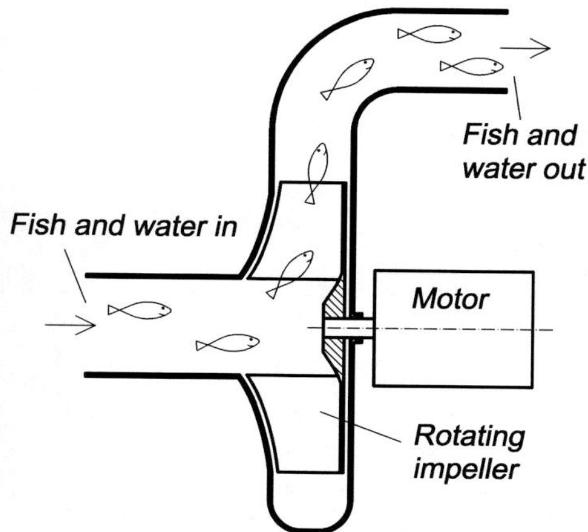

Figure 17.6 A centrifugal pump for pumping fish.

therefore deliver fish in a continuous flow of water. The pump has no moveable parts that can injure the fish. A suction head that is too high must be avoided; it is better to take a larger part of the lifting head on the pressure side. When fish have already lost scales, as occurs to young salmon during smoltification, they can easily lose more scales in these pumps, especially if the ejector is badly adjusted.

Different sized ejectors are available, adjusted to fry and to on-growing fish. The smallest size is portable and easy to move around the farm. Harvesting of mussels has also been performed with this type of pump. Some pumps have ejectors at both ends of the pipes so it is possible to change the flow direction; these are of interest on well boats for pumping fish in and out of the well.

Airlift pump: An airlift pump can be used for pumping fish (Fig. 17.8). Air is added to create bubbles that rise to the surface through a vertical, water-filled pipe suspended in the fish cage or tank. The bubbles cause drag on the water particles near them and create an upward flow of water inside the pipe. This moves the fish that are in the water up through the pipe. The capacity and the lift height in such a pump depend on the depth of the water, at which depth the air is supplied and the amount of air supplied. If using airlift pumps for fish transport, there are always possibilities for supersaturation of the water with nitrogen gas. Because of the short fish retention time in the pump, this will not normally result in any problems, but if the fish stay in the water for a longer period, problems may occur.

More generally, fish pumps can be used for different fish sizes from juvenile through to fish ready for harvesting. The difference is the size of the tubes or pipes. Pumps may also be used for harvesting shells and mussels from the bottom culture, especially airlift and ejector pumps.

It is very important that pumps are used according to the supplier's recommendations. Several problems have occurred when transporting fish due to incorrect pump use; examples include fish that have been cut by valves, eyes that come out because of incorrect pressure conditions, and scale loss; they all show the importance of correct pump use. Awareness of the correct suction head is especially important.

The great advantage of pumps over other handling methods is their large capacity. Today pumps are being used increasingly in intensive aquaculture as well as in slaughterhouses.

Fish screw: The fish screw, or pescalator, is based on the Archimedes screw which was used for lifting water in ancient times (Fig. 17.9). A screw is fixed inside a pipe which has a belt around its circumference. The belt is further connected to a small electric motor; when the motor starts the belt will rotate as will the pipe and therefore the screw relative to

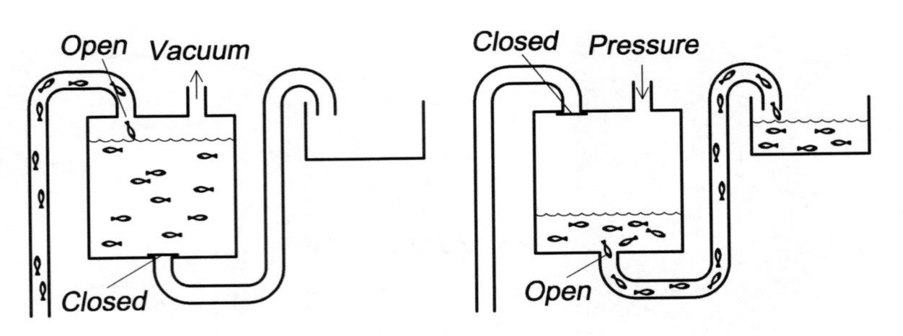

Figure 17.7 A vacuum–pressure pump for pumping fish.

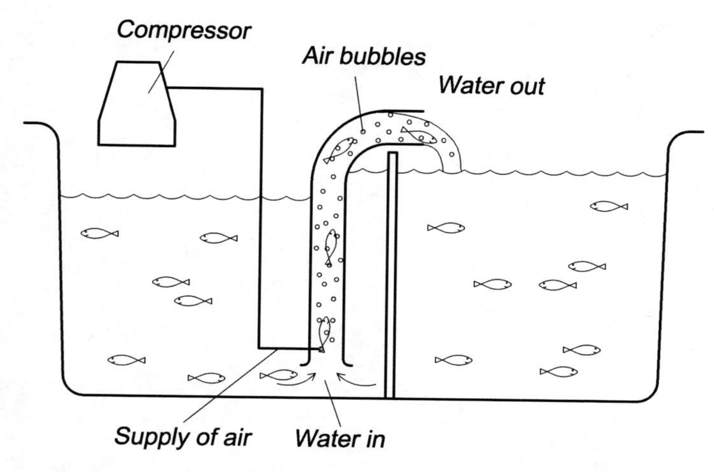

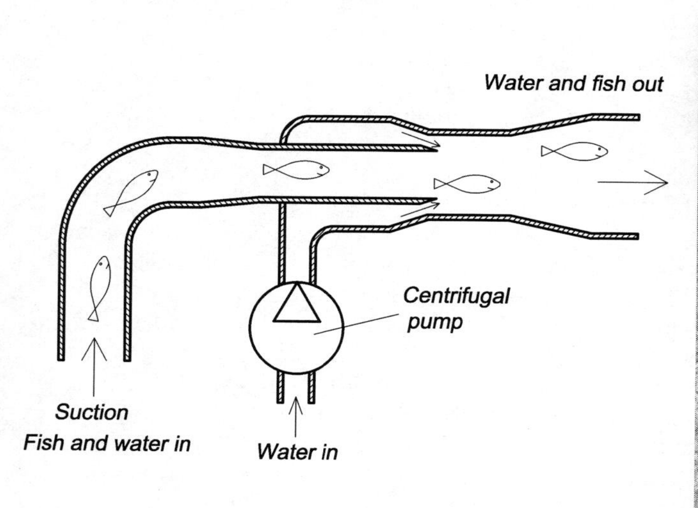

Figure 17.8 Airlift (top) and ejector (bottom) pumps for pumping fish.

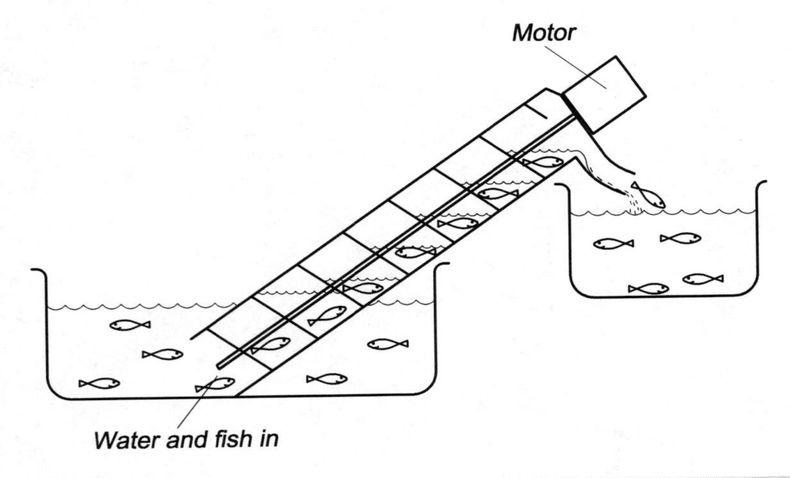

Figure 17.9 A fish screw used to move fish.

it. This is not a traditional screw which rotates; the pipe in which the screw is fixed rotates. Rotation of the pipe will result in lifting of water and the fish within it. The fish appear to be lying with water in small basins. Fish must be crowded around the inlet of the pipe for the screw to be filled with fish. Hence the screw may be placed in a well where the fish are automatically crowded. Another solution is to use a special perforated tank around the inlet, which again stands inside the fish basin or pond. A dip net may be used to lift the fish up and into a perforated tank. The lift height achieved depends on the length of the screw. To avoid damage to the fish it is not recommended to have an angle towards the horizontal plane of more than 40°. Usually the screw length is between 3 and 6 m, while the diameter of the pipe varies between 30 and 45 cm. An area around the tank or pond where the screw is to be used is required, so that the screw can be repositioned. Such screws are commonly used in tanks with low water levels and in ponds.

Transport tanks: The transport tank is specially designed with smooth surfaces and angles to avoid wounding the fish during transport (Fig. 17.10). The fish have first to be transferred from their ordinary production unit to the transport tank, for instance by use of a dip net or by pumping the fish. It is necessary to have mechanical equipment, such as a forklift truck or a tractor with a front loader, to lift the transport tank because of its heavy load of water and fish. The fish can stay in the transport tank at quite a high density; it is not usual to add extra oxygen into the tank during internal transport on the farm. The period that the fish can stay in the tank without a supply of oxygen is, however, limited.

The size of tank selected depends on the size of the fish to be transported. Tanks of 300–800 l are normal. Fibreglass or aluminium are the usual construction materials for transport tanks, and the design is very similar to that of tanks used for ordinary fish transport (Chapter 18), but smaller. Tanks for internal transport may therefore be used for transporting small amounts of fish out from the farm (external transport). The tanks are not, however, insulated like many of the long distance transport tanks. To enable the fish to be tapped out of the transport tank there may be a hatch close to the bottom which makes emptying of the tank quick and easy. The tank may also be used for vertical transport at the farm, being lifted with either a forklift truck or a tractor.

Figure 17.10 A tank used for internal transport of fish.

If tanks are to be used for fish transport, quite a large transport area is required. This must be as level as possible, to prevent the tractor or forklift truck tilting when lifting the tank. Use of a transport tank to move juvenile fish is physically undemanding. The capacity of this internal transport method depends on the size of the tank, the fish density and the duration. Fish density is species related; for example, over 500 kg in 1000 l of water can, easily result in lack of oxygen. Transport with such high densities must therefore only last for a short period of time.

Horizontal transport between units

Pipes are commonly used for internal transport of fish. The method is also often combined with pumping, for instance in cage farming. Here the pump transports the fish vertically and the horizontal transport is done through pipes. A certain excess pressure (head) is necessary for fish transport in pipes. If not using a pump to create this pressure, one possibility is to create a magazine of fish and water from which the fish can be tapped by gravity; alternatively, the pipe must slope downwards and use gravity.

If there is a natural slope in the building area, a fish farm may be built in terraces with start feeding in the upper part and on-growing in the lower parts (Fig. 17.11). Fish transport can then be performed

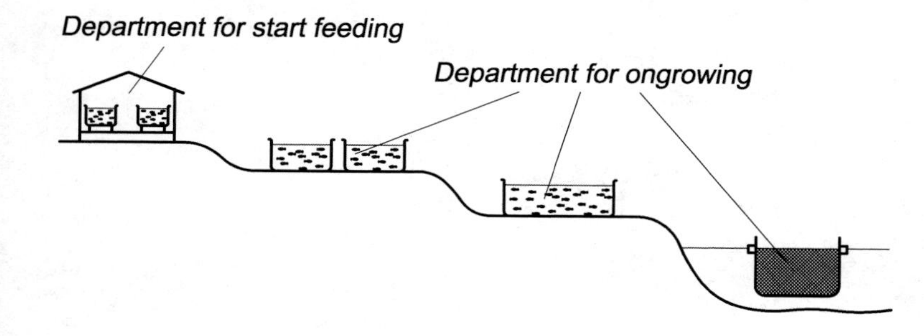

Figure 17.11 The farm may be terraced to utilize gravity for internal transport.

through pipes utilizing the natural slope. Another method is to use a common tapping centre to which tapping pipes from the fish tanks are connected. This can either be a common tapping pipe system, or there can be individual pipes from the separate tanks. From the tapping centre, vertical transport can be performed by some of the methods mentioned previously, a fish pump or a fish screw, for example. It must be possible to reduce the water level in the tapping centre to crowd the fish before vertical transport (Fig. 17.12); this can, for instance, be used to transport the fish back to other tanks or to a grader.

The same principle may also be applied in ponds, where the fish are tapped out to the harvesting tank. This requires the pond to be constructed to allow this, with a self-tapping pipe through the pond levee.

When selecting the diameter of the tapping pipe, experiments have shown that it should be large enough to enable the fish to turn inside.[27] A diameter of at least half a fish length is a good start, but this will of coarse also be species dependent. The flow velocity inside the pipe must be sufficiently high that the fish understand it is not possible to fight against the water flow and will only be dragged along with the water flow. If the velocity is too low the fish will fight against the water flow and be exposed to more stress. The correct velocity will, of course, depend on the species and its swimming performance; for salmonids an appropriate velocity in the pipe is three to four fish lengths per second[27] (Fig. 17.13).

The capacity of the system depends on the velocity of the water, the diameter of the pipe and the size of the fish; the distribution of fish through the pipe depends on the diameter of the pipe and the size of the fish. Normally, completely water filled pipes, such as from pumps, are recommended; otherwise quite large slopes on the pipes are nec-

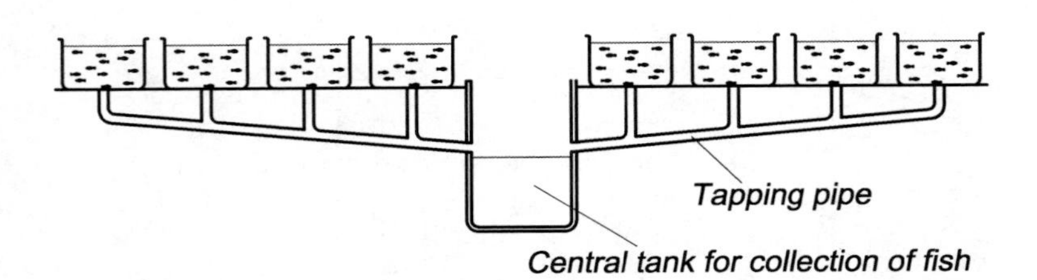

Figure 17.12 A common tapping centre may be used for collecting fish from several tanks.

Figure 17.13 Use of pipes for fish transport.

essary. If using partially filled pipes, after-flushing with additional water must be carried out to empty the pipe completely of fish. If after-flushing is started before the pipes are empty of fish a more even distribution of fish through the pipes will result. If the farm is constructed with a good system for tapping of fish there is minimal requirement for human work to operate the handling system. The capacity is large both in total and per man-hour.

One question may be raised regarding the use of a common tapping centre where pipes from several tanks are connected, because there is the possibility of transferring diseases between fish tanks through the pipes and the common tapping centre. When a farm is going to be constructed or reconstructed, much effort must be put into the disinfection systems, drainage and insulation of single tanks, when a tapping centre is to be used.

If the fish are to be tapped through pipelines, all the valves in the system must be of the ball or throttle type; with these valves the pipe diameter can be completely open through the total cross section. This is important to avoid narrowing in the pipes and by this possible damage to the fish.

17.4.2 Methods for moving of fish without the need for external energy

Today there are no commercial systems available based only on voluntary movement of the fish. Some experiments have, however, been performed where the aim has been to move fish without supplying energy. If stress, wounds and labour are considered, such methods are, of course, of interest.

To get the fish to move voluntarily, they need a signal or a stimulus that tells them to go. This can either be a positive signal that will attract the fish or it can be a negative signal from which the fish will swim away. The fish may also be trained so that a signal gives a positive or negative response. What type of signal or stimuli that can be used depends on the species and age. Practical observation has shown that the fish will keep away from areas with low oxygen levels. It is, however, important that environmental conditions are varied in a positive and not a negative way, which again may stress the fish. Flow of water is interpreted as positive stimulus for species that naturally prefer to stay in a water flow, such as salmon. Here juvenile and on-growing fish will swim against the water flow while smolts will drift with the flow.[28] One way to achieve voluntary transport is to equip the tanks with hatches, and have a channel where the fish can swim through and into the next tank (see Chapter 21). Water flow and light conditions can be manipulated to improve voluntary transport.[29,30]

The addition of chemical substances to the water can be interpreted as a positive stimulus by the fish.[31–33] Experiments have shown that certain types of amino acids can attract fish[34] (Fig. 17.14). Manipulation of the light conditions may also have a positive effect on the voluntary movement of fish; dark zones may attract them.[35]

The fish may also be trained to respond to a stimulus voluntarily. The simplest way is to teach the fish to associate a signal with a positive stimulus, i.e. to condition their behaviour. The Russian physiologist Pavlov demonstrated this with his famous dog trials: every time the dog was fed a bell rang and after a while the dog salivated in anticipation of food merely when it heard the bell ring. In fish farming, similar experiments have been carried out in connection with collecting fish.[36,37] In the author's experiments fish were trained to associate a flashing light with feeding[38,39] (Fig. 17.15); after a period the fish crowded together around the flashing light, even if they were not fed. Similar principles have been used in the sea. Sound has been used as a signal when feeding fish in a cage. After a period of training the net bag was removed and the fish

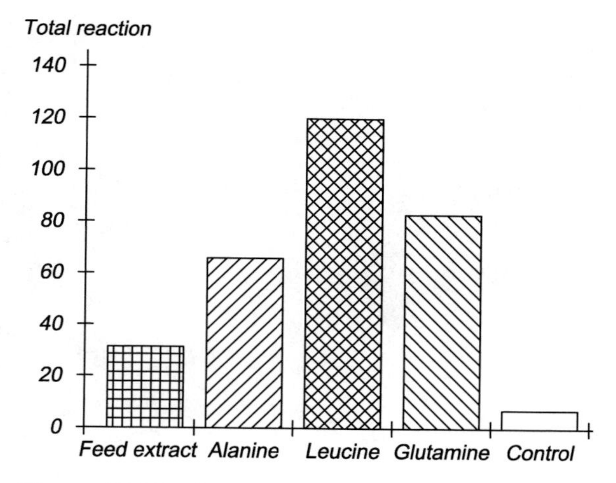

Figure 17.14 Chemicals can be used to attract the fish.

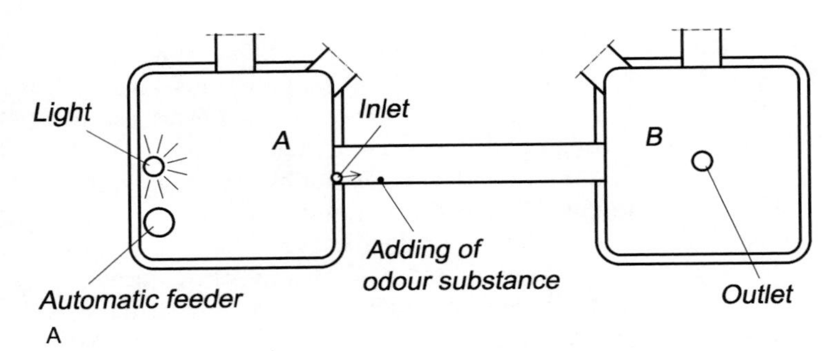

A

Figure 17.15 (A–C) If fish are trained to associate feeding with a flashing light, they will come to the light when it is flashed. (C) A rig was developed for doing experiments that allowed the fish to move between tanks.

released;[40] they could now swim and search for natural feed over a larger area. By making a sound again the fish crowded around the feeder and collection for harvesting was possible. In tanks a sound signal is difficult to use because of echoes from the walls which make it impossible for the fish to locate the sound source exactly.

The main problem with voluntary transport is its effectiveness and the time needed when the fish are to be moved. Not all the fish respond to the chosen stimulus, which also represents a problem.

17.5 Methods and equipment for size grading of fish

Similar to the methods and equipment for internal transport, the size grading equipment may also be divided into systems that do and do not require addition of extra energy. It may not be necessary to supply energy to the grader, but the equipment may be of a design that requires potential energy; the fish must be lifted to a higher level before grading and sent from there into the grader.

All size grading will stress the fish, even if there are variations from species to species. For this reason, it is important that the equipment used and methods employed are implemented correctly to minimize the stress response of the fish. Wounds may also occur as a result of incorrectly adjusted grading equipment. Therefore graders must be used according to the supplier's recommendations.

Several methods are used for grading of fish. Equipment can be separated into that needing a supply of energy and equipment where the fish voluntarily grade themselves. The first method is totally dominant. The effectiveness of the different methods is to some degree dependent on the species to be graded. The latter method is mainly used in research activities. A brief survey of the methods most used follows.

17.5.1 Equipment for grading that requires an energy supply

Methods where the fish are taken out of the water

Manual: Fish can, of course, be size graded manually. Each fish is taken on a table, visually graded by hand and sent to the different size classes. This method can be used to grade very small amounts of fish, but it is labour intensive. Therefore some types of automatic grader are used.

Fish cradle: A grading cradle is simple in construction and cheap to buy (Fig. 17.16). It is quite common to use a cradle, especially for the first grading, or for small species. The same is the case for smaller farms or on more extensive drift farms that seldom grade their fish. A cradle is basically a box with ribs or bars in the bottom. When using a traditional cradle, fish of different sizes are crowded into the cradle which is placed inside the fish tank. The cradle is then lifted up and shaken. Small fish will now fall between the ribs in the bottom of the cradle and remain in the fish tank. The cradle

Figure 17.16 Using a fish cradle to grade fish.

containing the larger fish is then lifted out of the tank and swung into another tank where the large fish are released. In relation to capacity the use of cradle for grading is labour intensive. A cradle that is lifted manually will impose large loads on the operator's back. The advantage with using a cradle is that it is simple, inexpensive and does not require any additional equipment.

Grading box: A grading box is based on the same principle as the grading cradle. The fish are lifted out of the fish tank, for instance using a dip net, and poured into a box with a grading grid in the bottom. The smallest fish will fall through the bottom of the grading box under gravity. The bottom consists of a grid made of bars; the distance between the bars is adjustable. The box normally slides on a rail system, and when the small fish have fallen out the box is moved over a new tank using the rail system. Here the distance between the bars in the bottom of the box is increased and fish of a new determined size then fall through the grid. Afterwards the sliding box can be moved above yet another tank and the distance between the bars increased once more, so the largest fish will fall out. The equipment has low capacity and is only recommended for grading small fish.

Tilt grader: A tilt grader is based on a similar principle. Fish are poured over a grid system, normally with two or more sections on top of each other (Fig. 17.17). The fish are poured into the middle section and the smallest fish fall through all the grids and into a tank below. Then the top grading grid, which has the largest distance between the bars, is tilted to one side and the largest fish will follow and fall into a tank. Then the grader tilts the intermediate grade to the other side and the medium sized fish will fall into another tank. If a small amount of fish is being graded, the grid may be tilted manually. For larger fish and larger quantities, hydraulic cylinders may be used to tilt the grids. This type of grader normally divides the fish into three size classes. It has quite low capacity and is also labour intensive.

Figure 17.17 A tilt grader for grading fish.

Another method based on the same principle is also used on large fish. The fish are crowded into a grading box fixed to a sledge. When a reasonable number of fish have come into the box, hydraulic cylinders lift it. The smallest fish then fall through the grids in the bottom and down into the tank fixed below. When the sledge reaches the top of the rail system, the fish are dropped into different size groups separate from each other. In this system the vertical transport is part of the grader.

Grading grids

Design of a grading grid: There are a number of ways to design a grading grid, which may be placed horizontally or vertically. The grading grid can either have a fixed or variable distance between the bars. If the distance between the bars is fixed separate sizes of grading grid must be available on the farm according to the size of the fish to be graded. To achieve a variable distance between the bars several methods are available (Fig. 17.18). One method is to set the bars in a frame where the two opposite sides can be displaced parallel to each other, so changing the distance between the ribs. The scissor principle may also be used. Here the separate bars are placed in the centre of a scissor construction; by opening this out, the distance between the bars is changed. It is also possible to place removable knots of various sizes between the single bars to obtain a grading grid with variable distances between the bars, but changing the spacing between the bars will require more time.

The same bar construction may also be used in grading grids which are to be placed vertically in the water, in a net cage, pond or sea cage. The grid

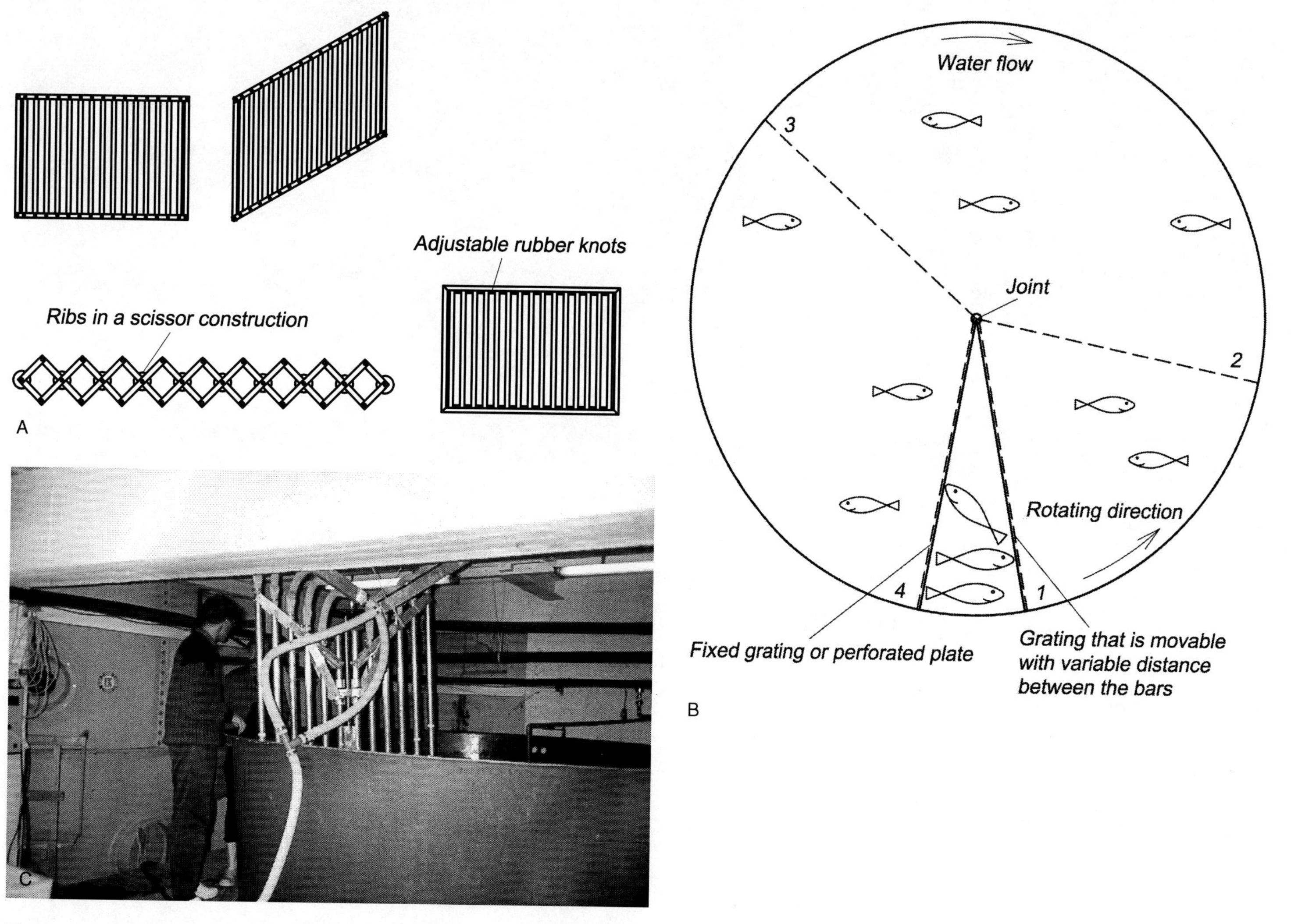

Figure 17.18 Several methods are used to achieve a variable distance between the bars in a grading grid: (A) grid construction; (B, C) the grid is rotated and the smaller fish swim through while the larger fish are retained.

may either stay fixed in one place or dragged through the production unit as an integral part of the seine net.

Distance between the bars: The distance between the grading bars determines the size of fish to be placed in the different classes. This size, or actually the thickness of the fish, is related to its weight. It is, however, difficult to give exact values for the distances that should be used to grade out fixed sizes of fish. This of course varies with the species, because they have different body shapes. However, it also varies with the condition of the fish within the same species. A fish in good condition will be thicker than one in poor condition. Normally a fish is thickest just behind the gills, but this may vary from species to species.

To obtain a rough estimate of the distance between the bars, salmonids can be taken as an example. A rough estimate says that The width of the fish i.e. the thickness (T), is around 1/10 of the length of the fish. The condition factor of the fish (C) and the weight of the fish (W) in grams may also be used to estimate in mm (and hence the distance between the bars). The following formula may be employed:

$$C = (W \times 100)/T^3$$

$$T = C/(W \times 100)^{1/3}$$

where:

T = thickness of the fish (mm)
C = condition factor of the fish
W = weight of the fish (g).

Grading machines (graders): A number of principles are used to determine the design of grading machines or grader. For all machines described the fish must be lifted out of the water; it is also necessary to lift the fish up to the grader, and the lifting height depends on the principle used for grading. Because the fish are graded in air, it is usual to spray them with water to prevent them drying out. After grading, the different fish sizes are delivered into different tanks through a pipe system, each housing a different size of fish, or the different sizes can return directly to the different fish tanks, depending on the total handling system. To get effective grading the grader must be fed continuously; the fish must not come in batches. Depending on construction, the grading machines are to various degrees adapted to take fish with different body shapes, for instance flatfish. Before choosing a grader it is therefore important to ensure that the grader is appropriate for the species.

Bar graders: In a bar grader the fish slide down on a slightly sloped 'rib table' constructed of beams or bars (Fig. 17.19). The distance between the bars is smallest close to where the fish enter the table and then gradually increases. Therefore the smallest fish will fall through the bars first into the tank underneath. Since the distance between the bars gradually increases larger fish will fall through when they have advanced some distance from the entry point. This type of grader is usually used to divide the fish

Figure 17.19 A bar grader.

group into two or three size classes. The advantage with the bar grader is that there are no movable parts, but it has limited capacity compared to the other graders. What decides the capacity is the slope angle on the grading board. However, if the slope is too large sub-optimal grading will result; many fish will enter the wrong weight groups. In each case there will be an optimal slope, and this must be tested on the site.

Roller grader: The design of the roller grader is similar to the bar grader, but the bars are replaced by rotating rollers. This system can be utilized for all size classes; the size of the graders is the only difference. Roller graders are normally installed on land-based farms, but may also be installed on boats or rafts for grading fish in sea cages. A machine located on a raft can be partly submerged to reduce the required lifting height.

The principle of 'dry placed' roller graders for juvenile and on-growing fish involves lifting and pouring the fish, so that they flow over the 'table' with the rotating rollers (Fig. 17.20). Two pairs of rollers rotate away from each other so that the fish

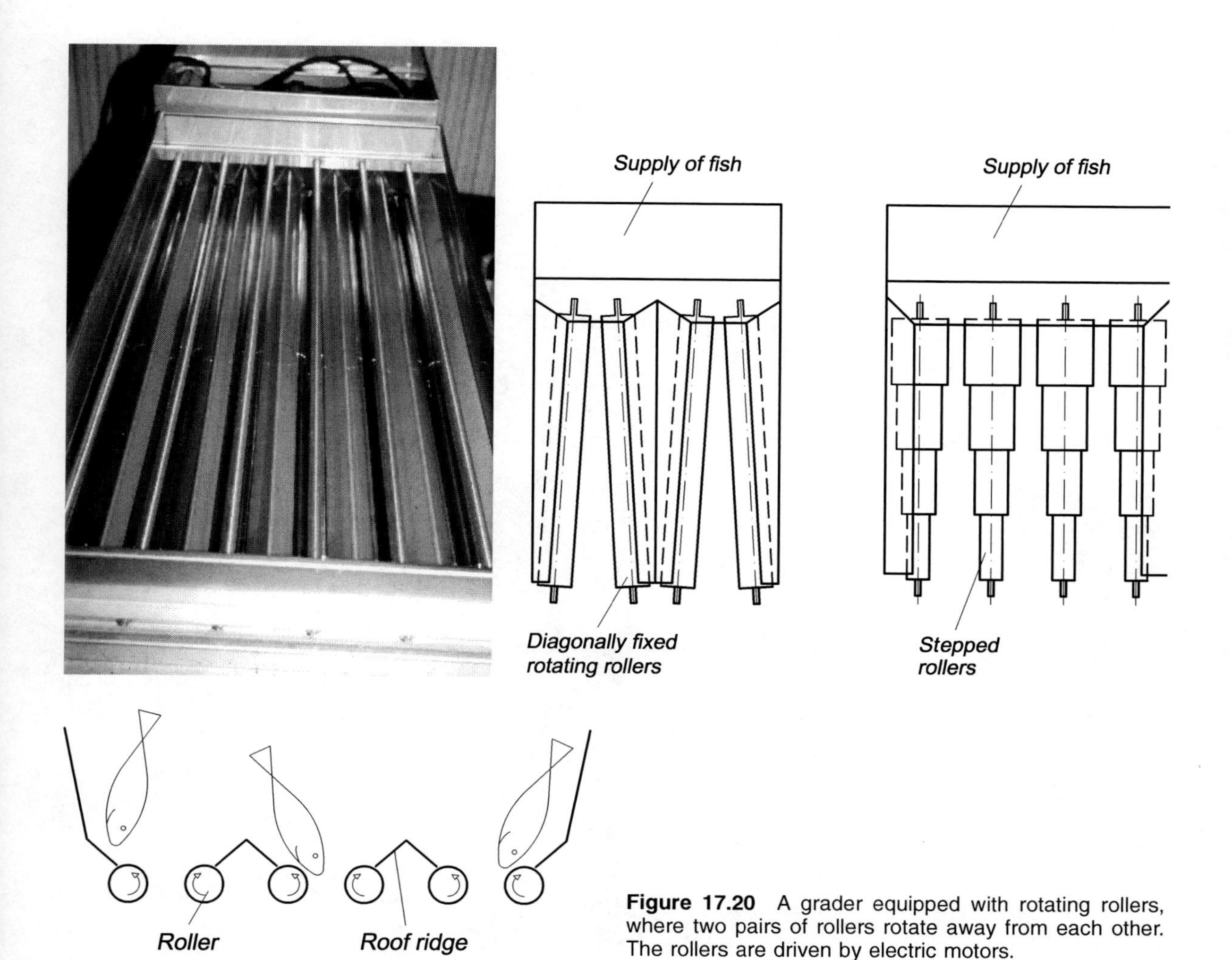

Figure 17.20 A grader equipped with rotating rollers, where two pairs of rollers rotate away from each other. The rollers are driven by electric motors.

are not squeezed between the rollers, but lifted up so that they fall through the grading table in the correct place. Between the pairs of rollers there is a ridge. The distance between the rollers increases from the start to the end of the grading table. Two different types of rollers are used: the first type of roller is the same diameter along the whole length and is installed with a fixed angle between the pair of rollers; in this way an increasing distance between the two rollers is achieved. The other type of roller has a diameter that decreases in steps along the direction of movement of the fish, because of which there will be an increase the distance between the rollers; this ensures separation of the fish based on thickness. A roller grader has an electric motor to drive the rollers via gear wheels. Normally this grader will divide the fish group into three to four different sizes. Roller graders have higher capacity than bar graders. The capacity depends on the number of rollers, or actually the width of the machine, the length and the slope of the rollers.

To use this type of machine, the fish must be lifted up to the grading table. The head loss over the machine is quite low (about 50 cm H_2O). Normally it is therefore possible for the fish to fall directly from the machine back into ordinary fish tanks through pipelines. This type of machine is normally equipped with wheels so can be easily moved around the farm and stored when not in use.

Belt grader: In a belt grader the fish slide between two rotating belts that are positioned obliquely to form a V-shaped channel with no bottom (Fig. 17.21). The rotation of the belts helps to drive the fish forward in the channel. To rotate, each belt is equipped with its own electric motor. From the point where the fish are poured into the machine, the distance between the belts gradually increases. When the distance between the belts is large

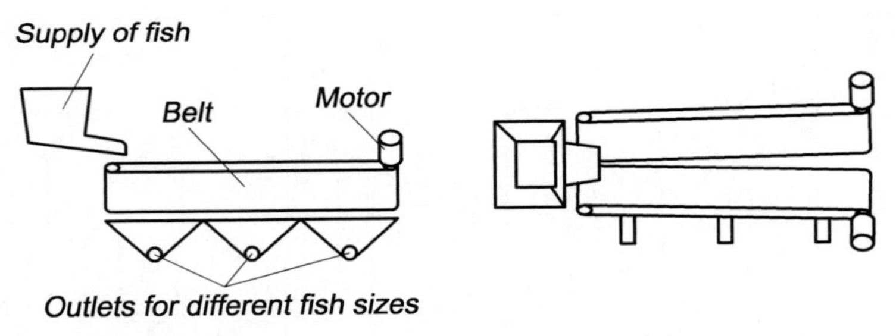

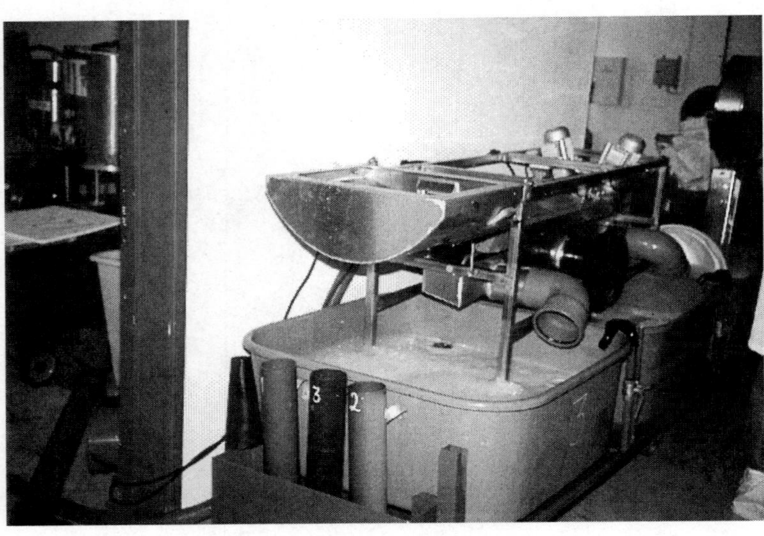

Figure 17.21 In a belt grader the distance between two rotating belts gradually increases.

enough the fish will fall through into tanks underneath, the smallest fish first followed by the other sizes. Normally this grader is used to grade into three size classes or more. These graders have a very low head loss, so only a low lifting height is required for the fish to enter the grader. This type of grader is also equipped with wheels for easy movement inside the farm; it is long and narrow (3–5 m) and so requires a long space.

It is important not to overload the grader with a too many fish. Sub-optimal grading will result, because the small fish may stay on top of the larger fish and in this way drop into the wrong size class. This problem may also occur with roller and beam graders, but in these cases a wider grading table can be used to increase the capacity and in this way reduce the problem.

Band grader: This is a fairly new grading principle which combines the belt and the roller grader (Fig. 17.22). A tilted rotating belt or band of PVC into which the fish are lead is used. The fish lie on their side and move in the direction of rotation of the band. A roller is sited above the belt. The distance between the band and the roller gradually increases in the direction of transport. When the distance is large enough the fish will slide under the roller. In this way division into groups is achieved. It is normal to grade in groups of three sizes with this machine. The machine has also proved adequate for grading flatfish such as turbot.

Level grader: On a level grader the fish are poured onto the top and gradually slide down through tilted grids that form a 'grading table' similar to the bar grader (Fig. 17.23). The larger fish will not go through the grid and will therefore be removed on the first grid. Smaller fish will continue to fall through onto a new tilted grid where the next fish size is removed, and then the same process may continue with new grids. Normally this grader will also divide fish into three size groups, but the machine may quite simply be adjusted to grade several sizes. The advantage with this machine is that there are no moveable parts and that the largest fish are removed first. The machine has a quite a high head loss and the fish must be lifted in order to enter the machine. Such graders are not normally mobile, but lightweight versions may be produced so that moving is possible. This type of machine is recommended to stay centrally placed in the farm. If the system for getting fish in and out

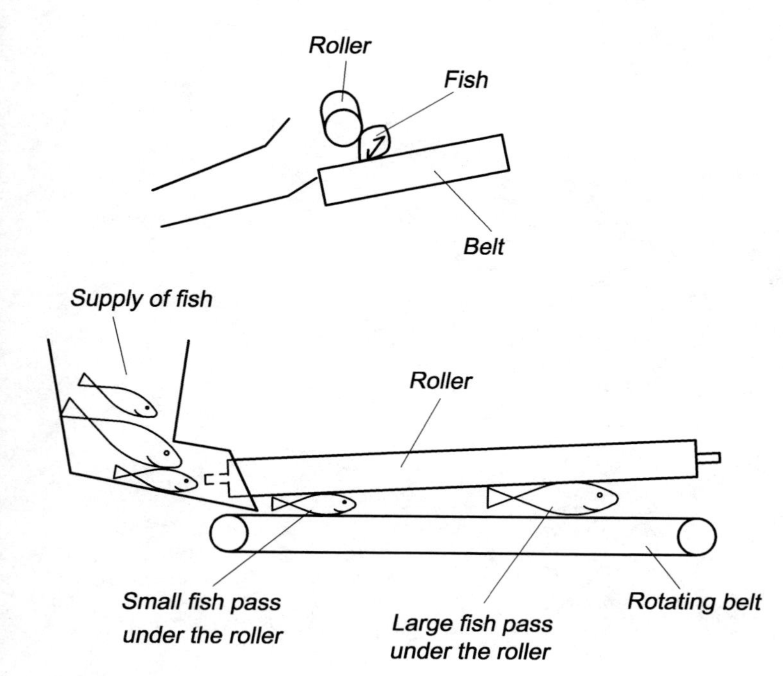

Figure 17.22 A band grader utilizes a principle that is also adequate for grading flat fish species.

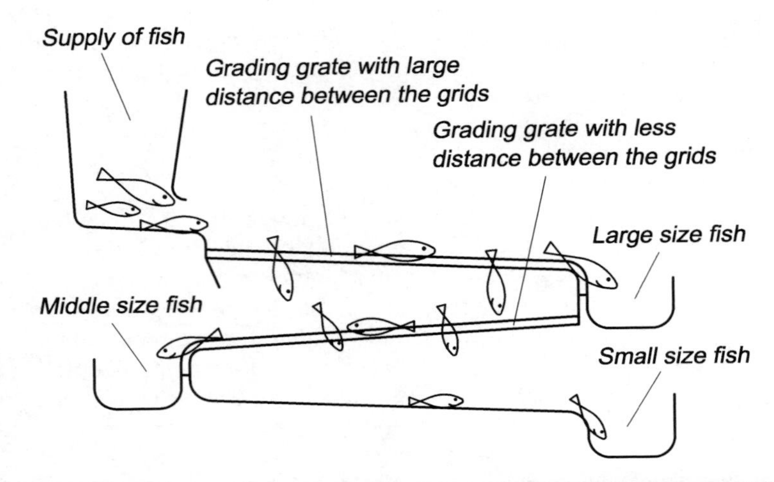

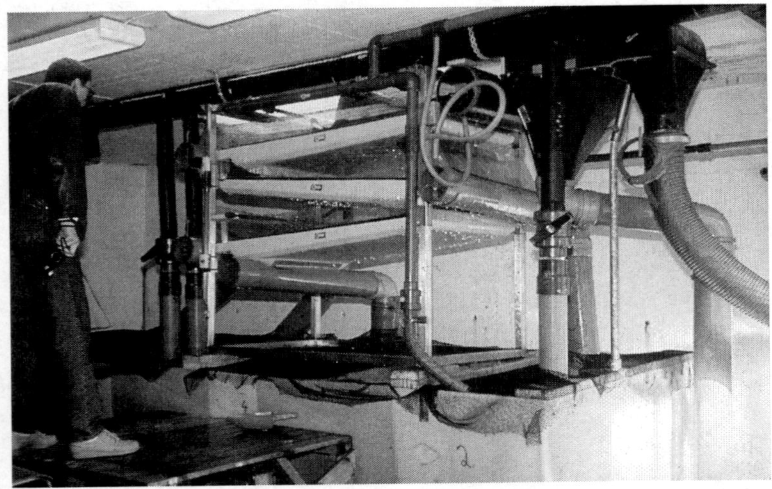

Figure 17.23 In a level grader the largest fish are separated away first, which prevents the smaller fish staying on the back of larger fish.

from the grader is well designed, a large grading capacity can be achieved with this system, for instance in connection with a tapping centre.

Other types of grader: There are also a number of other grading systems and principles that can be used. These are not described here where focus is on the most general types and basic principles. For instance, many fish farmers have constructed their own graders adapted to their specific needs.

Methods for grading the fish in the water

Methods where the fish are maintained in the water throughout the entire grading process have been developed for raceways, ponds, sea cages and grading channels.[5,41,42] In raceways this is possible by using movable vertical grading grids (Fig. 17.24). When using the grid it is moved towards the fish which are forced to swim through. It is often used in combination with a water flow towards the fish. The smallest fish will pass through whilst the larger fish will not be able to do so and will gradually be crowded together.

Pulling two cages together and placing the grading grid in the middle gives a similar system.

Alternatively, a seine net with a grading grid included can be dragged through a cage or pond (Fig. 17.24). The smaller fish will swim through the grid and remain in the cage or pond, while the larger fish will be crowded together and can be removed. In circular tanks a similar method can be used. A vertically placed grading grid is dragged through the tank like the hand on a watch. Another dense grid is fixed in the tank. Both the grading grid and the dense grid are fixed on a fitting in the centre of the tank; this may, for instance, be a centre drain in the tank.[43]

17.5.2 Methods for voluntary grading (self grading)

The same stimuli that are used for voluntary fish transport have been used for self grading[34,42,44] (Fig. 17.25). Water flow towards the fish, scented substances, manipulation of light conditions and behaviour training have all had some effect, but full size grading of an entire fish group has been shown

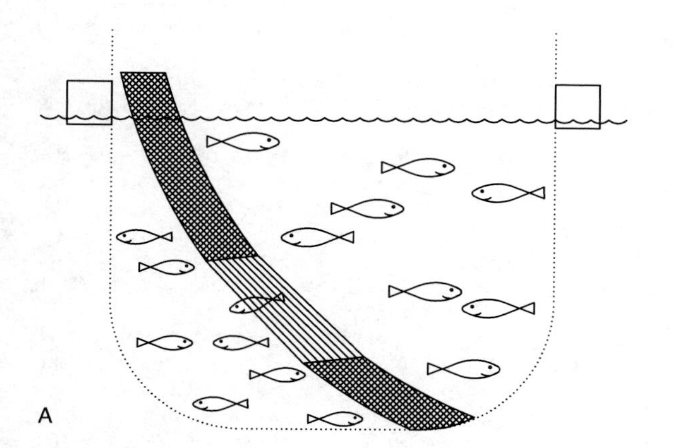

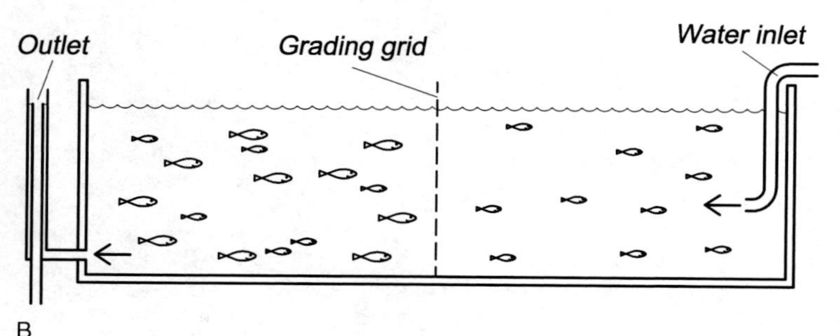

Figure 17.24 Moveable grids can be used for grading in the water in: (A) sea cages and (B) raceways.

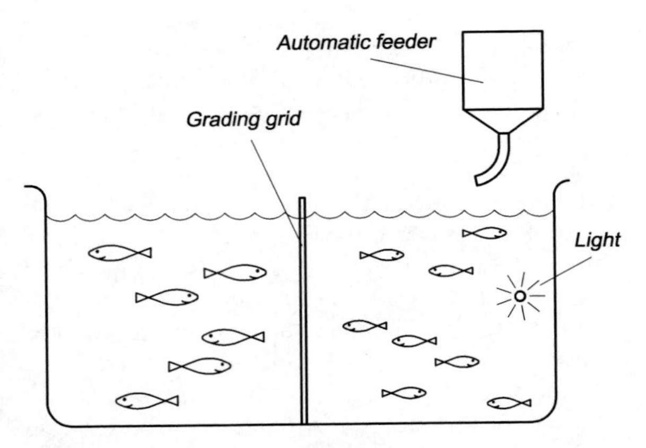

Figure 17.25 Use of stimuli to attract the fish through the grading grid.

to be impossible. In addition, it is necessary to have a slightly larger distance between the ribs when using stimuli compared to traditional methods for grading because the fish will not go through small passages voluntarily. In particular, when the fish feel the ribs on their sides they will stop swimming voluntarily. In practice, where time is limited, stimuli are not useful for voluntary grading of fish.

References

1. Lekang, O.I., Fjæra S.O. (1993) Fish handling in land based fish farms. In: *Fish farming technology. Proceedings of the first international conference on fish farming technology* (eds H. Reinertsen, L.A. Dahle, L. Jørgensen, K. Tvinnereim). A.A. Balkema.
2. Kjartansson, H., Fivelstad, S., Thomassen, J., Smith, M.J. (1988) Effects of different stocking densities on physiological parameters and growth of adult Atlantic salmon (*Salmo salar* L.) reared in circular tanks. *Aquaculture*, 73: 261–274.
3. Lekang, O.I., Fjæra, S.O., Skjervold, P.O. (1992) En lønnsomhetsvurdering av interntransport og flytting av fisk. *Norsk Fiskeoppdrett*, 1a: 22–23 (in Norwegian).
4. Lekang, O.I. (1994) *Logistikksystemer i landbaserte oppdrettsanlegg*. ITF rapport nr 53. Norwegian University of Life Science (in Norwegian, English summary).
5. Parker, R. (2002) *Aquaculture science*. Thomson Learning.
6. Gunnes, K. (1976) Effect of size grading young Atlantic salmon (*Salmo salar*) on subsequent growth. *Aquaculture*, 9: 381–386.
7. Fjæra, S.O., Lekang, O.I. (1991) *Betydning av størrelsessortering av lakseyngel på tilvekst, dødelighet og størrelsesvariasjon*. ITF-rapport nr 22. Norwegian University of Life Science (in Norwegian, English summary).
8. Lekang, O.I., Kittelsen, A. (1989) *Spørreundersøkelse blant settefiskprodusenter om utstyr og rutiner for handtering*. LTI-trykk nr. 101. Norwegian University of Life Science (in Norwegian).
9. Willougby, S. (1999) *Manual of salmonid farming*. Fishing News Books, Blackwell Science.
10. Lekang, O.I., Fjæra, S.O., Skjervold, P.O. (1991) Størrelsessorter ofte, unngå størrelsesspredning. *Norsk Fiskeoppdrett*, 7: 24–25 (in Norwegian).
11. Lekang, O.I., Fjæra, S.O. (1992) Erfaringer med uttak av fisk til veieprøver. *Norsk Fiskeoppdrett*, 1: 26–27 (in Norwegian).
12. Mazeaud, M.M., Mazeaud, F., Donaldson, E.M. (1977) Primary and secondary effects of stress in fish: some new data with a general review. *Transactions of the American Fisheries Society*, 106: 201–212.
13. Pickering, A.D. (1981) *Stress and fish*. Academic Press.
14. Lekang, O.I. (1989) *Effects of handling on juvenile fish*. PhD thesis, Norwegian University of Life Science.
15. Fjæra, S.O., Lekang, O.I. (1991) *Oksygenforbruk etter handtering for tre ulike fiskestørrelser*. ITF-rapport nr. 20. Norwegian University of Life Science (in Norwegian, English summary).
16. Fjæra, S.O., Lekang, O.I. (1991) *Additiv stressrespons ved trenging, transport i fiskeskrue, sortering og vaksinering av smolt*. ITF-rapport nr. 21. Norwegian University of Life Science (in Norwegian, English summary).
17. Barton, B.A. (2000) Stress. In: *Encyclopedia of aquaculture* (ed. R.R. Sickney). John Wiley & Sons.
18. Sniesko, S.O. (1974) The effects of environmental stress on outbreaks of infectious diseases of fishes. *Journal of Fish Biology*, 6: 197–208.
19. Pickering, A.D. (1993) Growth and stress in fish production. *Aquaculture*, 111: 51–63.
20. Wendelaar Bonga, S.E. (1997) The stress response in fish. *Physiological Review*, 77: 591–625.
21. Jentoft, S., Aastvit, A.H., Torjesen, P.A., Andersen, Ø. (2005) Effects of stress on growth, cortisol and glucose levels in non-domesticated Eurasian perch (*Perca fluviatilis*) and domesticated rainbow trout (*Oncorhynchus mykiss*). *Comparative Biochemistry and Physiology*, 141: 353–358.
22. Skjervold, P.O. (2002) *Live chilling and pre-rigor filleting of salmonids: technology affecting physiology and product quality*. Dr.Agric. thesis. Norwegian University of Life Science.
23. Pottinger, T.G., Moran, T.A., Morgan, J.A.W. (1994) Primary and secondary indices of stress in the progeny of rainbow trout (*Onchorhynchus mykiss*) selected for high and low responsiveness to stress. *Journal of Fish Biology*, 44: 149–163.
24. Pottinger, T.G., Carrick, T.R. (1999) Modification of plasma cortisol response to stress in rainbow trout by

selctive breeding. *General and Comparative Endocrinology*, 116: 122–132.
25. Gjedrem, T. (2005) *Selection and breeding programs in aquaculture*. Springer-Verlag.
26. Fjæra, S.O., Lekang, O.I. (1991) *Effekt av ulikt håvmateriale*. ITF-rapport nr. 19, Norwegian University of Life Science (in Norwegian, English summary).
27. Ravn Larsen, H.I. (1990) *Rørtransport av fisk*. Norwegian University of Life Science (in Norwegian).
28. Holm, M., Knutsson, S. (1977) Sorteringsforsøk med laksesmolt. *Norsk Fiskeoppdrett*, 5: 4–6 (in Norwegian).
29. Lekang, O.I., Fjæra, S.O. (1995) Effect of light conditions on voluntary fish transport. *Aquacultural Engineering*, 14: 101–106.
30. Lekang, O.I., Fjæra, S.O., Thommassen, J.M. (1996) Voluntary fish transport in land based fish farms. *Aquacultural Engineering*, 15: 13–25.
31. Hara, T.J. (1973) Olfactory responses to amino acids in Rainbow trout (*Salmo gairneri*). *Comparative Biochemistry and Physiology*, 44: 407–416.
32. Døving, K.B., Selset, R., Thomesen, G. (1980) Olfactory sensitivity to bile acids in salmonid fishes. *Acta Physiological Scandinavica*, 108: 23–31.
33. Mearns, K.J. (1986) Sensitivity of brown trout (*Salmo trutta* L.) and Atlantic salmon (*Salmo salar* L.) fry to amino acids at the start of exogenous feeding. *Aquaculture*, 55: 191–200.
34. Ness, G., Lekang, O.I., Mearns, C., Skjervold, P.O. (1991) Som man lokker på fisken får man svar. *Norsk Fiskeoppdrett*, 9: 24–25 (in Norwegian).
35. Claussen, O. (1992) *Kanalforhold i landbaserte anlegg*. Norwegian University of Life Science (in Norwegian).
36. Øiestad, V., Pedersen, T., Folkvord, A., Bjordal, Å., Kvenseth, P.G. (1986) Automatic feeding and harvesting of juvenile Atlantic cod (*Gadus morhua* L.) in a pond. *Modeling, Identification and Control*, 8: 39–46.
37. Ivanov, V. (1988) Tame trout: Soviet scientist develop new method. *Freshwater Catch*, 35: 12–13.
38. Lekang, O.I., Skjervold, P.O., Fjæra, S.O. (1991) Ny metode for interntransport av fisk. *Norsk Fiskeoppdrett*, 2: 30–31 (in Norwegian).
39. Ness, G. (1990) *Selvsortering av fisk ved hjelp av lokemidler*. Norwegian University of Life Science (in Norwegian).
40. Midling, K.Ø., Øiestad, V. (1987) *Fjordranching with conditioned cod*, ICES-CM F: 29. International Council for Exploration of the Sea.
41. Gessel, M.H., Farr, W.E., Long, C.V. (1985) Underwater separation of juvenile salmonids by size. *Marine Fish Review*, 3: 38–42.
42. Fjæra, S.O., Skogesal, G. (1993) Sub surface size-grading of fish. In: *Fish farming technology. Proceedings of the first international confernce on fish farming technology* (eds H. Reinertsen, L.A. Dahle, L. Jørgensen, K. Tvinnereim). A.A. Balkema.
43. Hovda, J. (1992) Effekt av økende tetthet på sortering av laks. Master thesis. Norwegian University of Life Science (in Norwegian).
44. Lekang, O.I., Skjervold, P.O. (1990) *Selvsorteringsmetoder for fisk*. ITF notat nr 28/90. Norwegian University of Life Science (in Norwegian).

18
Transport of Live Fish

18.1 Introduction

Because juvenile farms, on-growing farms and slaughterhouses can be located in different places, it is necessary to transport live fish and other aquatic organisms. Live fish may be transported as fry or juveniles to on-growing farms, and the adult fish may be transported to the slaughterhouse. There is also some transport of fry and juvenile fish associated with restocking in the wild. Transport of fish can be classified as external transport (normally known as transport) and internal transport of fish inside the farm area (see Chapter 17). The differences are in the distance and duration of the transport.

The equipment used for transporting fry/juvenile and adult fish is similar in design. The main difference concerns the size of the tank, which must be of sufficient volume, and fitted with large enough hatches and/or valves for filling and tapping out the fish.

All procedures will vary depending on the species to be transported. However, all transport will result in extra stress for the fish, possibly leading to death;[1,2] this will not necessarily occur during transport, but can do so over several days after transport. Good preparation before transport and good routines during transport and reception are therefore important. There may also be government regulations concerning the transport of live fish and other aquatic animals, based on animal welfare needs. Acceptable fish densities and requirements for adding new water, or water exchange, serve as examples. These regulations may also include requirements for design of the transport equipment, so this must be checked beforehand. Several methods are employed for transport of fish, and a survey of those used follows.

18.2 Preparation for transport

Since the duration of transport is normally quite long, several hours at least, it is important to prepare the fish beforehand. The fish must be in good conditions and should be starved before being transported to empty the stomach and digestive system and hence reduce the release of waste metabolic products that cause the quality of the transport water to deteriorate. Starving the fish will also reduce the metabolic rate and hence oxygen consumption and secretion of ammonia and carbon dioxide during transport. The length of the starvation period needed depends on the water temperature and the fish species, but is 24 hours or more.

Internal transport, size grading and other handling must be carried out in anticipation of transport; the fish must be fully recovered from the stress that these actions involve before being transported. Loading the fish into the transport container must be performed in a manner that affects the fish as little as possible and minimizes stress. After loading it is also important to expose the fish to as little stress as possible to keep recovery times short.

The water used in the transport equipment must have similar characteristics to the water quality that the fish are used to so that they are not exposed to a new stressor. Variations between the temperature of the farming water and that of the transport water must also be avoided to minimize stress to the fish.[3] Normally a reduced water temperature during fish transport is beneficial because the metabolic rate, and hence oxygen consumption and the release of metabolic products such as ammonia and carbon dioxide, will be reduced. Furthermore, the content of available oxygen in the water will increase. However, this may stress the fish unnecessarily, and some warm water species will also die if the temperature is too low. Whether it is appropriate for the transport water to be chilled by adding ice, for instance, must be checked with regard to the transported species.

Fish may be more vulnerable during some life stages than others; for instance, transport of salmon smolt during smoltification, when the scales are loose. Special care must be taken when transporting fish in such situations.

18.3 Land transport

Land transport with trucks or smaller vehicles is commonly used for live fish and also for other aquatic organisms. The method is especially suitable for fry, juvenile and small sized species (<1 kg), because the weight is limited. When moving large adult fish, the total cargo weight of the vehicle is a limiting factor. The same is the case for the size of the fish transport tanks.

18.3.1 Land vehicles

There are several types of land vehicle used for fish transport, including vans, wagons and trucks. Normally the vans and trucks have two axles, but larger trucks with several axles and trucks with a trailer or semi trailer may also be used. Calculations have shown that the unit transport cost decreases with increasing size of truck: in going from a two to three axle truck, and a tank volume of $4\,m^3$ to one of $8\,m^3$, the trucking costs will increase by 30% while the transport volume is doubled.[4] A good journey on the truck is an advantage for both the fish and the equipment.

Fish transport trucks can be completely specialized. Combined trucks that are also used for other purposes, such as general cargo, may also be used; the advantage with the second option is that it might be difficult to keep the truck fully booked all year round with fish transport alone.

18.3.2 The tank

Two main types of tank are used for fish transport (Fig. 18.1): either a separate removable tank is placed on a platform on the truck, or the tank can be attached directly to the truck chassis. Removable tanks are normally prefabricated in sizes ranging from 200 to 4000 l capacity. Fibreglass is commonly used for tanks, but stainless steel and aluminium may also be used. If the truck is to be used for other purposes the fish tanks can easily be removed. If the tank is specially designed to be fixed directly to the truck chassis more time will be needed to remove it before the truck can be used for other purposes. Fixed tanks are normally made of stainless steel or aluminium; inside they are divided into separate compartments. The principle is the same as that used for tanks for transporting other liquids such as milk or fuel. These tanks are also divided into separate compartments inside, even if this is not apparent from the outside.

Figure 18.1 Trucks for fish transport either have removable tanks placed on a platform body or larger tanks fixed directly to the chassis frame of the truck (lower photograph).

The fish are loaded through hatches on the top of the tanks and unloaded through special tapping valves in the bottom. It is important that there are smooth connections so the fish are not wounded. To avoid temperature variations during transport tanks may be constructed with a double wall with insulation in between. However, if insulated cabinet trucks are used for transport of single tanks, this is not necessary.

Regular cleaning and disinfection of the transport equipment is necessary; this requires the tank to be designed in a way that makes it easy to clean, with smooth surfaces for instance. 'Dead areas' where cleaning is difficult must be avoided. The inner tank surface area must tolerate standard disinfection liquids. The material must be of a closed structure to avoid water being forced into it so that later disinfection is difficult.

18.3.3 Supply of oxygen

The fish consume oxygen during transportation and additional oxygen must be supplied if the transport lasts for some time. The amount of oxygen needed can easily be calculated by looking at the available oxygen in relation to the consumption by the fish. Oxygen is usually supplied from bottles attached to the truck and is added to the water in the fish tanks through diffusers which lie on the tank bottom. Air may also be used as an oxygen source for the fish; in this case the truck must be equipped with an air blower. If transporting fish at high density and only adding air through diffusers in the bottom of the tank, large numbers of bubbles in the water can result, which is not recommended because it will stress the fish. When adding air, supersaturation with nitrogen must be avoided (see Chapter 8). It is important to get a good distribution of the added air or oxygen throughout the tank;[5] an airlift pump may be used to create a flow inside the tank and hence an improved distribution.

When fish are transported for long periods of time (>12 h), depending on the density, problems may result from accumulation of carbon dioxide in the tanks. Airlift pumps may be used to aerate the water and will also remove some of the excess carbon dioxide. They will also cause the water to circulate inside the tank (Fig. 18.2). When using such a pump, the air is supplied via a perpendicular closed pipe which runs from a few centimetres above the tank bottom to a few centimetres above the water surface. The addition of air to the water inside the pipe creates bubbles which will drag the water towards the surface and in this way generate a water flow inside the tank. The air blower is fixed to the frame of the truck. To get proper water circulation, an airlift pump can be placed near the centre of the tank; however, it must not inhibit filling of the tanks with fish through the hatches on the top.

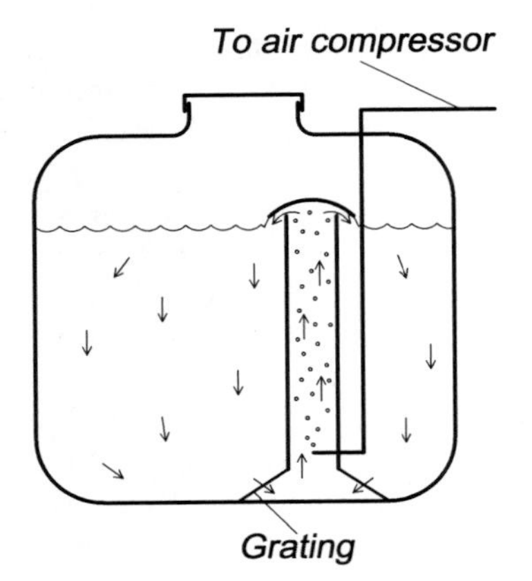

Figure 18.2 An airlift pump may be used to add oxygen and at the same time remove excess carbon dioxide. The photograph shows the installation in a transport tank.

Near the bottom of the tank the airlift pump will create an area of low pressure and the fish may get stuck to the inlet of the perpendicular pipe. To avoid this, a wide perforated screen should be placed around the inlet to reduce the water velocity and possibilities for the fish to get stuck.

When using the airlift pump it is important to be aware that when carbon dioxide is removed from the water there will be an increase in pH and hence an increase in the concentration of ammonia relative to ammonium ion. Ammonia is rather more toxic to the fish than ionic ammonium and could cause problems in high concentrations. The reason for this is that the equilibrium $NH_3 \leftrightharpoons NH_4^+$ is pH dependent (see Chapter 9).

18.3.4 Changing the water

In addition to high concentrations of carbon dioxide it is possible to get high concentrations of ammonia (total ammonia nitrogen, TAN) depending on fish density, water temperature, transported species and duration of transport. To avoid fish mortalities it is therefore necessary to have the following:

- Proper water exchange routines
- A system for cleaning and re-use of the transport water or dosing with additives.

Few trucks are equipped with cleaning and water re-use systems, and few operators use additives (see below), so water exchange is therefore the most common method. Because of this fish transport must be properly planned, taking into consideration the duration of the trip, the oxygen requirements of the fish and the necessity of water exchange. When exchanging water, the incoming water must be of satisfactory quality, so that a stable water environment is maintained for the fish. To avoid additional stress it must have a similar quality to that of the production water. When exchanging water, mixed zone problems must be taken into consideration,[6] In addition the water must not contain any micro-organisms which are harmful to the transported fish. The outlet water must be treated and released in a secure place, because it might contain micro-organisms that could harm the natural fish stocks in the area. It is therefore important to know the government regulations regarding intake and release of transport water.

The length of time during which fish can be transported without water exchange depends on water temperature, fish density and tolerance of the fish species to the decreasing water quality. The water quality gradually deteriorates as a result of fish metabolism until it is toxic to the fish. Normal water exchange intervals are 10–16h, but it is possible to have longer intervals between water exchanges: low water temperature will reduce the metabolic rate and therefore the fish may be transported for a longer period without water exchange.

18.3.5 Density

The density of the fish during transport is of importance for transport economy. Hence the amount of water transported must be as low as possible in relation to the amount of fish. An accurate limit of the correct density for transport is difficult to calculate and may vary greatly among species, size and development stage. For instance, Tilapia can be transported at higher densities than rainbow trout. If the water temperature increases, the density must normally be reduced. For salmon, which is a species that has very high requirements regarding water quality, a rough estimate says that for temperatures below 8°C a density of up to 90–100kg/m^3 may be supported;[7,8] if the temperature is above 8°C, a density of up to 50kg/m^3 is recommended. If the temperature is very high, transport is difficult, because both the oxygen requirements of the fish will increase and the oxygen content in the water will decrease. Another way to express density is as the percentage of fish in the total transport water volume. At a water temperature of 18°C, densities for catfish of up to 38%, *Tilapia* 32%, carp 15% and drum 4% have been reported for transports of duration 8–16h for fish of various sizes.[9]

Some insurance companies have their own guidelines regarding density of fish under transport. These guidelines must be known before transportation, if the fish are to be insured. There may also be government regulations governing fish density.

18.3.6 Instrumentation and stopping procedures

The temperature and oxygen level should be monitored during fish transportation to control and avoid critical situations. Therefore the truck must possess the necessary instruments such as a thermometer and an oxygen meter. These may be simple portable devices or fixed so that they can be read straight from the driver's cab. An alarm system may also be installed to detect and respond to variations in these parameters.

When transporting fish one should not, however, depend entirely on the instruments installed in the tank. Visual control is also necessary. Programmed stops are recommended for observation of fish behaviour. The frequency of these stops may vary: a stop quite early in the trip is recommended, for instance 15 min after the start, and subsequently each hour or every two hours thereafter.

It is necessary to keep a logbook during the journey in which parameter values and observations are registered. It is also necessary to record fish behaviour and any special occurrences during transportation. If something happens to the fish during or directly after transportation, the logbook could be used to find a possible explanation and to claim on the insurance policy. The economic value of the transported fish can be quite high: if transporting 100 g salmon smolt, each valued at 1 €, in a $10\,m^3$ tank with a fish density of $50\,kg/m^3$, the total value of the cargo will be 5000 €.

18.4 Sea transport

Large quantities of fish can be effectively transported by sea. This may be carried out by special boats, well boats or by hauling the sea cages or other floating installations with boats. Compared to trucks on shore, the advantage with using well boats is that the well is much larger than the truck tanks so the amount of transported fish can be increased and therefore the costs per fish transported reduced. Larger tanks are also better for larger fish, for instance for transporting fish to the slaughterhouse because the available volume for the fish to swim in is greater.

The fish can also be transported by hauling the net cage or another closed construction. However, the recommended drag velocity is quite low; this is of course also species dependent, but is never recommended to be above 1 or 2 knots. Such velocities will also impose large forces on the dragged constructions.

18.4.1 Well boats

Boats used for live fish transport are equipped with wells (internal tanks) for holding the fish (Fig. 18.3). Well boats have for many decades been used for transporting various species, such as herring and pollack, from traditional fisheries. The same boats may today be used for transporting farmed fish. Special systems for loading and unloading the fish may be installed. The normal size of well boats varies from 20 to 60 m and they may accommodate up to 150 t of live fish.[10] Traditional V-shaped hulls are usually used, but catamaran types that are faster

Figure 18.3 Well boat for live fish transport.

have also been tried. The wheel house is placed either in the bow or the stern. On new boats it is typically placed in the bow.

18.4.2 The well

In the well boat there is a constant flow of new water through the well to ensure enough oxygen is supplied to the fish and the waste products are removed. Special well valves are placed in the front and in the back of the well (Fig. 18.4). If the boat is moving these valves are opened and new water will flow through the well from front to back. The well valves are operated directly from the bridge. If the well boat is in dock, large circulation pumps ensure the renewal of water in the well and by this the supply of necessary oxygen to the fish and removal of waste products.

When the well boat is transporting fish it may be necessary to close the well valves in some areas (special zones) and continue with closed valves. Reasons for this are to avoid possible transfer of disease either from surroundings to the fish in the well, or from the fish in the well to the surroundings. Special zones can be areas near fish farms or important natural fish stocks. On these occasions it is necessary to ensure that the fish get enough oxygen even if the circulation pumps are inoperative. An oxygen supply system or air blower is therefore necessary, together with the necessary distribution system, for example diffusers located in the bottom of the well. Most modern well boats are equipped with systems for adding oxygen and may carry the fish for a long period of time without opening the well valves. The wells function similarly to those used on truck transport. Similarly the tanks can be equipped with systems for individual circulation of the water and also water purification (re-use) systems. The well boat may also be equipped with a refrigerated seawater (RSW) system for cooling the seawater, to keep a low temperature in the well and also to chill the live fish before slaughter.

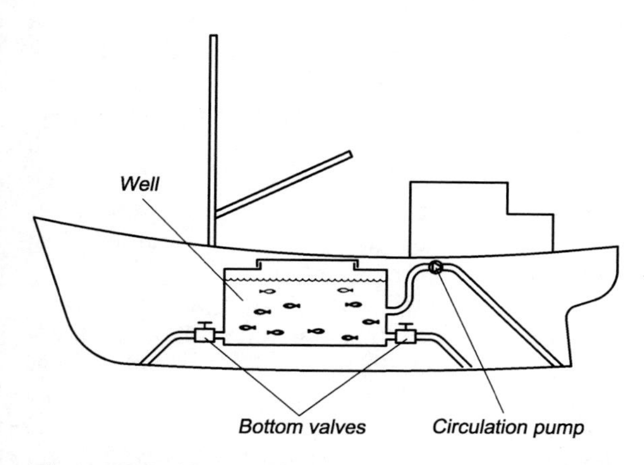

Figure 18.4 Valves that are open in the front and back of the wells ensure water flow through the well during transport.

Well boat size commonly varies from 50 to 1000 m^3. Inside the well is normally divided into several tanks separated by fixed walls. The construction and surface of the sidewalls and bottom must be designed not to damage the fish. Use of nets for dividing the well into additional compartments is therefore not satisfactory. It is also important that the well is designed in a way that ensures good water distribution and exchange of the total water volume. Well boats with circular tanks have also been constructed. The well will then function in the same way as a circular tank with the same flow pattern, and a very good distribution of the incoming water is ensured.

As for tanks on trucks, it is important that the wells are easy to clean and disinfect. It is also important to have control of the real well volume when adding disinfectant.[11]

18.4.3 Density

As for truck transport, the transport density will vary with the species, but here it is also important from an economic perspective. For Atlantic salmon, the density is normally around 35–50 kg/m^3, and for adult fish to slaughterhouse densities of between 150 and 180 kg/m^3 have been reported.[12] When transporting large amounts of fish to the processing plant, the cargoes are very valuable. A well capacity of 1000 m^3 and a fish density of 150 kg/m^3 mean that 150 t of fish is being transported. If the price is 3 €/kg the total value of the cargo is 450 000 €. This also shows the necessity of control to avoid accidents.

18.4.4 Instrumentation

Instrumentation to control water temperature, oxygen content and salinity, when transporting in seawater, is essential in well boats. In all boat transport it is important to have proper records and to keep a logbook. This can either be done manually or in a computer log with automatic data-logging of the parameters.

When transporting fish by well boat, special attention must be given to variations in temperature and salinity in the sea because these changes may be very stressful for the fish. Large variations in temperature may occur when there are large supplies of freshwater, for example close to the outlet of large rivers. Variation in water quality during transport is not good; for instance, there may be reduced salinity just outside large rivers and also large organic burdens so that a mixing zone with negative water quality may occur.[6,13]

Boat transport in rough weather must also be avoided. The fish in the well boat may suffer from some kind of seasickness. Some species will attempt to compensate for the wave pressure and try to stay at the same depth in the well by taking air in and out of the swim bladder. This may exhaust the fish totally and they will sink to the bottom of the well where they will lie, sliding about on the bottom of the well so incurring wounds and losing scales. Therefore it is important to listen to the weather forecast before transport, and record the weather conditions and approximate wave height.

All well boats have some type of equipment for loading and unloading the fish. Some sort of fish pump, for instance an ejector or vacuum–pressure pump, is normally used today, but a wet net may also be employed (Fig. 18.5). There have also been some specially designed boats that allow the fish to move in and out of the well voluntarily. This is possible by lifting the stern so that the well is opened. Sliding walls can be used to crowd the fish without pumping the water out from the well and reducing the water level. It is thus possible to reduce the volume in the well by moving one wall.

Many well boats are also equipped with grading equipment so that they can be used to grade fish in sea cages in addition to transporting fish.

Figure 18.5 Detail of the deck on a well boat with the well hatch and wet net for lifting the fish.

18.5 Air transport

Live fish may also be transported by helicopter or by aircraft. Because of the high cost this option is seldom used. Reasons for using air transport are mainly remote geographical conditions or bad infrastructure (roads, bridges), that make sea or land transport difficult and expensive. For restocking of lakes transport of live fish with small seaplanes is quite normal, but because of the low loading capacity (some hundreds of kilograms) this is not very useful in commercial aquaculture. Larger multi-engine aircraft can be used; in such cases seaplanes are equipped with large internal tanks for storing the fish, and pure oxygen gas is used. However, it is fry or small fish that are most often transported by air.

The use of helicopters is quite a new method for live fish transport. In one method, a large barrel of 400–500 l capacity filled with fish and water hangs on a wire below the helicopter.[14,15] This is basically the same equipment as when using helicopters to fight fire. To fill the barrel, the helicopter lowers it and the fish and water are lifted in; to empty the barrel, the helicopter lowers it into the water and the fish are released by tilting the barrel. The helicopter is maintained in flight throughout the entire loading and unloading procedure. This method may be a solution for fry/juvenile transport in areas with low quality roads and where the geographical conditions made long distance sea transport necessary.

18.6 Other transport methods

Plastic bags may be used to transport small amounts of fish, for instance to transport fish for lake cultivation and for transport of ornamental fish. A common size of transport bags is 85 cm × 60 cm. Thick polyethylene (PE) is used to make the bags resistant to puncturing. The bag is normally filled with between one-quarter and one-third of its volume with water before the fish are loaded. Air is then removed from the bag by pressing it together, which is then filled with pure oxygen until it assumes a balloon like form. The atmosphere over the water in the bag will then be pure oxygen. Cooling of the plastic bag minimizes the oxygen consumption of the fish and maximizes the duration of oxygen supply. Using ice on top of the bag, and not exposing the bag to direct sunlight may help with this. During transportation it is recommended that the bag is shaken from time to time to mix the oxygen that is in the atmosphere above into the water. The duration of the journey using this method depends directly on the fish density and their oxygen consumption. For example, a density of 1 kg salmon smolt per 3–4 l of water and 10 l of pure oxygen in plastic bags at 5°C has been used for air transportation for a couple of hours.[16]

An alternative to plastic bags is plastic containers. Here also the containers are filled with approximately one-third of water and fish, above which there is a pure oxygen atmosphere. Special equipment is required to remove the air from the can and replace it with pure oxygen.

Special types of tractor-trailers designed for fish transport can be used for short journeys (Fig. 18.6). Goods wagons equipped with tanks for live fish transport have also been tried for railway transport.[17]

18.7 Cleaning and re-use of water

In some situations it can be necessary to transport live fish over large distances and for longer periods of time (12–16 h). In these circumstances there must be either a water exchange or continuous flow-through of water as in well boats, depending of the species being transported. It is, however, also possible to design a transport tank system that includes a circuit for cleaning and re-using the water (Fig. 18.7) so that high concentrations of carbon dioxide and ammonia in the water can be avoided. Such transport tanks really function as a recycling plant with 100% re-use of water. Complete removal of metabolic waste products is impossible from an economic view, so even here the water quality will gradually decrease, but over a much longer time period. Compared to a re-use plant, the degeneration of the water quality will be slower because fish respiration is reduced due to prior starvation.

Figure 18.6 Using a tractor trailer for fish transport.

Airlift pumps may be used for removal of carbon dioxide in addition to adding oxygen and creating water flow. Biofilters are not efficient systems for removal of TAN during transport because these are biodynamic systems that require a start-up time to establish a culture of bacteria before use. However, an ion exchanger can be used to remove ammonia. When the water passes through the ion exchanger the ammonia is removed (NH_4^+ exchanged with Na^+); after use the ion exchanger must be regenerated, and this can be done between journeys.

Long journeys and high degrees of water circulation may also cause the water temperature to increase as a result of the pumping required for water re-use and fish metabolism. Installation of a cooling system on the transport tank can therefore be advantageous. The oxygen supplied to the water

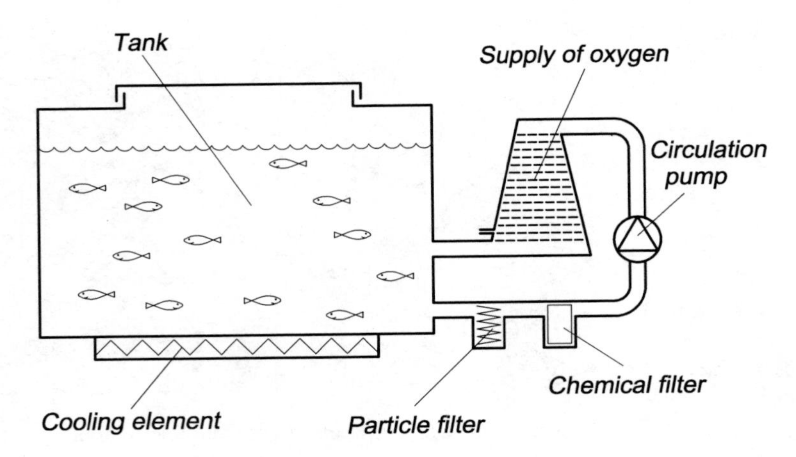

Figure 18.7 It is possible to design a transport tank system for long journeys that includes a circuit for cleaning and re-using the water.

may be either produced by generators or by bottles of oxygen gas attached to the truck.

In Norway salmon smolt and juvenile turbot have been transported for up to 5 days without water exchange using a specially designed tank with a water re-use system.[18]

18.8 Use of additives

Many additives are available that can be added to the transport water to improve the transport results and allow fish transport over longer periods.

Instead of using filters and water re-use systems, additives can be applied to the water to give similar effects. Addition of seawater to freshwater will have a positive pH regulating effect because of its buffering capacity. The same will be the case with addition of other buffering substances. Sodium chloride can be added to freshwater as a defoaming agent. Addition of clinoptilolite to the water (as used in an ion exchanger) will reduce the concentration of ammonia.

Addition of salt may reduce osmoregulatory disturbance, and by this the total stress when transporting freshwater fish.[19,20] Antibiotic can be added to reduce the development of bacteria in the transported water.[21]

Addition of sedatives, such as MS222 and clove oil, to the transport water may also be done to calm the fish down and reduce the metabolic rate.[22–26] This will reduce oxygen consumption and decrease the excess of carbon dioxide and ammonium ion; the fish will also tolerate more stress.

All use of additives will, however, require accurate control of the water quality during transport; they may not always function or the effect may be minor. It is best therefore normally to avoid using additives. Adding of antibiotics and sedatives ought to be avoided unless there are very special conditions pertaining to the live transport. These may not be allowed before the fish go to processing plants because they may leave residues in the flesh.

References

1. Barton, B.A., Haukenes, A.H., Parsons, B.G., Reed, J.R. (2003) Plasma cortisol and chloride stress response in juvenile walleyes during capture, transport, and stocking procedures. *North American Journal of Aquaculture*, 65: 210–219.
2. Iversen, M., Finstad, B., McKinley, R.S., Eliassen, R.A., Carlsen, C.T., Evjen, T. (2005) Stress responses in Atlantic salmon (*Salmo salar* L.) during commercial well boat transport, and effects on survival after transfer to sea. *Aquaculture*, 243: 373–382.
3. Strange, R.J., Schreck, C.B., Golden, J.T. (1977) Corticoid stress responses to handling and temperature in salmonids. *Transactions of the American Fisheries Society*, 106: 213–218.
4. Wahlberg, B. (1977) *Aktuelle fisktransportsfrågor*. Rapport Vattenfall (in Swedish).
5. Børjesson, H. (1987) Synspunkter for transportutrustning for fisk. In: *Fisktransport handbok*. Mittnordekommitten før vattenbruk (in Swedish).
6. Krogelund, F., Teien, H-C., Rosseland, B.O., Salbu, B. (2001) Time and pH-dependent detoxification of aluminium in mixing zones between acid and non-acid rivers. *Water, Air and Soil Pollution*, 130: 905–910.

7. Mittnordenkomitten før vattenbruk (1987) *Fisketransporthandboken*. Mittnordenkomitten før vattenbruk (in Swedish).
8. Pursher, J., Forteath, N. (2003) Salmonids. In: *Aquaculture, farming aquatic animals and plants* (eds J.S. Lucas, P.C. Southgate). Fishing News Books, Blackwell Publishing.
9. Johnsen, S.K. (2000) Live transport. In: *Encyclopedia of aquaculture* (ed. R.R. Sickney). John Wiley & Sons.
10. Willougby. S. (1999) *Manual of salmonid farming*. Fishing News Books, Blackwell Publishing.
11. Johnsen, S., Simolin, P. (2002) *Volumbergning av vannmengde i brønn og rørsystem i brønnbåt ved bruk av sporstoff.* VESO rapport (in Norwegian).
12. NIVA (2003) *Transportkvalitet av fisk I brønnbåt.* Prosjektfakta, NIVA (in Norwegian).
13. Rosseland, B.O., Blakar, I.A., Bulger, A., Kroglund, F., Kvellestad, A., Lydersen, E., Oughton, D.H., Salbu, B., Staurnes, M., Vogt, R. (1992) The mixing zones between limed and acidic river waters: complex aluminium chemistry and extreme toxicity for salmonids. *Environmental Pollution*, 78: 3–8.
14. Sheperd, B.G., Bérézay, G.F. (1987) *Fish transport techniques in common use at salmonid enhanchement facilities in British Columbia.* Canadian manuscript report of fisheries and aquatic sciences 1946. Canadian Department of Fisheries and Oceans.
15. Anon. (1988) Up, up and away. Air-lifting smolts is a job for helicopter specialists. *Fish Farmer*, 6: 37.
16. Gjedrem, T. (1986) *Fiskeoppdrett med framtid.* Landbruksforlaget (in Norwegian).
17. Berka, R. (1986) *The transport of live fish: a review.* EIFAC Technical Report 48, FAO.
18. Lekang, O.I. (1992) Avansert levendefisktransport. *Nordisk Akvakultur*, 6: 32–33 (in Norwegian).
19. Carneiro, P.C.F., Urbinati, E.C. (2001) Salt as a stress response mitigator of matrinxã; *Brucon cephalus* (Günter) during transport. *Aquaculture Research*, 32: 297–304.
20. Johnson, D.L., Metcalf, M.T. (1982) Causes and controls of freshwater drum mortality during transport. *Transactions of the American Fisheries Society*, 111: 58–62.
21. Amed, D.F., Croy, T.R., Beverly, A.G., Johnson, K.A., McCarthy, D.H. (1982) Transportation of fish in closed systems. *Transactions of the American Fisheries Society*, 111: 603–611.
22. Mishra, B.K., Kumar, D., Mishra, R. (1983) Observations on the use of carbon acid anaesthesia in fish fry transport. *Aquaculture*, 33: 405–408.
23. Prinsloo, J.F., Schoonebee, H.J. (1985) Note on procedures for the large scale transportation of juvenile fish. *Water SA*, 11: 215–218.
24. Stickney, R.R. (1994) *Principles of aquaculture*. John Wiley & Sons.
25. Swanson, C., Mager, R.C., Doroshow, S.I., Cech, Jr., J.J. (1996) Use of salt, anaesthetics and polymers to minimize handling and transport mortality in delta smelt. *Transactions of the American Fisheries Society*, 125: 326–329.
26. Cooke, S.J., Suski, C.D., Ostrand, K.G., Tufts, B.L., Wahl, D.H. (2004) Behaviour and physiological assessment of low concentration of clove oil anaesthetic for handling and transport of largemouth bass (*Micropterus salmoides*). *Aquaculture*, 239: 509–529.

19
Instrumentation and Monitoring

19.1 Introduction

Equipment for measuring and recording of various parameters is more and more commonly used in aquaculture, especially in intensive aquaculture. Such equipment controls and adjusts the environmental conditions to obtain optimal production results. Until now several of the measurements have been taken manually, which is normally more time consuming and labour intensive, and therefore results in fewer measurements. During the past few years, there has been rapid development in the automation of instruments and monitoring systems that can also be used in the aquaculture industry, mainly based on developments in electronics and computer science. Therefore many of the trivial manual measurements are now carried out by specially designed instruments, releasing manpower for more important intellectual tasks and to improve the production results, especially in intensive aquaculture.

One reason for using instruments is to automate the management of fish farming as much as possible. For example, video cameras and image analysis can be used to monitor fish and give alarm signals if odd behaviour is observed. The biological processes underlying fish production are, however, both complex and difficult, unlike the production of nails. Even with today's knowledge, it is only a dream to believe that it is possible to fully replace the fish farmer with instruments and robots.

When buying and installing instruments, the requirements for maintenance and calibration, adjusted for special circumstances must be taken into account. The values read from the measuring equipment must be reliable; otherwise the result can be more damaging than if no measuring equipment were used at all. This implies that maintenance and running costs must be included in the price of an instrument, not just the purchase cost. Extra effort must be given to maintenance of instruments used to monitor water quality. This also includes frequent calibration according to the manufacturer's instructions so that the values shown are reliable. Depending on the type of instrument, the sensors may have a limited duration, so must be exchanged at fixed intervals.

Measurement of biological performance has also increased during the past few years as a result of the increased focus on profitability in intensive fish farming. By automatically measuring development in terms of weight and total fish biomass, it is possible to control the development and intervene if something does not correspond to the production plans.

Due to the large expense involved and the amount of technical equipment that can fail, it is increasingly common to have a total monitoring system on the farm, which also includes a significant use of computer tools.[1–4] On land-based farms using pumps for the water supply or in farms with re-use of water such systems are essential.[5]

There is much general literature available describing measurements, instruments and sensors (see, for example, refs 6–8). This chapter gives a general description of the construction of instruments, with some thorough investigation of those used in aquaculture. In addition there is a brief review of methods used to count fish, measure fish size and estimate biomass. The chapter ends with a description of monitoring systems of which measuring instruments constitute a major part.

19.2 Construction of measuring instruments

The construction of measuring instruments depends on the measuring principles used and the signal transfer. One classification is mechanical, hydraulic, pneumatic, electrical and electronic, where the last is being increasingly used.[9]

A measuring instrument often comprises three major parts (Fig. 19.1):

(1) A sensor or probe
(2) A transmitter to transfer the signal
(3) A display or another type of indicator (connected to the transmitter).

In some instruments the three major parts are connected within the same unit, while in other instruments the parts are separate and connected via cables for electric signals or another principle for transfer of the measured values. Measuring equipment can either give continuous signals (analogues), or on/off signals (digital). An example of the first case is an oxygen meter that shows the concentration of oxygen. Flow indicators that register if there is water flow or not (on/off) are an example of the latter case.

The sensor in the unit is used to record the physical conditions in the medium, such as the probe in an oxygen meter. The transmitter can either be electrical or mechanical and translates the signal coming from the sensor to a scaled signal that as is further transported to the display or indicator where the results are shown in an understandable way. In the display the physical conditions of the medium are shown. Signals may also go directly to a recording unit such as a computer for storing the results, or can be used to control a regulator.

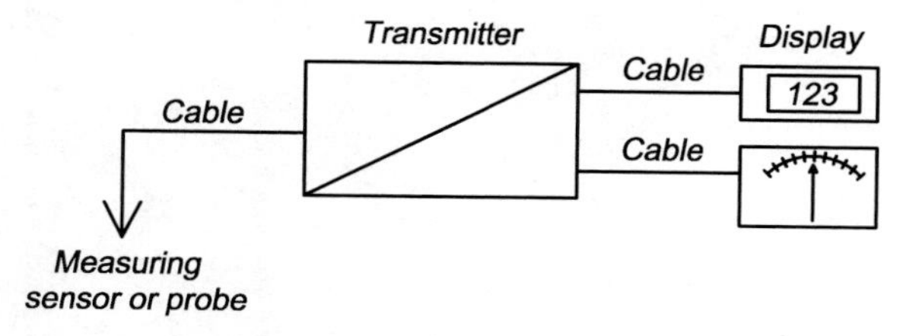

Figure 19.1 An electronic instrument normally comprises three major components either linked by cables or integrated.

A short description of the measuring instruments most used in aquaculture facilities is given below. Equipment is separated into that used for measuring water quality and that used to measure physical conditions.

19.3 Instruments for measuring water quality

Both chemical and physical parameters are used to characterize water quality, and many instruments based on different principles are used. Instruments for measuring water quality can be divided into on-line and off-line instruments based on their construction. On-line instruments that can stay in the water carry out continuous monitoring and are always a type of electrode or sensor. The same instrument may, however, also be used off-line for individual measurements. Off-line instruments are more closely related to those in laboratories, and normally more work is needed to perform an analysis, together with experience in performing laboratory work in some cases.

On-line instruments or chemical analysers can be separated into those that use sensors but no reagents, and wet-chemistry analysers.[10] Use of instruments with a sensor is advised, because they reduce the manual input. Wet-chemistry analysis can be classified on the principle of the analyses: typical examples are colorimetric, titrimetric and ion-selective electrodes. Colorimetric measurements are much used, and function by adding a reagent to the water sample and monitoring the colour change with a detector. A typical detector is a spectrophotometer that measures the light absorption of the sample. The great disadvantage with wet-chemistry analysers is that a sample from the water flow needs to be taken to which a reagent is added. Stopped flow/batch mode analysers and flow injection analysers are available.[10] In the latter pumps and valves ensure that samples are automatically taken from the water flow and transported to the instrument for monitoring before being released so that a new sample can be taken. Some common devices for water analyses are described briefly below; more information can be obtained from the literature, for example, refs 2 and 9.

19.3.1 Measuring temperature

In fish farming, temperature needs to be measured in several situations, for instance, the farming water or inside slaughtered fish. One principle utilized to measure temperature is the expansion of certain substances with temperature as in a mercury thermometer. The construction of the thermometer is such that even quite a small expansion of the mercury causes a noticeable change in the level in the thin pipe where the reading is taken. Thermometer manufactures calibrate the scale according to known values: a bath with freshwater and ice is 0°C at normal atmospheric pressure and is the boiling point of freshwater is 100°C under the same conditions. The Celsius scale divides the interval between 0 and 100 into equal degrees. Other substances besides mercury can be used in thermometers using this principle.

That the electrical resistance in materials changes with temperature may also be used to measuring temperature. What actually is measured is the electric current going through a circuit of which the resistance material is an integral part; if such material is exposed to temperature variation its resistance will change together with the electric current passing through the circuit. Platinum is often used in digital thermometers which measure electrical resistance.

The difference in voltage occurring between two different materials may also be used for measuring temperature. If two different metals are soldered together, for instance copper and constantan, a current will flow, the size of which will be proportional to the temperature. If the point where the two materials are soldered together is affected by a temperature gradient, the electrical current will also vary. This equipment is called a thermocouple; it is simple but not as accurate as other devices.

A thermistor is also a temperature dependant resistance, but is more complex and includes semiconductor technology. Higher accuracy is achieved with this principle than with the other methods described. The device is simple and the price reasonable; thermistors are therefore widely used for temperature measurement.

19.3.2 Measuring oxygen content of the water

The oxygen content may be measured either chemically or electronically. The normal chemical method is the so-called Winkler method which is a titration method, consisting of adding certain chemicals to the water and observing a colour change which will be directly related to the oxygen concentration.

When measuring oxygen concentration electronically, a sensor or probe is used. This can be constructed using a positively charged conductor (anode) and a negatively charged conductor (cathode) separated by an insulator (Fig. 19.2). In one design, an electrolyte is located around the electrodes (anode and cathode). A special membrane covers the electrolyte, keeping it in place and protecting it. The membrane is permeable to oxygen molecules. As there is a positive and a negative electrode, electrolysis will occur, and electrons will pass between the two electrodes. The magnitude of the electron transport is affected by the amount of oxygen in the electrolyte; this again

Figure 19.2 Measuring oxygen content: detail of the sensor for measuring oxygen content.

depends on the amount of oxygen transported through the membrane, which is related to the amount of dissolved oxygen in the water.

It is important to calibrate this type of instrument accurately. For some instruments it is necessary to calibrate for air pressure and temperature, while in others this is integrated. The electrolyte and the membrane or the complete sensor, must be changed at defined intervals. Since the membrane is very thin it is easy to break, care must be taken when handling it. Ageing of the membrane can be detected, because it becomes difficult to get stable results and the values fluctuate constantly. The membrane is also exposed to fouling, and must regularly be visually inspected and cleaned. If fouling occurs the oxygen values will normally drop in relation to the correct values.

19.3.3 Measuring pH

Several methods are used to measure pH to prevent it dropping to critical values. A simple but inaccurate way is to observe the colour change of pH paper, also known as litmus paper, when it is dipped into water. This paper contains a chemical, the colour of which is pH dependent. Similarly, a small water sample can be taken, a specially prepared liquid added to it and the colour change observed.

The pH can also be measured by a standard chemical analysis with titration and observing the change in colour, which is pH dependent. This system may be implemented for daily measurements on a fish farm and has the advantages of requiring low maintenance and being reliable. The disadvantages are, however, that this takes some time and experience of working with chemical equipment, and measurements in a laboratory are necessary.

A pH meter is constructed on the same principle as the oxygen meter. Here the probe is constructed with two electrodes, one for measuring the concentration of H^+ and the other as a reference electrode. In this case the pH electrode consists of a membrane permeable to hydrogen ions. Here also the membrane creates a cell that encloses the electrolyte. Between the electrodes a current will pass that depends on the concentration of H^+. The voltage between the pH electrode and the reference electrode is then measured, and the pH calculated.

Instruments for measuring pH must be calibrated before use. This is performed by placing the probe in different solutions of known pH and then adjusting the instrument. The pH probe is compact, in contrast to the O_2 probe; therefore it must be completely replaced if it ceases to function. A disadvantage of the pH probe is its limited duration, normally from 3 months up to 1 year in special cases. Practical experience with the use of pH meters in fish farming shows that maintenance is of extreme importance. Both calibration and changing of electrodes must be done regularly to achieve reliable measurements.

19.3.4 Measuring conductivity and salinity

Conductivity is a measure of the ability of water to conduct an electric current. In fish farms this is important in order to evaluate the ability of the water to inhibit pH fluctuations, i.e. the buffering capacity. In seawater, Na^+ and Cl^- ions dominate and here the instrument is used to measure the salinity.

The probe consists of two electrodes and is lowered into the water. A small electric potential (voltage) is applied across the electrodes. An electric current will occur between the electrodes, the size of which depends on the ion concentration in the water. To prevent the establishment of a layer on the electrodes which affects the current, it is necessary to use an alternating current as pre-voltage.

Conductivity is affected by temperature, so it is important to compensate for this parameter when taking measurements. Each instrument should have a special table setting out the effect of temperature on the conductivity. Advanced instruments incorporate automatic temperature compensation.

19.3.5 Measuring total gas pressure and nitrogen saturation

The total gas pressure in the water is measured mainly to find not only the total pressure, but also the amount and saturation of dissolved nitrogen gas (N_2). If the saturation of nitrogen in the water is above 100%, the fish may suffer from gas bubble disease. This is more critical in fry stage fish than in adult fish. In salmonids problems have been observed when saturation is over 102%, but it is recommended that saturation be maintained below

100.5%. Marine fish fry has been shown to be very sensitive for supersaturation of nitrogen. Problems may also occur if the total gas pressure is too high and there are some indications that above 100% total pressure may be detrimental.

One method to measure the total gas pressure in the water is to use a saturometer (saturation meter) (Fig. 19.3). The main part of the instrument is a small silicon tube into which the dissolved gases from the water pass (the silicon acts as a membrane) and become enclosed. A pressure meter is attached to the tube and the measurement is carried out by determining the difference in pressure between the local atmospheric pressure and the pressure inside the silicon tube.

Before using the saturometer, the pressure inside the cylinder must be equalized to the surrounding environmental pressure. This is carried out by placing the probe in the air for some minutes before placing it in the water. A perforated cover surrounds the cylinders and a manual pump is used to ensure water flow past the cylinder. The instrument has to stay in the water for 5–10 minutes before a reading can be taken.

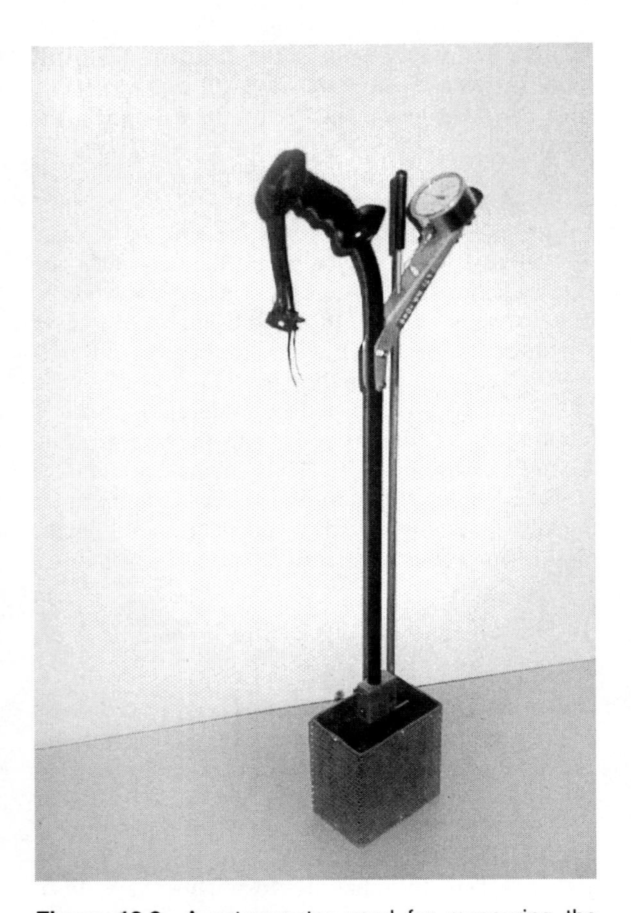

Figure 19.3 A saturometer used for measuring the total gas pressure in the water.

In order to determine the total gas pressure when using a saturometer the following equation must be employed:

$$\mathrm{TGP}(\%) = \left(\frac{\mathrm{BP} + \Delta P}{\mathrm{BP}}\right) \times 100$$

Where:

TGP = total gas pressure of dissolved gases
BP = local barometric pressure (normally read in mmHg (mercury))
ΔP = pressure difference (read from the saturometer) between total gas pressure in the water and local barometric pressure (mmHg).

When calculating the nitrogen pressure, which is a critical factor for avoiding gas bubble disease, the following equation may be used:

$$N_2\,(\%) = [\mathrm{BP} + \Delta P - ((O_2/\beta_{O_2}) \times 0.5318 - P_{H_2O})/((\mathrm{BP} - P_{H_2O}) \times 0.7902)] \times 100$$

Where:

N_2 = partial pressure of nitrogen gas in the water (percentage nitrogen saturation)
BP = local barometric pressure (mmHg)
ΔP = difference between total gas pressure and local barometric pressure, measured by a saturometer (mmHg)
O_2 = oxygen concentration in the water measured by an oxygen meter (mg/L)
β_{O_2} = Bunsen's coefficient for oxygen (see Appendix 8.2)
P_{H_2O} = partial pressure of the water vapour (mmHg)

The two numbers in the equation are conversion factors.

There are also other instruments available for monitoring the nitrogen gas saturation which are simpler to use.

19.3.6 Other

As a result of the rapid development of instruments for online measurements, sensors for online measuring of carbon dioxide, ammonia and nitrate are available. The sensors can either be single, or multi instruments comprising several sensors connected in a multiprobe (Fig. 19.4). The sensors in the multiprobe can be changed with sensors for other quality parameters, but the instrument remains the same.

Instruments using traditional chemical laboratory methods have also been developed for use under field conditions; for instance there are special types for aquaculture facilities. One commonly used instrument is based on a spectrophotometer, in which the sample is illuminated by light of a specific wavelength and the amount of light passing through is monitored. Before putting the sample in the spectrophotometer a chemical is added and so that a colour change will occur, the size of which will depend on the amount of substance to be measured in the sample. To avoid much work with weighing out of chemicals specially prepared ampoules, each with the correct amount of chemical for one water sample, are delivered with the instrument. Many chemical analyses can be run in such instruments which are quite simple to use.

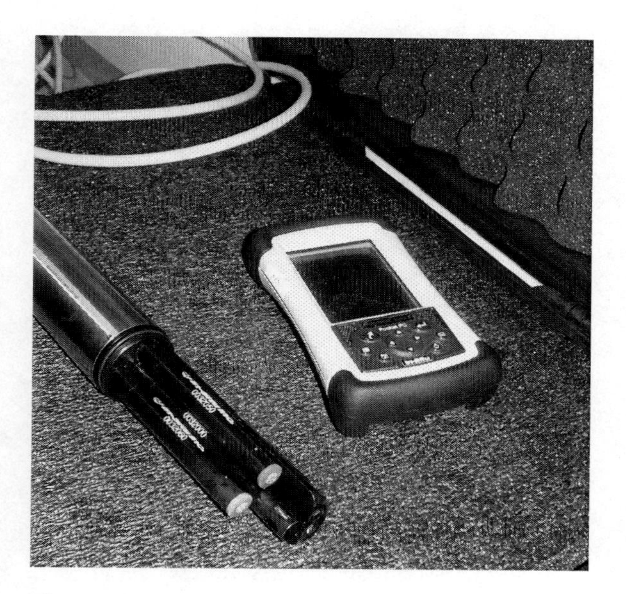

Figure 19.4 Sensors connected in a multiprobe.

19.4 Instruments for measuring physical conditions

There are many places were there is a need to control physical conditions. In aquaculture, the water condition is of major importance, particularly the following factors: water flow, water level and water pressure. Many methods can be used to measure these parameters; some can also be used to measure more than one. There are differences both in price and accuracy of available instruments. Methods used are reviewed below; for more information see, for instance, refs 9 and 11.

19.4.1 Measuring the water flow

It is a common practice to measure water flow to be sure that it is constant and that the correct amount of water passes through the pipes. Water flow measurements can be carried out at various places in a fish farm. Flow meters may be located in the main inlet pipe, in part flow pipes or in the inlet pipe to a single tank. Several principles are actually used, such water as velocity and head loss. The methods used to measured the flow in open channels and pipes are different. The main emphasis here is on measurement in pipes that are full of water, since this is the most common situation in fish farming. Some of the methods presented may, however, also be used in open channels.

Measuring water velocity

The water velocity is proportional to water flow (see Chapter 2). If the water velocity and the internal diameter of the pipe are known, it is easy to calculate the water flow.

A simple way to measure the water velocity in a pipe is to use a propeller, paddle wheel or turbine (Fig. 19.5). Propellers with a variety of designs are used. The working principle is that the water will move the propeller and the rotational velocity of the propeller will reflect the water velocity in the pipe. Either the propeller can be installed in an existing pipe system, known as an inset meter, or it can be a completely separate system in which the propeller and instruments are connected, forming a complete unit adapted to the pipe. Propeller systems are simple and inexpensive, but one of the disadvantages is that the head loss will increase.

Figure 19.5 Instruments for measuring water velocity: (A) propeller; (B) ultrasound; (C) electromagnetic.

However, the main disadvantage is that the propeller will be exposed to fouling because it is within the water flow. Normally, fouling will reduce the measured velocity compared with the actual velocity. Correct maintenance is therefore important in such a system. A further disadvantage is wear of the continuously moving propeller.

An electromagnetic flow meter can also be used to measure water velocity (Fig. 19.5). The principle utilized here is that an electromagnetic field changes when the water velocity varies. A conducting fluid, such as water, in a magnetic field will induce an electric voltage, which correlates with the water velocity. The water must have some electrical conductivity if this type of instrument is to function; in fish farming this will always be the case.

Ultrasound waves may also be used for measuring the water flow, as in an ultrasonic flow meter (Fig. 19.5). Two combined ultrasound source and receiver units are placed, one on each side of the pipe and crossed diagonally. Sound waves are sent from one side and travel with the water flow; from the opposite side of the pipe another sound wave is sent against direction of the water flow. The times taken for the sound waves to reach the receivers are compared; the water velocity can be calculated because the water flow deflects the sound waves and alters their velocity. A fixed unidirectional ultrasonic system can also be used which calculates the time difference during which the sound waves travel. The advantage of this system is that the instrument may be attached to the exterior of existing pipes; if the type of pipe is known the water flow inside can be found. Flow meters utilizing this principle will not give any head loss and are very accurate; however, they are quite expensive. The instruments may also be affected by air bubbles or high particle concentrations.

The various types of flow meter have different accuracies, and this must be checked before selection. Generally, greater accuracy incurs greater expenditure, so the meter must be fit for purpose.

Measuring head loss

Head loss in a pipe will occur when the water passes an obstruction in the pipe or any other object that reduce the water flow. Measurement of the head loss may therefore be used to calculate the water velocity, since the head loss depends on the velocity; an increase in velocity will increase the head loss. When this principle is used for flow measurements, a known obstruction is set inside the pipe. A very accurate plate with a hole in the centre, slightly smaller than the internal pipe diameter, known as an orifice plate, is usually used (Fig. 19.6). The difference in the pressure before and after the plate gives an indirect measurement of the water flow. This instrument is also called a differential pressure flow meter. The physical relation used when measuring the head loss is defined by the Bernoulli equation:

$$p + \frac{1}{2}\rho v^2 + \rho g h = \text{constant}$$

Where:

p = pressure
ρ = density
v = velocity
h = elevation (height)
g = acceleration due to gravity.

The disadvantage with this method is that a head loss occurs, especially with the orifice plate, so instead a venturi can be used which functions in the same way, but the head loss is reduced.

Another instrument based on measuring the differential pressure (actually the head loss) to calculate water flow is the pitot tube which measures the dynamic head (total head) and the static head (Fig. 19.6). Based on the difference between these two values, the water velocity can be calculated (from the Bernoulli equation). Correct use of this is important because there is a velocity profile inside the pipe that must be taken into consideration when locating the pitot tube.

A rotameter or hover velocity meter also utilizes the head loss measurement, but with a constant head loss. The construction of a rotameter includes a conical glass or plastic pipe, with a floating device inside (Fig. 19.6); this device is lifted when the water starts to flow. Since the pipe is conical the height by which the floating device is lifted will depend on the magnitude of the water flow, because the area where the waters is flowing is increased. A normal rotameter must be placed vertically to function. In a similar instrument the device is set on a spring with known characteristics, and this is compressed when the water flows by an amount that depends on the water flow. Vertical installation is not necessary for this flow meter.

19.4.2 Measuring water pressure

Water pressure is measured to control water levels in tanks or the pressure in pipes. If the pressure is too high or low a warning signal can be given.

Diaphragm manometers are often used to measure the water pressure. The principle is that pressure can alter the shape of a diaphragm (Fig. 19.7). An increase in pressure will change the form of the diaphragm and this change can be used to control an indicator. The manometer is fixed directly to the pipe where the measurements are taken.

A bourdon tube manometer may also be used for pressure measurements (Fig. 19.7). This system consists of a hollow bent pipe that is connected directly to the place where the pressure is measured. When the pressure increases, the bent pipe will try to straighten as the pressure is considerably higher on the outside of the bend than on the inside because

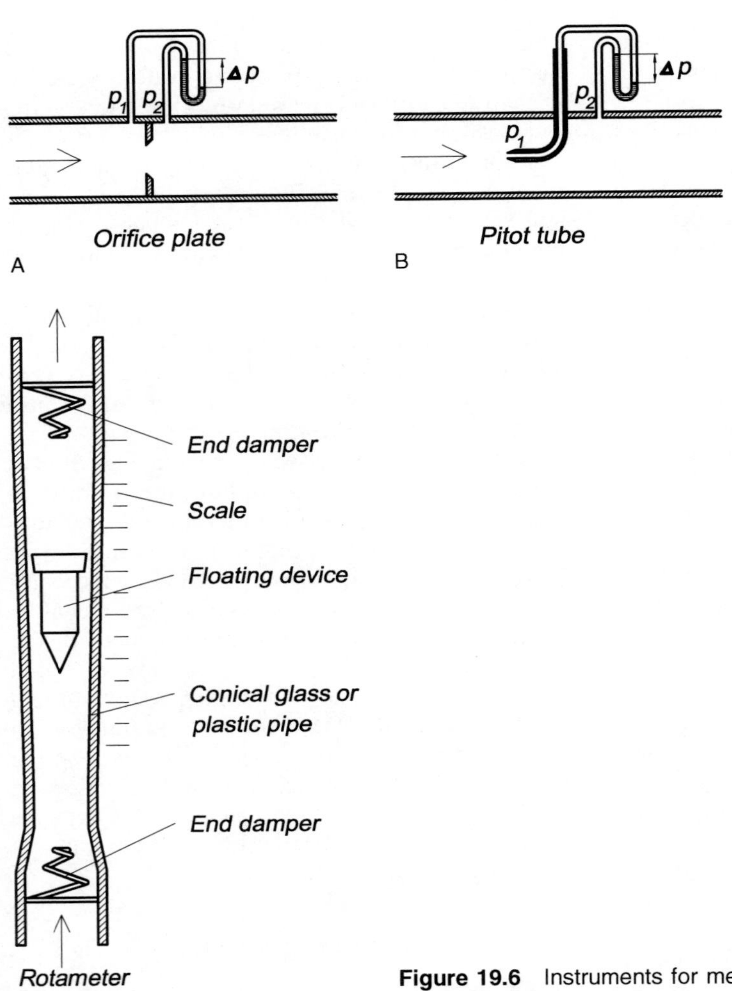

Figure 19.6 Instruments for measuring head loss: (A) orifice plate; (B) pitot tube; (C) rotameter.

the area where the force acts is larger; this movement will be registered on a scale calibrated to show the pressure. There are also other devices using the same principle, such as pressure capsules and bellows.

Today, however, the most commonly used instruments are electronic load cells or pressure transducers. They function because there is a small change in the electric resistance in an electric circuit related to the pressure. A pressure transducer may be installed on the bottom of a tank or inside a pipe.

19.4.3 Measuring water level

Measuring the water level is necessary at various places to avoid overflows and water shortages. This can be carried out in a head tank or directly in the individual fish tanks. Water level sensors are normally digital, sending a signal if the level of the water is below or above defined values. The system may also be analogue; this generates a signal that is proportional to the water level.

Different types of electronic floats may be used for controlling the water level. The float will lie on the surface and change position if the water level varies; this variation can be utilized to control the level. When this happens a signal will be sent and action taken, for instance switching the water supply on and off. A commonly used float is the level rocking sensor (Fig. 19.8) which floats on the surface; if the water level decreases it will gradually become more upright. The sensor is located in the

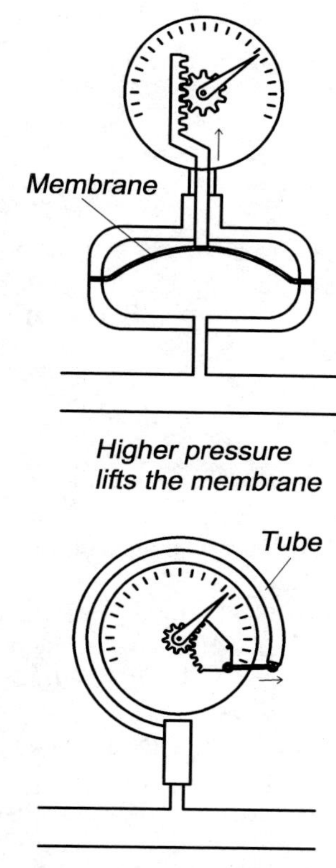

Figure 19.7 Membrane and bourdon manometers used to measure water pressure.

end of a cable and when it is in the floating position there may or may not be contact between the two conductors located inside the float. If the sensor is hanging vertically an automatic switch will take the opposite position (opened or closed as the case may be) to make or break the electric circuit. This can then be used to control the water level. On submerged pumps this is a common method for starting and stopping the pumping, depending on the water level.

Other methods are also used to measure the water level in fish farms, of which pressure sensors, as described above, are the most common.

Capacitance sensors are becoming popular for monitoring water levels (Fig. 19.9). These sensors detect how well a material keeps its electrical charge. Water may keep 80 times as much electrical charge compared to air and the sensor can therefore register changes in electrical charges associated with changes in water level. If the tank wall is thin, the sensor may be placed on the outside of the tank wall and will be capable of sensing whether the tank is filled with water or not. If the water level decreases to beneath the position of the sensor, an electrical circuit will be either opened or closed and a signal given.

Water level can be controlled very accurately by ultrasound devices A transmitter and receiver are placed above the water surface. By transmitting a sound wave and measuring the time of reflection with the receiver, it is possible to calculate the distance from the transmitter to the water surface and hence the water level. If the water level decreases, the time taken by the sound wave to travel between the transmitter and receiver will increase.

Figure 19.8 A floating sensor used for monitoring and controlling the water level.

19.5 Equipment for counting fish, measuring fish size and estimation of total biomass

19.5.1 Counting fish

To count fish manually on large aquaculture farms is very labour intensive; for example, a juvenile production plant delivering 1 million juveniles to an on-growing plant. Therefore it is beneficial to do this automatically. The problem when counting fish automatically is to separate the individuals and to avoid two or more fish coming together and

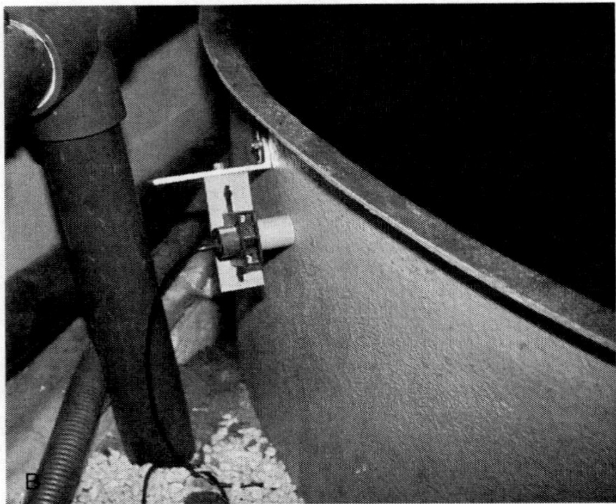

Figure 19.9 Capacitance sensors used for measuring water level (A) in a pipe and (B) in a tank.

being counted as one. One challenge when developing equipment for counting fish is therefore to find a way to separate the fish on an individual basis.

Counting can be done when the fish are in or out of the water. Self-counting of the fish when they are put into water has been very difficult in practice because the fish are reluctant to pass any obstacle voluntarily (see Chapter 17). Some of the equipment used for counting fish can also be used for measuring the size of the fish (see section 19.5.2).

A suitable place to count the fish is in connection with internal transport or grading. Here the fish are normally taken out of the water and it is quite easy to count them and carry out the other operations, most of which will also require counting to control the number of fish.

One fairly cheap and simple method used to separate the individual fish is to let them slide in a convex V-shaped channel (Fig. 19.10). By having a convex channel the velocity of the sliding fish will gradually increase. This will separate the individuals and it is now quite easy to count the single fish, for instance with the use of light sensitive cells. The light diodes create a beam that is broken by the sliding fish, which is thus counted.

Counting of the fish in water is normally done in connection with pumping or transport through pipes. Some kind of camera is commonly employed, either a video camera or a linear camera.

During the past few years camera technology has developed rapidly together with the use of image analysis, which has also been utilized for counting fish (Fig. 19.11). The challenge when using cameras is to achieve good pictures, with sufficient contrast to the background because black and white images are commonly used. A chamber with proper light conditions is therefore required. This may be included on the pump pipe as a separate unit. To avoid two fish coming together and being counted as one, image analysis, either with two cameras or one camera and a mirror to give two pictures, is employed. This system may also be used to calculate the size of the fish (see section 19.5.2) but then other algorithms are used.

The difference in conductivity between water and fish may also be used to count them, but here also it will be difficult to separate the individuals.

Figure 19.10 Counting fish with a convex V-shaped channel and light cells.

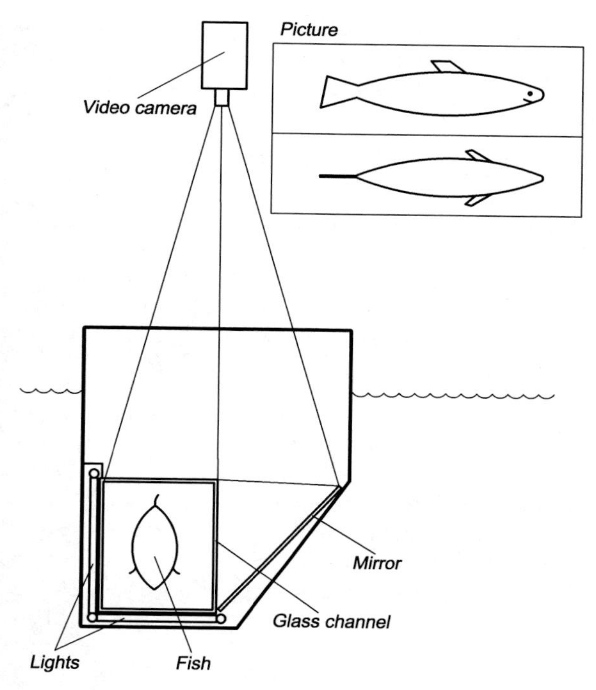

Figure 19.11 A video camera in a chamber used for counting fish and estimating their size.

19.5.2 Measuring fish size and total fish biomass

It is no longer necessary to lift fish out of the water and weigh them to obtain a traditional weight sample, or to lift them into a calibrated tank and use Archimedes' law to calculate the weight of the fish group. Archimedes law can be used because the fish will displace a volume of water equal to their own volume. If it is assumed that 1 l of water weighs 1 kg of water the volume displaced in litres will be equal to the weight of the fish in kilograms. By having a scale on the tank wall the increase in water level can be monitored and in this way the biomass found. Equipment for measuring fish size and total biomass is useful for production planning and control, and also for deciding when to size grade and for planning harvesting.

Today, various equipment is available for measuring fish size automatically whilst they remain in the water. The equipment used for automatic counting and weighing of the fish can be separated in two depending on whether: the fish stay in or are taken out of the production unit. The first category comprises equipment that is lowered into the production unit and does not disturb the behaviour of the fish; the second normally comprises measuring equipment connected to a fish pump.

Measurements where the fish stay in the production unit

Several systems have been developed, mainly for sea cages, to measure either the single fish size or the total biomass of the fish in the cage, or a combination of the two. The reason for the interest in measuring in sea cages is due to the large amount of fish and high feed consumption. Good control of fish weight and development are therefore important. In a cage it is impossible to control totally environmental factors that affect the appetite of the fish. Neither is it possible to foresee exactly changes in environmental factors (i.e. the weather) for prognosis.

Two principal methods dominate[12]:

- Remote sensing by hydroacoustics or submerged cameras
- Detailed measurements obtained by leading the fish through a sensor for individual registration, e.g. by optical, impedance, capacitance or ultrasounds techniques

One commonly used technique based on the second method is a measuring frame[13]. This is a rectangular frame that is lowered into the cage. In the walls of the frame there are a number of light diodes placed in rows so they create a linear camera (Fig. 19.12). When the fish swim through the frame the light diodes create a shadow picture. Using a computer, the approximate size of the fish is calculated, based on image analysis of the shadow picture. The system is based on a random sample of fish voluntarily swimming through the frame during a given period. This will occur in cages because the fish are swimming voluntarily around in the cage. The number of passages through the frame depends on the fish density. At fish densities between 10 and 20 kg/m^3 a typical swimming frequency is 30–200 fish per hour[13]. If two fish pass through the frame together, the image analyses will remove the picture as not being useful. Only the passages where a single fish passes through the frame and a good picture is achieved are used to calculate the average fish size. This is done based on a side view of the fish when it passes through the frame. The side area of the fish is highly correlated with its size. Based on the large number of measurements on single fish, the average fish weight with a standard deviation can be calculated for the total cage. Comparison of these data with the actual weight achieved at the slaughter house has shown good correlation. The standard deviation gives important information about the size variation of the fish in the cage and can be used to tell when size grading is necessary. If the average size of fish is known, it is possible to calculate the total biomass provided that the number of fish in the cage is known. When using such a system for calculating the total biomass, it is of course very important to account for the dead fish, and there must be no unregistered escape of fish from the cage. To use the frame it is necessary that the fish swim voluntarily around in the production unit; it is not possible to use the system in fish tanks with a water flow, because here the fish will stay still in the water flow.

Another method based on camera technology for finding the size of fish in a cage utilizes a stereo-video system[14,15]. Two video cameras sitting above each other in a rack with a known distance between them are used (Fig. 19.13). By taking pictures of the fish, with both cameras, it is possible to calculate

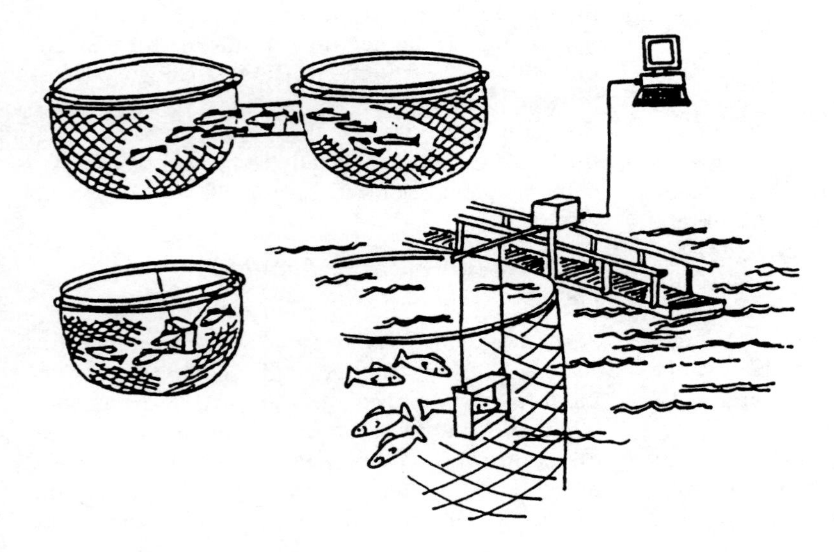

Figure 19.12 A measuring frame used to determine fish size in sea cages[13].

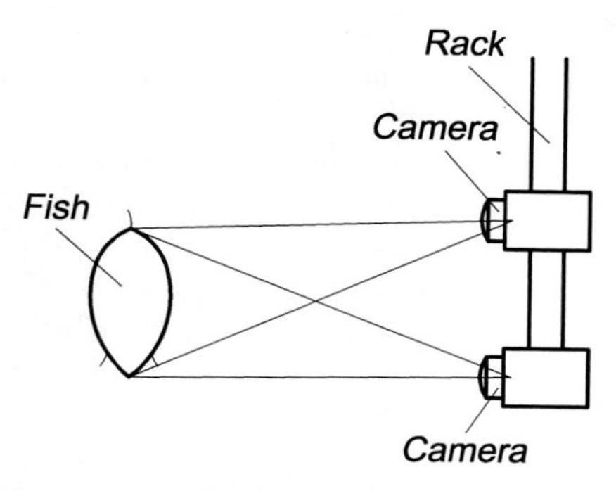

Figure 19.13 Use of two cameras and stereoscopy for calculating fish size.

the distance from each camera to every point on the fish; image analysis will give the length and height of the fish; there is a high correlation between these measurements and the weight of the fish. Experiments using three cameras simultaneously to count fish in cages have also been performed.[16]

Video cameras like those used for counting fish can also be used for biomass estimation.[17,18] This system is quite large for use in cages and it may be difficult to get the fish to swim through the unit or channel voluntarily. When using such a unit in connection with pumping, this does not present any problem because here the fish are moved by force. By taking a picture from the side when it passes the video camera and calculating the area of the fish, its weight can be found.

Hydroacoustics is a totally different principle used for measuring total fish biomass in a sea cage which employs an echo sounder.[19] This is actually the same method as used in traditional fishing to find where and at what depth the fish are swimming. It also gives some information about the amount of fish. An echo sounder contains a source that emits sound pulses at fixed intervals and a transducer which is normally placed below the cage. The source sends the sound pulses up to the water surface in the cage; this is the opposite direction to the echo sounders used for fishery which send sound pulses down to the bottom. When the sound pulse hits a fish it is reflected. A receiver fixed near the sound source beyond the cage receives this reflected signal (echo), which is travelling in the opposite direction to the sound pulse that was sent out (Fig. 19.14). The sound pulses are very short (1 ms is typical) and separated by longer intervals during which the receiver is listening for echoes before the next sound pulse is sent out. Therefore several echoes are received over the course of a minute. The size of the echo depends on the size of the individual fish or fish shoal. By using an echo sounder it is also possible to obtain information about the behaviour of the fish and the depth in the cage at which they are swimming under various environmental conditions. It is often the swim bladder of the fish that is the source of the echoes registered by the echo sounder; a good echo is

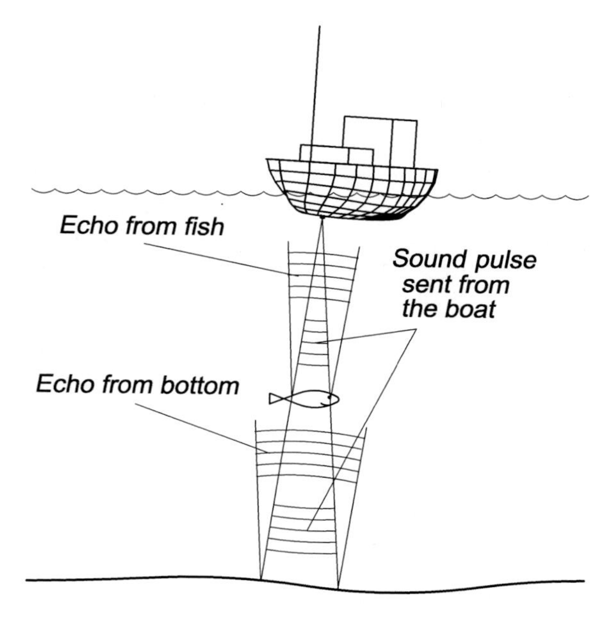

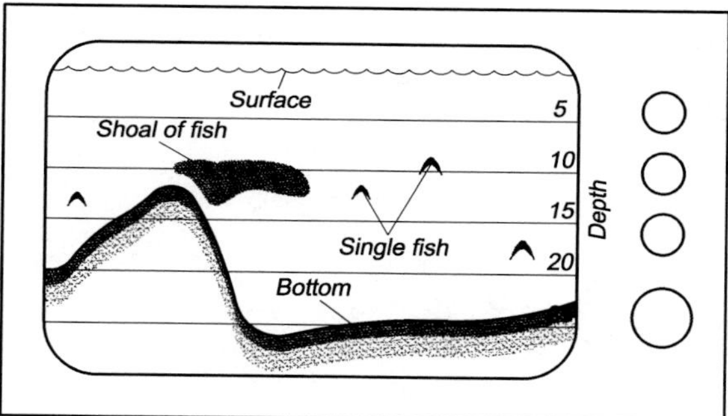

Figure 19.14 The principle used in an echo sounder for finding fish in the sea may also be used for finding the total biomass in a sea cage, but it is used in the opposite way with the receiver being placed under the cage.

generated here because the swim bladder is filled with air. Echo sounders are not used to measure the size of single fish with high accuracy because the magnitude of the echo is highly dependent on the direction of the swim bladder in relation to the sound source and will vary if the fish is swimming upwards, downwards or horizontally. For the whole fish group it will, however, give quite a good estimate. Echo sounders cannot so easily be used in tanks because sound reflections from the tank bottom and walls will interfere with the measurements.

Measurements where the fish are removed from the production unit

It is possible to install a video camera in a pipe used for pumping fish or an ordinary pipe. This takes pictures from the side and eventually above, and calculates the fish size from image analysis of the pictures (Fig. 19.15). The system is the same as described earlier for counting fish, the only difference being the image analysis program used to calculate fish size.

19.6 Monitoring systems

On an aquaculture facility it is necessary to control all the different factors that are involved in the management of the farm. This can for instance be a water quality factor such as the concentration of oxygen, ensuring sufficient water flow, or the correct water level in a head basin. Of course this can be done with manual measurements and observations, but is very time-consuming if done with high regularity. How regularly measurements must be taken depends on the economic consequences if something fails. In intensive fish farming with high stock density and much use of technical installations that can fail, the need for regular measurement and control is obvious. The time available for remedial action when something fails is also limited. Systems for automatic control are therefore used based on economic and practical consequences, especially in intensive fish farming. To illustrate this a land-based marine fish farm that requires pumping to supply water to the fish can be used as an example. The importance of continuous control of the water flow into the fish tanks is obvious.

A monitoring system comprises three major components (Fig. 19.16):

- Sensors and measuring equipment which control the conditions
- Monitoring centre which receives signals from the sensors and measuring equipment, interprets them and eventually sends out alarm signals or signals to regulators
- Equipment for warning when something is failing and emergency equipment that is started and stopped by regulators.

Figure 19.15 A well boat equipped with a system for automatically measuring fish size when they are pumped.

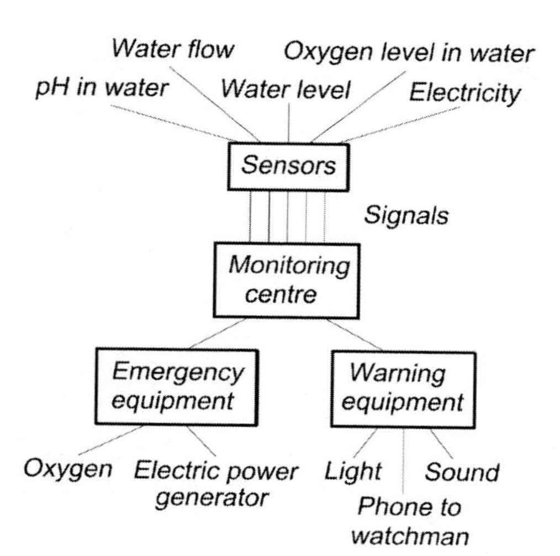

Figure 19.16 A monitoring system comprises three major parts: the sensors, the monitoring centre and the warning equipment.

Signals are transferred between the components. These are normally electrical signals via cables; wireless connections however, may also be used. A monitoring centre is always equipped with a battery back-up so that it can always function independently whether or not there is an electricity supply.

19.6.1 Sensors and measuring equipment

To control the conditions a number of sensors which function as monitoring points in the alarm system can be used.[20] Sensors can be used to control water level, water flows or water quality parameters, all described earlier in this chapter. Sensors are much used also to indicate whether the electricity supply is connected or not. The number of factors to be controlled determines the size of the monitoring system. From the sensors or measuring equipment electrical signals are sent to the monitoring centre: these can be digital signals, i.e. current or not current, for example from a level sensor where an electric circuit is broken and the signal to the monitoring centre changes from current to no current. The signal may also be analogues which means that the current coming into the monitoring centre varies depending on the value read from the sensor; an example of such an instrument is an oxygen meter. A current corresponding to the measured oxygen level is sent to the monitoring centre; thus if the oxygen concentration decreases, less current flows. Normally this standardizes low current signals of between 4 and 20 mA.

19.6.2 Monitoring centre

The construction of the monitoring centre that receives the signals, interprets them and eventually sends out signals to the warning equipment, depends on the complexity of the system. Even though the monitoring centre is usually built especially for aquaculture facilities, it is based either on a programmable logic controller (PLC) or a computer.

A PLC can be visualized as an electronically operated switch system. It includes a number of input channels and a number of output channels. Electric signals from the sensors come through the input channels. The electrical output signals (i.e. how the switches are functioning) can be programmed based on the input signals. Programming of the PLC is done using an external keyboard connected to the 'brain' of the PLC. Whether output signals will be sent, and through which of the output channels, depends on the input signals and the location of the switch system. In and out signals can be digital or analogue, or most commonly both.

To illustrate the functioning of a PLC, take a simple monitoring system on a farm as an example. The farm equipment includes a pump that delivers water to a head tank from which it flows under gravity to the fish tanks. Pure oxygen gas from an oxygen bottle can be added to the single fish tanks through diffusers at the bottom of the tanks. The following measuring instruments are included: a level sensor in the head tank and an oxygen sensor in each of the fish tanks. In addition there is a reserve pump on stand-by, a magnetic valve that controls the emergency oxygen supply, and a siren available on the farm.

The programme for the PLC might be as follows. If the water level in the head tank is too low a signal will go from the level sensor to the PLC. The PLC is then programmed to send an output signal to the reserve pump to start it. If the oxygen level in the tank is too low, the value input to the PLC is under a programmed value and a signal is sent out through the channel that controls the magnetic valve on the oxygen gas bottle. This will change from the closed to the open position. In addition, a signal is sent to the channel that starts the siren.

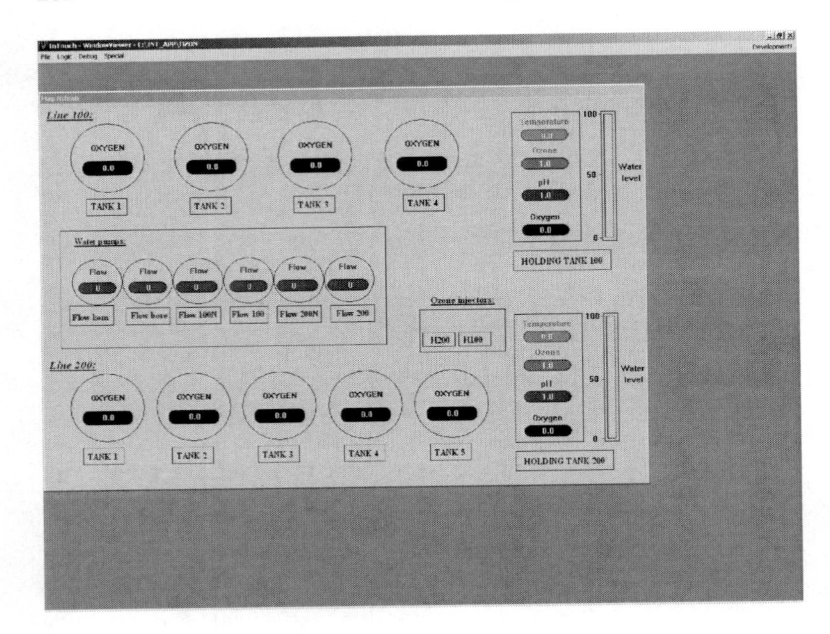

Figure 19.17 By using a PC as the monitoring centre it is easy to obtain a picture of the farm and the status of each sensor on the computer screen.

If a computer is used as a monitoring centre it is normally equipped with special cards known as I/O cards or data acquisition cards (DAQ) that make it possible to import signals, including analogue signals, and send out the same types of signals as the PLC. By using a computer the input signals can easily be recorded and stored. It is also possible to get a hard copy of the time points for alarms, including which of the sensors registered the failure. It is also possible to get a picture of the farm, including the sensors and their individual status, on the computer screen, which makes visual control of operations easier (Fig. 19.17). Today, both PLC and PC are used in combined systems.

Figure 19.18 A generator for providing electrical power in the event of a power failure on the farm.

19.6.3 Warning equipment

From the monitoring centre electric signals are sent through the output channels to external warning equipment. The warning equipment is installed on the farm and normally includes both equipment that creates light and sound, such as warning lights and sirens. A telephone may also be included in the system. This dials fixed programmed phone numbers, among others to the watchman's mobile phone. On more advanced systems, a message saying what is wrong appears on the telephone display.

19.6.4 Regulation equipment

The signals may also be used to start emergency equipment or to regulate the functioning of equipment. For instance, in the event of a power failure, a generator will start (Fig. 19.8). The equipment may also open emergency oxygen supplies to the fish tanks or start stand-by pumps. Often there is a combination of starting emergency equipment and starting external warning equipment.

19.6.5 Maintenance and control

For a monitoring system to be reliable it is very important to carry out proper maintenance and testing. The maintenance normally includes testing and calibration of the sensors. It is advisable to have a fixed routine for testing the monitoring system, for instance once a week. The sensors are then tested to ensure that the signals being emitted and also that the signals are being sent from the centre to the warning equipment. Emergency systems such as electric generators and systems for adding oxygen must also be tested at fixed intervals, to be sure that they will function if a real emergency situation occurs.

References

1. Lee, P.G. (1995) A review of automatic control systems for aquaculture and design criteria for their implementation. *Aquacultural Engineering*, 14: 205–227.
2. Hochheimer, J. (1999) Equipments and controls. In: Wheaton, F. (ed.) *CIGR handbook of agricultural engineering, part II. Aquaculture engineering*, pp. 281–307. American Society of Agricultural Engineers.
3. Ernest, D., Nath, S. (2000) Computer tools for siting, designing and managing aquaculture facilities. *Aquacultural Engineering*, 23: 1–78.
4. Lee, P.G. (2000) Process control and artificial intelligence software for aquaculture. *Aquacultural Engineering*, 23: 13–36.
5. Ebeling, J.M. (1994) Monitoring and control. In: *Aquaculture water reuse systems: engineering design and management* (eds M.B. Timmons, T.M. Losorod). Elsevier Science.
6. Webster, J.G. (1998) *The measurement, instrumentation and sensors handbook.* CRC Press.
7. Anderson, N.A. (1997) *Instrumentation for process measurement and control.* CRC Press.
8. Morris, A.S. (2001) *Measurement and instrumentation principles.* Butterworth-Heinemann.
9. Montgomery, J.M. (1985) *Water treatment, principles and design.* John Wiley & Sons.
10. Gibbs, R. (1990) Advances in on-line monitoring. In: *Aquaculture and water quality* (eds D.E. Brune, J.R. Thomasso) *Advances in world aquaculture, vol. 3.* World Aquaculture Society.
11. Wheaton, F. (1977) *Aquacultural engineering.* John Wiley and Sons.
12. Heyerdahl, P.H. (1995) Biomass estimation, a powerful tool in fish farming management. In: *Quality in aquaculture*, Special publication no. 23. European Aquaculture Society.
13. Heyerdahl, P.H. (1995) Optical biomass estimation, a technological approach. In: *Quality in aquaculture*, Special publication no. 23. European Aquaculture Society.
14. Nailberg, A., Petrell, R.J., Savage, C.R., Neufeld, T.P. (1993) A non-invasive fish assessment method for tanks and sea cages using stereo video. In: *Proceedings of an Aquaculture Engineering Conference, Spokane, Washington*. American Society of Agricultural Engineers.
15. Harvey, E., Cappo, M., Shortis, M., Robson, S., Buchanan, J., Speare, P. (2003) The accuracy and precision of underwater measurement of length and maximum body depth of southern bluefin tuna (*Thunnus maccoyii*) with a stereo-video camera system. *Aquacultural Engineering*, 63: 315–326.
16. Petrell, R.J., Neufeld, T.P., Savage, C.R. (1993) A video method for noninvasively counting fish in sea cages. In: *Proceedings of an Aquaculture Engineering Conference, Spokane, Washington*. American Society of Agricultural Engineers.
17. Boyle, W.A., Ásgeirsson, Á., Pigott, G.M. (1993) Advances in the development of computer vision fish biomass measurement procedure for use in aquaculture. In: *Proceedings of an Aquaculture Engineering Conference, Spokane, Washington*. American Society of Agricultural Engineers.
18. Eide, T. (1997) *Videoregistrering av fisk i elv.* Internpublikasjon, Norwegian University of Life Science, (in Norwegian).
19. Dunn, M., Dallard, K. (1993) Observing behaviour and growth using the Simrad FCM 160 fish cage system. In: *Fish farming technology* (eds H. Reinertsen, L.A. Dahle, L. Jørgensen, K. Tvinnereim), pp. 269–274. A.A. Balkema.
20. Huguenin, J.E., Colt, J. (2002) *Design and operating guide for aquaculture seawater systems.* Elsevier Science.

20
Buildings and Superstructures

20.1 Why use buildings?

Buildings or superstructure are used in aquaculture for several reasons. It will shelter the production facilities from environmental factors such as wind, sun, rain or snow and the working environment for the farmers will be improved. It also makes possible photomanipulation of the grown organisms, which is an important part of optimizing the farming conditions for different species, especially in the fry and juvenile stages and during maturation.

The buildings or superstructure can be used to cover the entire farm, part of the farm, or only the production units. Buildings are used mostly on land-based fish farms; on sea-based farms they can be used for protection of feed storage on a barge and for housing the mess room and WC. In a land-based farm there will always be a building housing the toilet, mess room and office; it might, however, be an integral part of a larger building that also covers the production area.

Building design and construction is a large specialist field and here only a very brief survey is given with special attention to aquaculture facilities. Many textbooks are available on the subject; see, for example, refs 1–4.

20.2 Types, shape and roof design

20.2.1 Types

Several types of buildings can be used for aquaculture facilities. On the basis of their construction they can be categorized into different types ranging from simple superstructures with or without walls to complete buildings (Fig. 20.1). Buildings can be insulated or uninsulated, and can be built for various lifetimes, either short or long.

The simplest superstructure is a shadow net. This is used to prevent the fish getting sunburned and to inhibit solar heating of the water. A slightly more advanced construction results when plastic sheeting or tarpaulin is substituted for the shadow net and a tent is created. Normally a framework of steel pipes keeps this upright. The tent can either be for one tank or larger for the complete farm. On single tanks a hemispherical shaped construction is commonly used. It is also possible to have an insulated tent made of two layers of plastic with insulation in between. A bowed shape is a cheap construction to use here.

A typical building is, however, based on weather-tight walls and a roof made of fixed materials. The duration of such buildings is much longer than of the simpler tent constructions. Buildings on the farm may have one or several storeys. If there are possibilities for utilizing the terrain and hence gravity for internal transport of fish and feed, it will be advantageous. For instance, if small fish can be grown on the first floor and on-growing takes place in the ground floor, tapping of the fish to the on-growing department is possible, which gives a simple system for internal transport of fish. The feeders may also be filled from the first floor if they hang in the roof of the ground floor over the fish tanks.

20.2.2 Shape

Buildings may have different shapes. When thinking of economy, the best is to have a square or especially a rectangular shape for the building. Angles in the 'body of' the building will increase the costs.

Figure 20.1 Different types of buildings used in aquaculture.

The reason for having angles must therefore be to improve the functionality of the building, for example, by improving the feed or fish handling.

20.2.3 Roof design

The roof may also have different constructions. A ridge roof is the most usual, but a sloping roof may also be used. Normally a ridge roof will be the most economic. A flat roof may also be used, but is much more difficult to build correctly and is especially difficult to get completely watertight. In areas with snow a flat roof is not recommended; it may be necessary to clear the snow, which is difficult. In addition the weight of the snow will press the roof down, so a large and costly frame is necessary. The chosen shape and roof of the building will also depend on available materials.

20.3 Load-carrying systems

A load-carrying system keeps the walls upright and the roof in position; it also takes the weight of the roof which, in colder regions also includes the load from snow. The cost of the load-carrying system is an important part of the total cost of the building.

Various constructions are used as load-carrying systems; these are based, among other things, on the material used (Fig. 20.2). Either separate constructions can be used for the wall and the roof, or combined systems can be employed, where the walls and roof are integral parts of the load-carrying system. In the first case the wall system carries the roof; different types are explained later.

When designing the load-carrying system, the necessary free span, i.e. the available floor area with

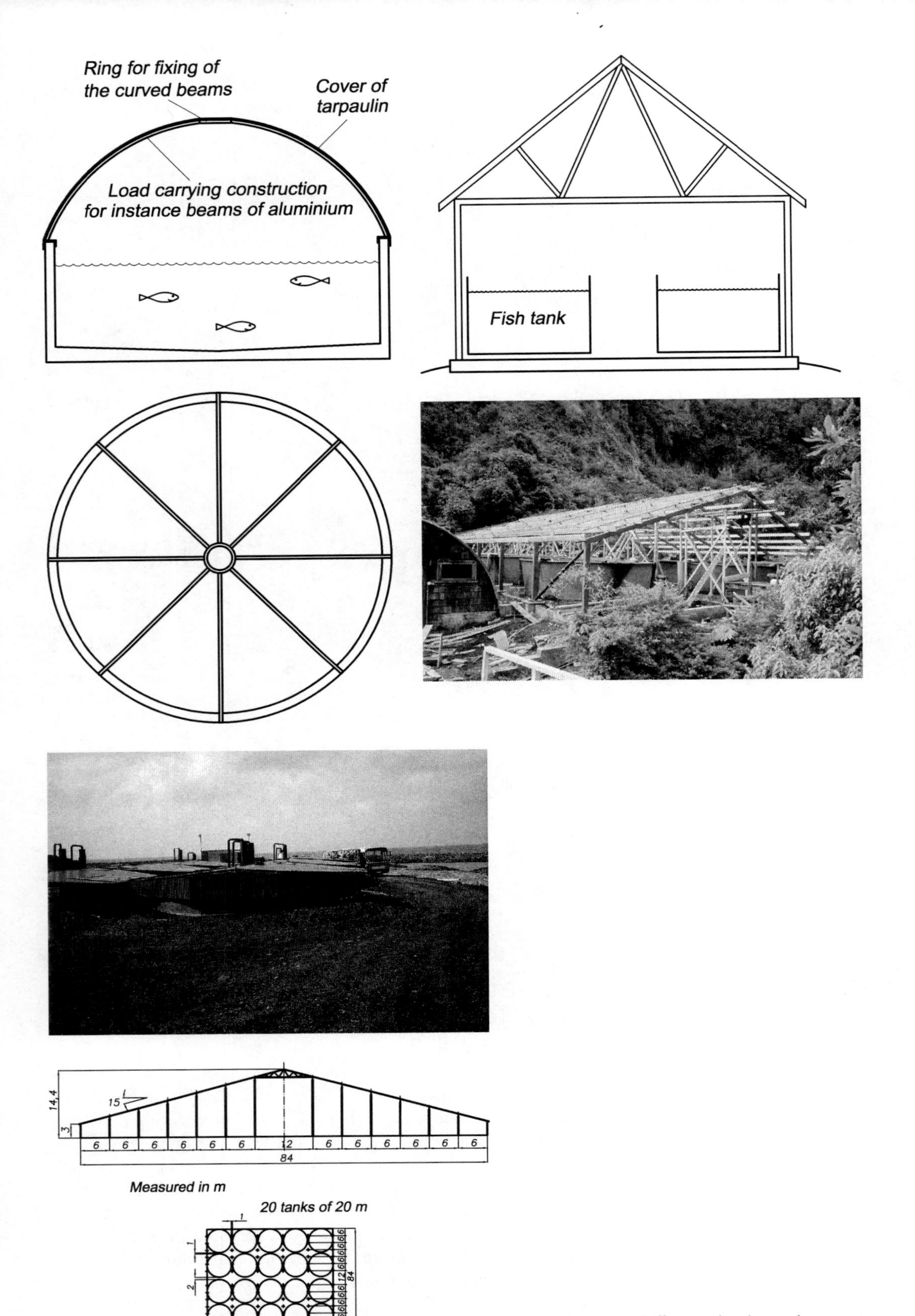

Figure 20.2 Different load-carrying systems can be used to keep the walls upright and the roof in position.

no pillars, is of great interest. The size of the free span determines the size of the construction; a larger free span results in a larger construction. To reduce the size of the free span, pillars or internal walls are used to carry the weight of the roof at several points. The use of pillars will reduce the cost of the load-carrying system, but inhibit the use of the building, constraining internal handling and space utilization.

Roof constructions that cover a large free span (above 10–15 m) are expensive. Long distances between the outside walls, without pillars or internal walls, will increase the building costs because the size of the roof construction is increased.

The timber lattice roof (W) truss is quite a commonly used roof load-carrying system. It may be built on site, but is normally prefabricated. A free span distance of up to 12–13 m width is easily covered with such a construction. If the building is wider, it is recommended that pillars be used to take the weight of the roof, if the aim is to keep the cost of construction low; a load carrying internal wall will have the same effect. This is a cheap and simple building construction.

Steel, concrete or wood beams or trussed beams are also much used. They can be used on large free spans in excess of 20 m, and even up to 100 m if the beams are bowed, as in a velodrome. Trussed beams are favoured for large constructions. However, for this type of construction also, the costs increase exponentially with the free span. Longitudinal beams of wood or steel in the building give quite a simple construction, especially when used in free spans below 20–25 m; however, pillars are recommended if the span is longer.

In combined systems, the load-carrying system for the roof and walls is integrated. A much-used construction here consists of large frames, either in steel or wood and often of a bowed shape, that are set into the foundations. The distance between the frames varies, but is normally in the range 3–5 m, or equal to a section of the building. The rest of the wall and roof structure is also built of steel or wood and connected directly to the frames. A great advantage of this construction is that it is quite fast to erect, and can be used on large buildings with a large free span. The method is fairly expensive and will normally require cranes to put up the building.

If the tanks are large, individual superstructures for separate tanks can be used. A cheap construction uses a hemisphere created with a frame of steel pipes and a cover of tarpaulin.

20.4 Materials

Various materials are used for constructing buildings (Fig. 20.3). Wood is simple to work with and simple to join together. Normally it is cheap, at least in smaller buildings and in areas with timber. It is used in walls, roofs and load-carrying constructions. To increase the strength and the length, glued beams can be used. Wood may also be part of constructions, for example in chipboard panels.

Metals such as steel or aluminium are also quite easy to work with; pieces are either welded or screwed together. Metals are much used in load-carrying constructions, such as beams or framework. Metal plates are used for covering interior and exterior surfaces on walls and roofs.

Concrete is a widely used building material. It is made of a mixture of sand and gravel with cement that functions as a glue, and water. After mixing followed by some hardening time this makes a permanent construction. The method of mixing and proportions of materials used will give concrete of different strengths.

Concrete has good compressive strength but poor tensile strength. Therefore iron is used, either as rods or mats, as reinforcement in concrete to enable it to withstand tension, while the concrete can withstand compressive forces. In small constructions, concrete is quite simple to handle and work with, and in addition it is fairly cheap. It can either be mixed on site or in a factory and delivered in special trucks ready mixed but not hardened. This latter method is most normal in larger constructions.

Concrete can also be delivered as prefabricated elements which are finished and hardened. Elements, such as beams, may also be pre-stressed to increase tolerance to higher forces without increasing the weight too much. Sizes of components vary from small blocks to bars to complete parts of buildings, such as wall elements or roof elements. Concrete may also be used for beams in load-carrying constructions.

A lightweight version of concrete is also available as blocks and bars. Here the gravel is replaced with a light material, for example, expanded clay products.

Figure 20.3 Different materials used in buildings on aquaculture plants: (A) wood panelling; (B) metal sheets; (C) concrete; (D) lightweight concrete blocks.

20.5 Prefabricate or build on site?

Buildings can be built on site from the foundations up or can be delivered as prefabricated parts that are put together on site. Various sizes of prefabricated elements can be delivered; options include the degree of finishing. These elements could be complete parts of the walls or the roof, where everything is finished. The time to establish a prefabricated building is, of course, much less than for buildings constructed on site. Whether prefabricated buildings are to be used or not depends on the price which will vary from case to case depending, among other factors, on the freight costs. The available building period will also influence the final choice.

20.6 Insulated or not?

In a cold climate, the building should be insulated if the air inside is to be temperate. The greatest amount of heat is lost through the roof; it is also lost through the walls and the floor. In addition, there are high heat losses through the windows and doors. The thickness of the insulation depends on the climate and winter temperature; lower winter temperatures mean that more insulation is required. In walls and roofs mineral wool made of glass or rock is used for insulation; expanded polystyrene (PS) is commonly used in the floor when this is of concrete. All these materials have a low k value (see Chapter 7). PS can also be used in walls or roofs providing it is covered by concrete.

This is because it can produce toxic gases if it catches fire. Taking Norway as an example, it is normal to use 15 cm mineral wool in the walls and 30 cm in the roof. The PS in the floor is 5–10 cm thick (equal to 20–25 cm of mineral wool which has a higher *k* value.

Whether insulation is to be used or not depends on the rooms and how they are utilized. The office, mess room and toilet are normal insulated in cold climates. Whether the production rooms are insulated or not depends on the desire to improve the working environment.

If the building is insulated, it will be of sealed construction so it will be necessary to use a ventilation system to ensure exchange of the air inside. If the walls are not insulated and airtight, the wall construction can be open so that a ventilation system is not required. Then the wall can be simply made with split panels or only a plastic grating; natural air exchange is thus ensured and there is no need for ventilation.

20.7 Foundations and ground conditions

When starting to build, proper foundations are very important to prevent part of the construction moving after the building is finished. Important components may break if the foundations move under the load from the building.

The ground conditions must be suitable for erecting a building. Rock, stone and sand/gravel form good building ground, while clay and silt are not as suitable because they are less stable. In colder regions, where the ground freezes in the winter season, clay and silt are not recommended for building ground, because the frozen ground will create movement in the building during freezing and thawing. The ground must carry the weight of the building, which makes marshy areas unfit for normal building constructions unless special precautions are taken, such as having raft construction as the foundation for the building.

Foundations are normally laid as concrete slabs. In areas with frozen ground, insulation under the slabs is necessary to avoid problems with heave. Alternatively, a ring foundation wall that goes down to frost-free ground must be used to ensure that the frozen ground not does affect the building. Normally this is recommended to go below 1.5 m depth, but this varies with the depth of the frozen ground. It is important to use drainage pipes to ensure that water is removed from the proximity of the walls to avoid possible movement.

20.8 Design of major parts

20.8.1 Floors

In aquaculture facilities concrete is normally used for the floor (Fig. 20.4). Iron mats or rods are used to reinforce the concrete. The normal slab thickness is 10–15 cm, but it might be greater under the walls and pillars because the floor needs to withstand the additional weight transferred from the wall or pillars. In cold climates is it necessary to use PS as insulation under the floor. To avoid frost heave either flank insulation or a ring foundation can be used for the building. If using flank insulation, it is laid under the surface, extending 1.5–2 m outside the flanks of the building. In this way frost inhibit from going down in the ground close to the building; it will go some distance out, so does not reach into the building and the problem with frozen ground is eliminated. If using a ring foundation this goes down to below the frozen ground and ensures that the building remains independent of the frozen ground. The thickness of a ring foundation wall is 10–20 cm, and typical depths are 1.5–2 m below ground level.

In the production room, at least, it is necessary to have a smooth surface on the floor that is easy to clean. It is normal to seal the floor with some

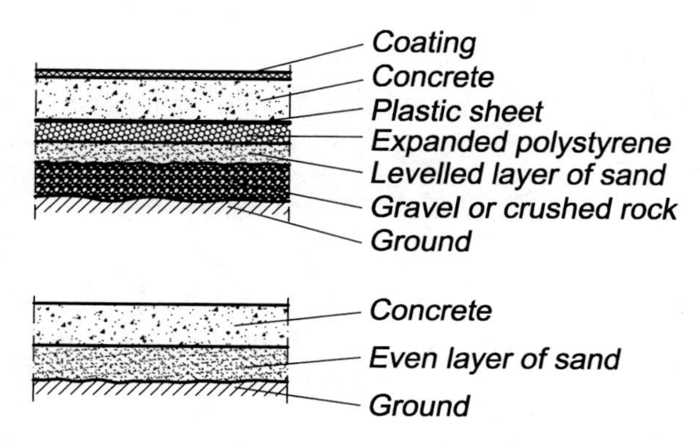

Figure 20.4 Normal method for constructing a concrete floor in a production room.

type of epoxy two component paint, especially suitable for concrete surfaces, to increase the smoothness of the floor so that it is easier to clean. However, it must not be too slippery; in the walkways, sand grains can be included in the epoxy paint.

To keep the floor clean and dry, it is important to have sufficient slope on the floor and enough gullies to collect the water; the recommended slope is from 0.1 to 1%. It is better to have too great a slope than too little. Farms have also used heat cables in the concrete floor to keep the floor dry, but this is very expensive.

20.8.2 Walls

The walls must be constructed to stay upright and keep the roof in position, keep the building stable, and prevent wind damage to or deformation of the building. How the wall is constructed depends whether it is insulated or not, the material used, and the load carrying system employed.

If the wall is not insulated, an open construction is recommended, because then no ventilation system is necessary and the air inside will be the same as the air around the building. It is, however, an advantage to have some walls to shield the inside of the building from the wind.

A closed or insulated wall construction, if this is necessary, consists of the following three major parts:

- Exterior covering to shelter and protect the load-carrying system
- Wall construction including the load-carrying system and insulation layer
- Internal covering to screen the inside and protect the load-carrying system.

These parts can be separate, or all the parts can be included in one construction, depending on the material used.

Several materials can be used for construction of the wall (Fig. 20.5). One major material could be used or the wall can be a mix of different materials. The major materials are wood, steel and concrete. A normal weathertight wall (the same principle is also used in the roof) of wood or steel can be constructed as described below.

The exterior covering that protects the wall construction and prevents ingress of rain and wind can be wooden boarding or metal plates (steel or aluminium). Under these boards or plates, sheathing cardboard or plates can be used to stop strong wind and driving rain penetrating into the wall construction.

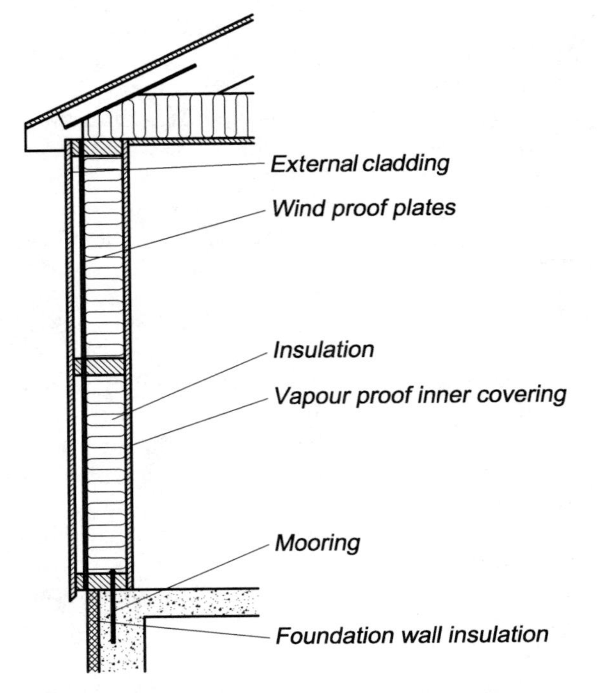

Figure 20.5 Construction of a wall.

In the farming room, plastic-covered chipboard sheets or metal plates can be used for the internal wall covering. To avoid humidity from the production room penetrating into the wall, a damp course is used under the plates and on the wall construction. This is a thin clear plastic sheet, and is of great importance in protecting the wall against high humidity and possible decay.

Timber framework is widely used; typical plank sizes are 5 cm × 10 cm, or 5 cm × 15 cm, which gives walls of thickness 10 cm or 15 cm, respectively. The planks are joined to form a frame that is torsion stable in every direction. The frame is typically covered on both sides with metal plates or wood panelling.

If using concrete, the thickness of the walls is 10–30 cm, and iron bars are used as reinforcement. If the building is insulated, PS can be an integral part of the wall. Because the PS is enclosed in concrete, problems with toxic gases from the PS in the event of fire are avoided. When building concrete walls, formwork must be used on both sides before the concrete is poured in and the hardening process can take place. Concrete blocks or light concrete blocks may also be used. This is a simple way to build a wall, where no formwork is needed. The blocks are laid with cement mortar. Blocks are also delivered with integral insulation. The concrete or concrete blocks represent the total wall construction, and will keep the roof in position, so they function as the load-carrying system.

The exterior and internal coverings are easy to apply to concrete walls. The concrete can be plastered and painted so a smooth surface that is easy to clean can be established on both sides (Fig. 20.6).

When using concrete and light concrete, a weathertight construction is achieved. Some kind of ventilation is therefore necessary, even if the walls are not insulated, to avoid excessive humidity.

20.9 Ventilation and climatization

In a building with a weathertight wall construction, there is no exchange of air. After working there for a period the air quality will gradually decline because, no new oxygen is added and no carbon dioxide removed. Exchange of air (ventilation) is therefore necessary.

Both in areas for water treatment and in the fish production rooms there are large free water surfaces. This will increase the humidity, which means that there are large amounts of water vapour in the air. The material and equipment that is going to be used in the room must be able to withstand high humidity.

The amount of water that the air can take up depends on the air temperature (Fig. 20.7); if this drops the amount of water that can be kept in the air is reduced and condensation will result. Typical surfaces that have a lower temperature and where condensation will accumulate are exterior walls, windows and floors. Excessive humidity can cause problems with the feed. Especially problematic is when the feed is taken from a cold room; because

Figure 20.6 It is important that surfaces are easy to clean.

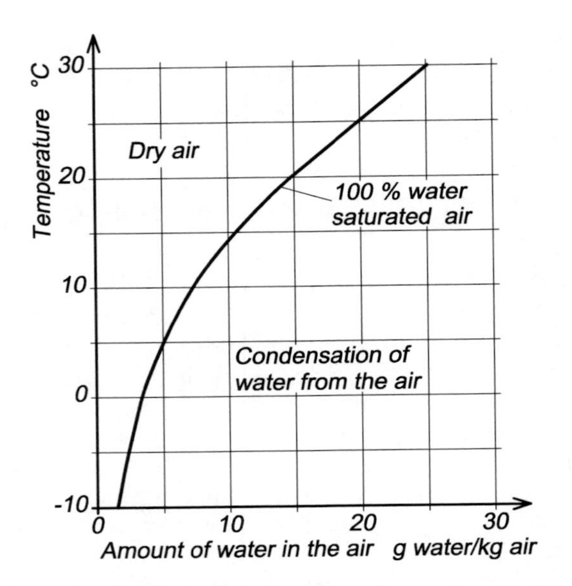

Figure 20.7 The amount of water that can be taken up by air depends on the temperature.

Figure 20.8 A ventilation system containing an air-to-air heat exchanger.

the feed is colder than the new surroundings, water will condense on its surface. Problems with humidity occur when the water temperature is high and it is cold outside the building. To avoid these, the humidity of the air must be reduced; this is known as air conditioning.

Two things must therefore be done: exchange of air to improve the air quality and reduce the humidity. These problems can be solved in combination or separately, and a number of methods and solutions are available.

One method of reducing the humidity in the air is to use a dehumidifier. This functions in the same way as a heat pump or a refrigerator. A fan blows the air from the room over a cold surface (the evaporator); the water in the air condenses on the surface and is collected in a tray. Since the temperature of the air is reduced the amount of water that the air can contain will be reduced. Afterwards the air is transported over a heated plate (the condenser) and its temperature is increased; because of this the humidity is decreased even more.

Another method to reduce the humidity is to increase the air temperature 2–3°C above that of the water. It is, however, difficult to achieve a higher temperature in all parts of the building; this shows the importance of good insulation in all cold areas of the building. In practice, this is difficult, because there will always be parts of the building where the temperature is lower and condensation will occur. Otherwise the room temperature must be very high, which is very expensive. Good circulation of air in a room is also very important when using such methods to smooth out temperature differences.

To combine reduction of humidity with ventilation is also a solution. Air from outside normally has a lower humidity than air inside a building. By bringing air in from outside to replace the air inside, humidity is reduced and ventilation enhanced.

The simplest way is to achieve ventilation is to create a small vacuum inside the building. By using a fan that blows air from inside the building reduces the air pressure. The air from outside will then flow in through air valves on the walls due to the partial vacuum inside the building. In this way good air exchange can be achieved. Alter-natively, air from outside can be forced into the building with a fan and inside air forced out through valves in the walls. This is not recommended, because humid air will be forced into equipment and the fabric of the building.

To recover the energy from the air that is taken out, heat exchange with the inflowing air can be used. An air-to-air heat exchanger is shown in Fig. 20.8. Air is withdrawn with a fan, and an equal amount of new air is dragged in via a pipe system. A neutral pressure is then created inside the room.

Another method is to use a so-called breathing intermediate ceiling. The phenomenon utilized is that hot air will go upwards because it is less dense than cold air. A special open ceiling through which the hot air can go is then used to create a small vacuum in the building and new cold air can be taken in from outside through valves in the wall. Of course, a combination of the different methods may also be used.

To avoid the need for ventilation, as mentioned previously, an open wall and roof constructions can be used. This requires free circulation of air from the outside to the inside. Split panels or plastic gratings are examples of this. In addition an open roof ridge can be used, so that hot air can escape. This is a specially designed roof part.

References

1. Ching, F.D.K., Ching, F.D. (1996) *Architecture: form, space and order.* John Wiley & Sons.
2. Ching, F.D.K., Adams, C. (2000) *Building construction illustrated.* John Wiley & Sons.
3. Allen, E., Iano, J. (2003) *Fundamentals of building construction: materials and methods.* John Wiley & Sons.
4. Chudly, R., Greeno, R. (2004) *Building construction handbook.* Elsevier Science.

21
Design and Construction of Aquaculture Facilities

21.1 Introduction

The design and construction of a production plant for aquaculture depends on a number of factors, including the intensity and size of the production. Production plants can be classified according to whether the production is based on freshwater or seawater, and if it is onshore or offshore. The development stage of the organism may also be used as a basis for classification. A farm may produce eggs, fry, juvenile or on-growing fish ready for market, or it can have a complete production system with all life stages, from eggs to harvesting, on the farm.

In this chapter some general examples of farm design are given. The focus is on intensive fish farms because such farms use technology to the greatest extent. A land-based farm for hatching and juvenile production, and a sea-based farm for on-growing are used as examples.

21.2 Land-based hatchery, juvenile and on-growing production plant

21.2.1 General

Land-based production may be classified according to the production units used: tanks, ponds or net pens in lakes. The tanks can either have circular water flow or be raceways. Here the focus is on intensive farms having tanks with circular flow, because this system can give the largest production per unit surface area.

The design of an intensive land-based farm depends on whether a flow-through system or a re-use system is used. Use of a re-use system will reduce the amount of new inlet water necessary, but the need for equipment for water treatment increases. The design will also depend on whether the re-use units are placed on single tanks, or whether one re-use unit is centrally placed to serve several tanks. In a plant with a water re-use system, the amount and size of the equipment for water treatment is increased, and the same will be the case for the required area.

A land-based farm for hatching and juvenile production based on intensive production and a flow-through system can be separated into the following sections (Fig. 21.1):

- Water intake and water transfer to the farm
- Water treatment
- Hatchery
- First feeding
- On-growing
- Feed storage
- Workshop, staff room
- Wastewater treatment
- Equipment for feeding and handling that could be integral parts of the farm construction.

21.2.2 Water intake and transfer

The design of the water intake and transfer system depends on whether the water source is at a higher altitude than the farm, so the water flows under gravity into the farm, or whether the source is on the same or at a lower level than the farm, so that pumping of the inlet water is necessary. If possible, the first alternative is of course recommended (Fig. 21.2).

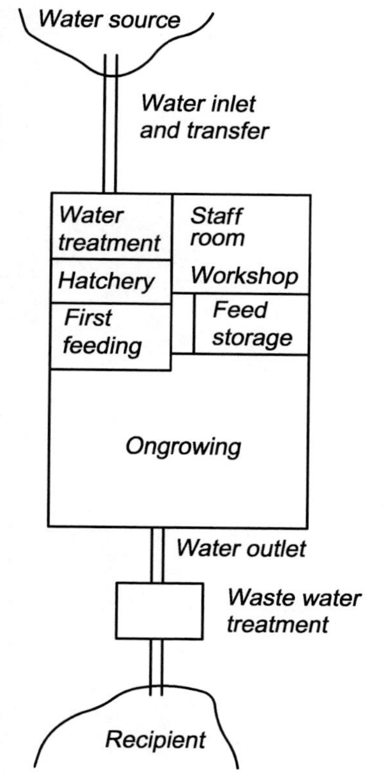

Figure 21.1 Main parts in an intensive drifted land-based farm for hatching and juvenile production.

Transfer pipes can be quite large, and therefore expensive, so it is recommended to have as short a distance from the water source to the farm as possible. If the distance is more than 500 m the cost of the inlet pipe will be considerable. Hence land-based fish farms should be located as close to the water source as possible. The same is of course the case for the distance between the farm and the recipient water body, if the water is to be transported through a pipeline.

If the water is to be pumped into the farm, a high lift head is not recommended because this increases the pumping costs; heads 10–15 m will result in quite large costs. If such conditions prevail, the case for increased oxygenation and use of recycling systems must be evaluated, to reduce the amount of water needed and hence reduce the operating costs of the farm.

Water inlet

The inlet design depends on the source: lake, river, groundwater or the sea.

Lake: In lakes deeper than 10–20 m there is normally thermal stratification. The ratio between the water temperature and the water density causes this stratification. Because water is most dense at 4°C, water at this temperature will sink to the bottom and warmer water will float on top. In areas where the water temperature falls below 4°C in the winter, there will be mixing of water during spring and autumn. In the autumn, the surface temperature is reduced; when it reaches 4°C mixing will occur because all the water will have the same density and only a slight breeze on the surface will ensure mixing. Strong winds may also result in mixing of water in the column at other times of the year.

It is normal to divide the water column into three layers, the surface layer (epilimnion), a layer with a steep temperature gradient where a large difference in temperature occurs (metalimnion), and the bottom layer (hypolimnion) (Fig. 21.3). The depth at which the steepest temperature increase/decrease occurs is known as a thermocline. Water collected from below this depth is said to have been collected below the thermocline. If there is such a layer in the lake it is recommended that the inlet be sited below the thermocline to avoid temperature variations that are large and difficult to control. Two intakes can be installed, one above and one below the thermocline, so that the warm surface water can be used to increase fish growth, but then possible problems with fouling must be taken into account.

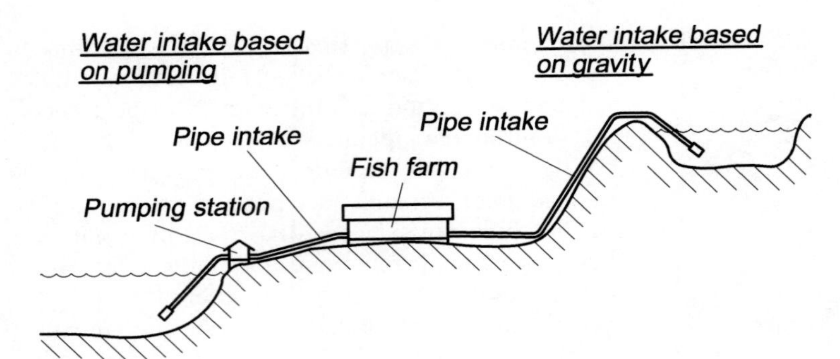

Figure 21.2 Water sources supplying the farm under gravity are recommended to avoid pumping.

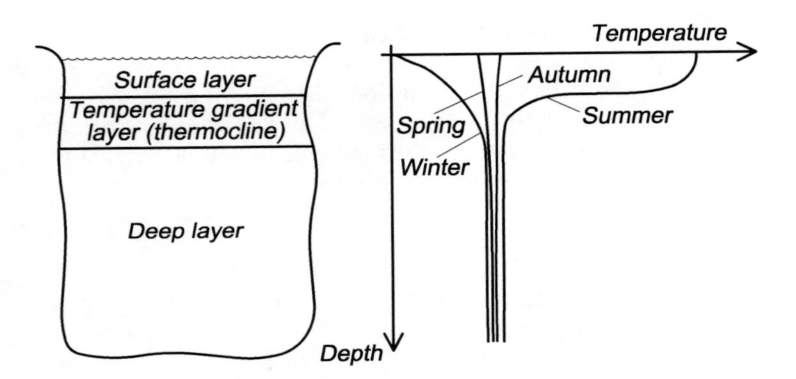

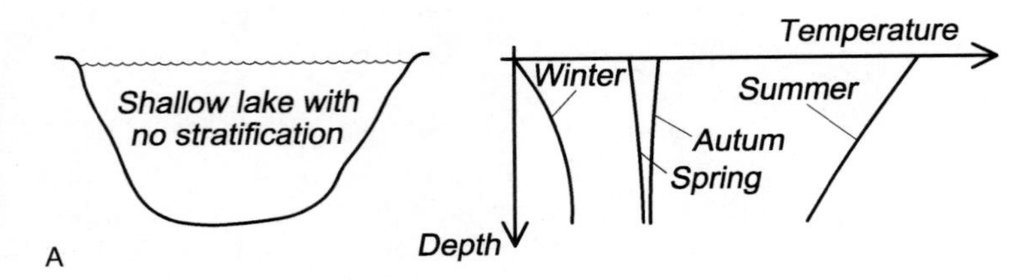

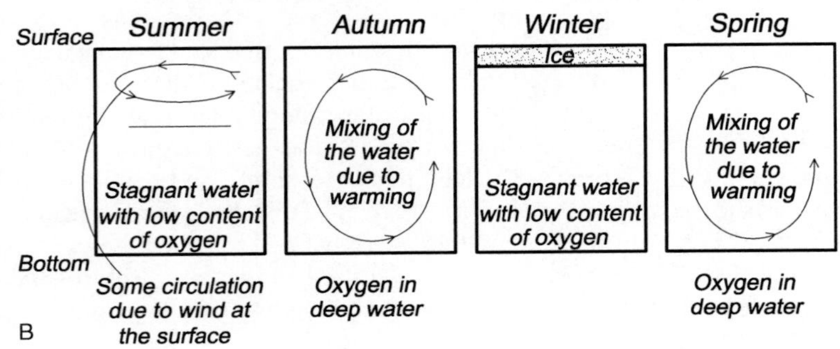

Figure 21.3 (A) The water in a deep lake can be divided into three layers, but a shallow lake may show no stratification. (B) Circulation of water by season due to water temperature gradients (highest density at 4°C).

In a lake the inlet pipe is normally laid on the bottom. The pipe must be correctly aligned, and large stones and deep cracks avoided; regular sloping of the bottom is advantageous. Under watercameras or an echo sounder should be used to align the pipe before it is fixed in position with weights. These are normally concrete blocks clamped to the pipe; the distance between them depends on the pipe size and water flow.

The actual inlet ought to be placed some distance from the bottom to avoid mud and small stones being sucked into the inlet pipe. This can be achieved in different ways, for instance by adding a float at the end of the pipe or by using an elbow that directs the pipe upwards in the water column (Fig. 21.4). To avoid fish and other large substances being dragged into the inlet pipe, a screen is used at the orifice. The water velocity through the screen must not be too large to prevent small fish and other objects getting sucked onto the surface of the screen and blocking the inlet. Water inlet velocities of less than 0.1 m/s are recommended, while in the inlet pipe itself flows of 1–1.5 m/s are used. Polyethylene (PE) has proved to be a suitable material for inlet pipelines, because it is both reasonably priced and to some extent will follow the contours of the terrain.

Sea: In the sea the same principles for an inlet in a lake are used. However, the thermocline is normally located deeper in the sea, normally from 30 to 70 m. Here also it is advantageous to have the inlet below the thermocline. The water temperatures will then be predictable, at least when the farm has been in use for some years, or if much historical data are available. For future production planning predictable temperatures are a great advantage but predicting surface water temperature is impossible. Another great disadvantage with having the water inlet above the thermocline and close to the surface is the problem of fouling inside the pipe. With high water temperatures, the inlet pipe will become totally fouled quite quickly and the water flow will be dramatically reduced. Facilities to clean the inlet pipe must, in such cases, be an integral part of the construction.

For water inlets at great depths, it will be difficult to clean the grating if it becomes blocked, so self-cleaning inlet gratings should be used (Fig. 21.5). Another, and probably a better solution, is to place the grating in the pumping station near the surface instead of end of the inlet pipe. The inlet pipe is now completely open at the bottom; objects will be swept in with the water but are stopped by the grating in the pumping station. A funnel at the start of the inlet pipe may also be used. This will reduce the water velocity and at the same time the resistance head (see Chapter 2). In the pumping station it will be quite easy to inspect the grating visually and remove accumulated debris.

An infiltration intake may also be used if ground conditions are suitable. As for infiltration inlets from rivers described below (Fig. 21.6), various designs of the inlet are possible.[1] For example, a well can be built on the beach but the beach material must be sufficiently permeable.

The cost of inlet pipes in the sea can be considerable, because they must be long enough to reach adequate depth and be below the thermocline; a good site therefore has a short distance to the acceptable depth, so deep water close to the shore is advantageous. A site on a long shallow coastline is normally poor.

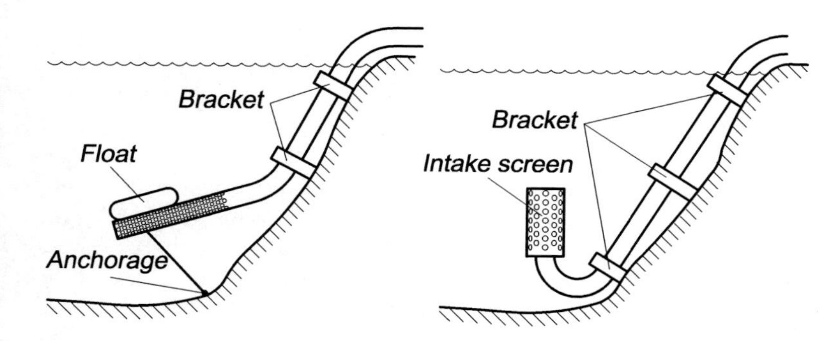

Figure 21.4 Methods for installing the inlet pipe to avoid bottom mud being dragged into the inlet pipe.

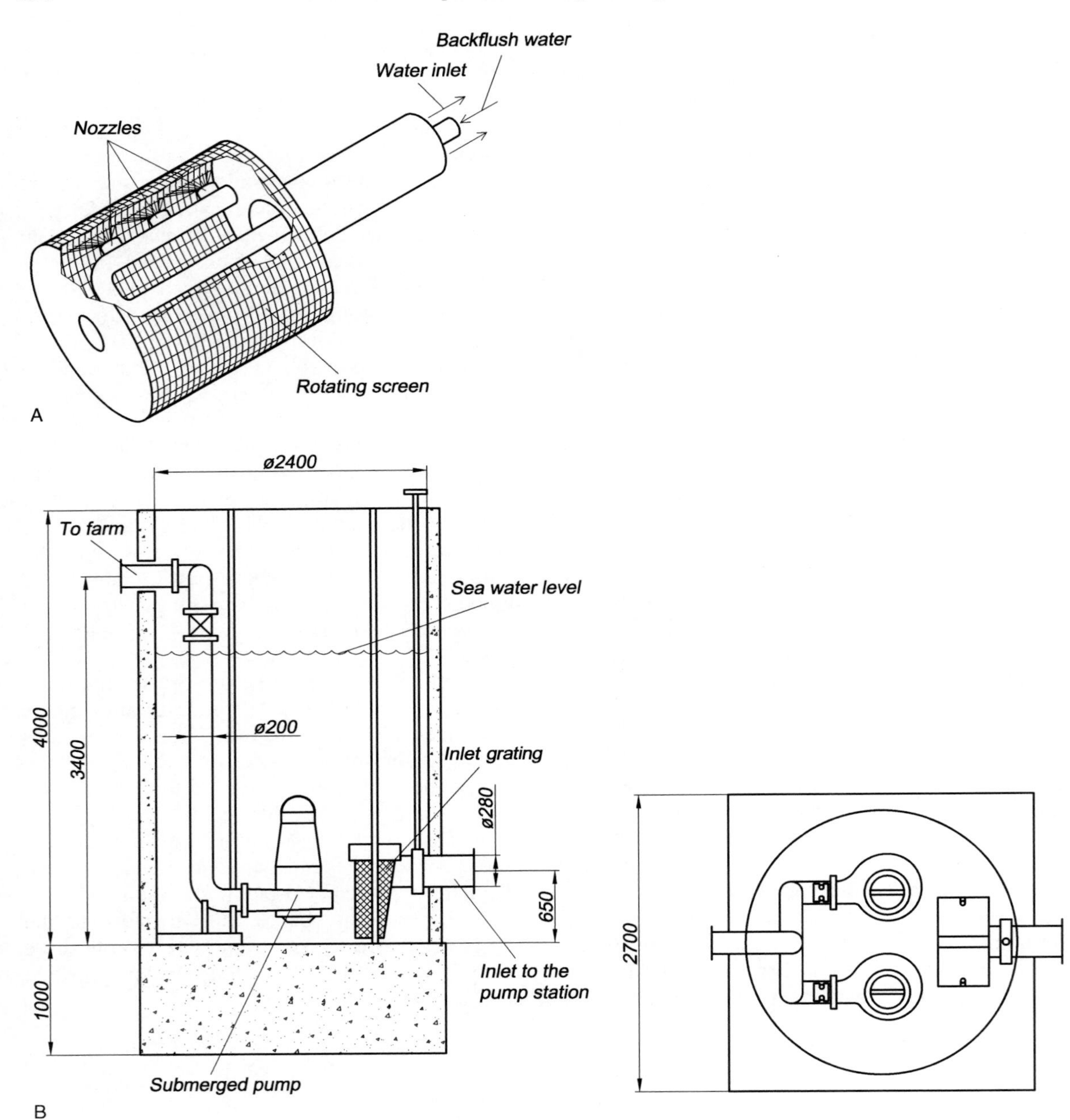

Figure 21.5 The inlet grating can (A) be made self-cleaning, or (B) placed within the pumping station so that it is close to the surface and easily available for cleaning.

River: When the inlet source is a river, the location of the inlet is very important. Rivers normally transport a lot of suspended particles and larger objects that are unwanted in the inlet water to the fish farm. The autumn period is especially critical when a lot of fallen leaves are transported in the water. Flood situations will also be critical because the amount of suspended particles in the water increases due to erosion. The inlet ought therefore to be laid in an area were the velocity of the river

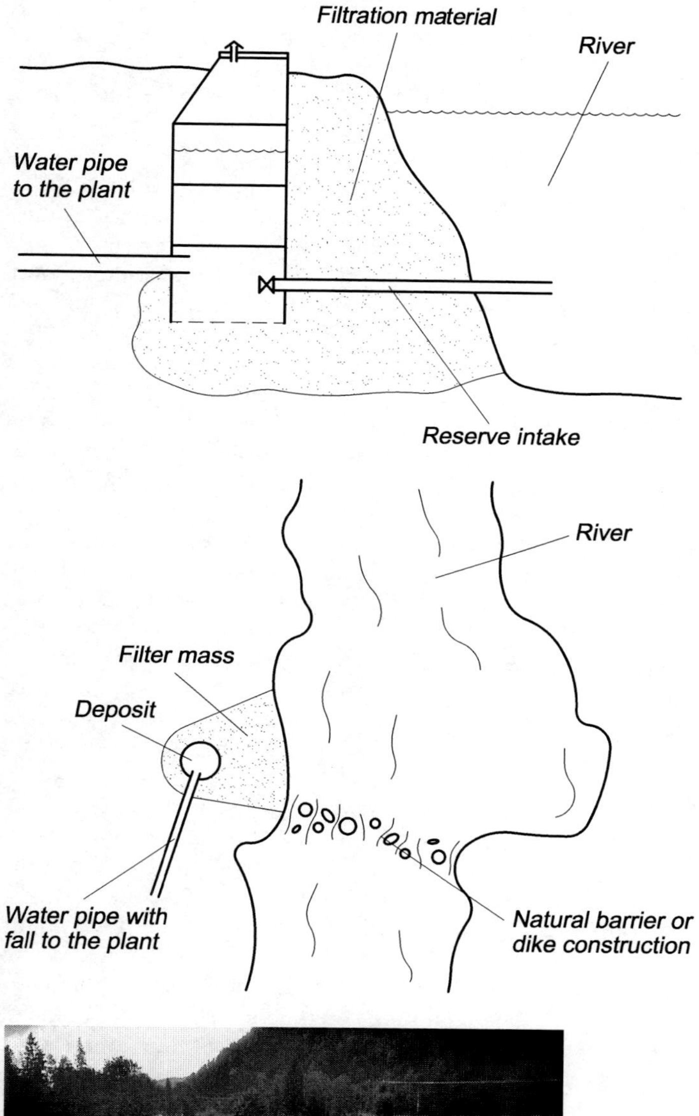

Figure 21.6 An infiltration intake ensures that the inlet water is purified before entering the farm.

is low and there might be some settling of particles before the water enters the inlet. This will for instance occur in areas were the river enlarges in depth or width. To avoid larger objects entering the inlet pipeline, a solution is to cover the orifice with a screen; this may also be of the self-cleaning type, a rotating screen or equivalent, (see Chapter 5). Another solution is to use an infiltration intake, if adequate material such as sand or gravel is available (Fig. 21.6). A basin may then be excavated close to the river; the river water is not then taken directly from the river but from the basin close to where the water enters by infiltration through the loose material. Purification of the water is then achieved during infiltration.

Groundwater: Groundwater whether fresh or salty could be used with advantage for aquaculture facilities, because of its stable temperature and low content of micro-organisms such as parasites, bacteria and viruses. The quality of the water should be checked before use because it may contain excessive concentrations of metal ions such as iron and manganese.[11] Proper aeration is also important because the oxygen concentration is always low and the concentration of carbon dioxide can be very high. In addition, groundwater may contain toxic concentration of hydrogen sulphide (H_2S) due to anaerobic decomposition in the ground. If drilling for groundwater on the shore or even some distance inland, the water may be salty (Fig. 21.7) because seawater may be forced in, depending on the ground conditions.

Groundwater must always be pumped; artificial water supplies will seldom be of a size necessary for aquaculture facilities. Even with traditional groundwater it is difficult, because the amount is normally quite low. Groundwater may either be found in mountains or in loose materials. If drilling wells in rock it is rare to get a flow above 50 l/min; in suitable loose materials, for instance in old glacial areas, the flow can be up to several hundred litres per minute.

Groundwater is taken from wells (Fig. 21.8). To establish a well in the mountains a special drilling rig is used; such wells can be very deep (more than 100 m). A specially designed small diameter pump, normally a multistage centrifugal pump, is lowered into the well.

Normally wells are vertical, but horizontal wells can also be used in loose materials. Wells can be either excavated or drilled. The choice depends on the water table in the area; if it is deep down, wells are drilled using special drilling rigs like those commonly used when drilling for oil. The inside of the well shaft is lined with a riser pipe to prevent sand and other loose material falling into the well and filling it up. When using excavated wells, excavators are used to make a hole that goes below the groundwater table; the depth that the excavator can reach limits the depth of the well. A basin built of concrete rings, for example, is lowered into the excavated hole. The bottom of the basin is open and filled with a layer of gravel that filters the groundwater forced into the well. A submerged pump can now be lowered into the well. A dry placed self-priming pump can also be used; it is set above the surface with a pipe down into the well.

The shaft of a horizontal well is set horizontally in the ground. One possible solution is to put drainpipe in it and collect the water from the pipe. Another solution could be to use a combination of an excavated vertical well and horizontal well. From the vertical well horizontal drainpipes are

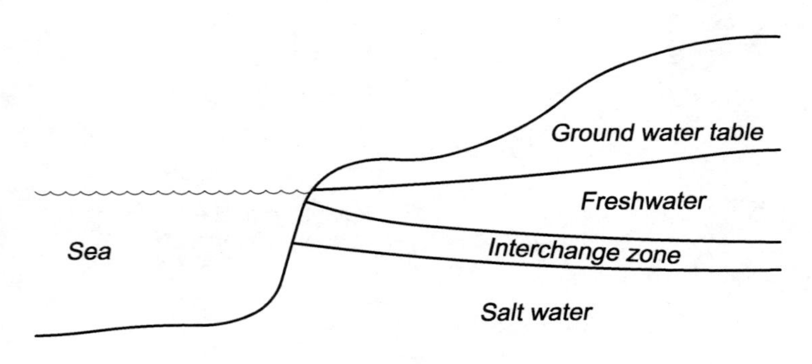

Figure 21.7 When drilling for groundwater close to the shore, salt water may be found instead.

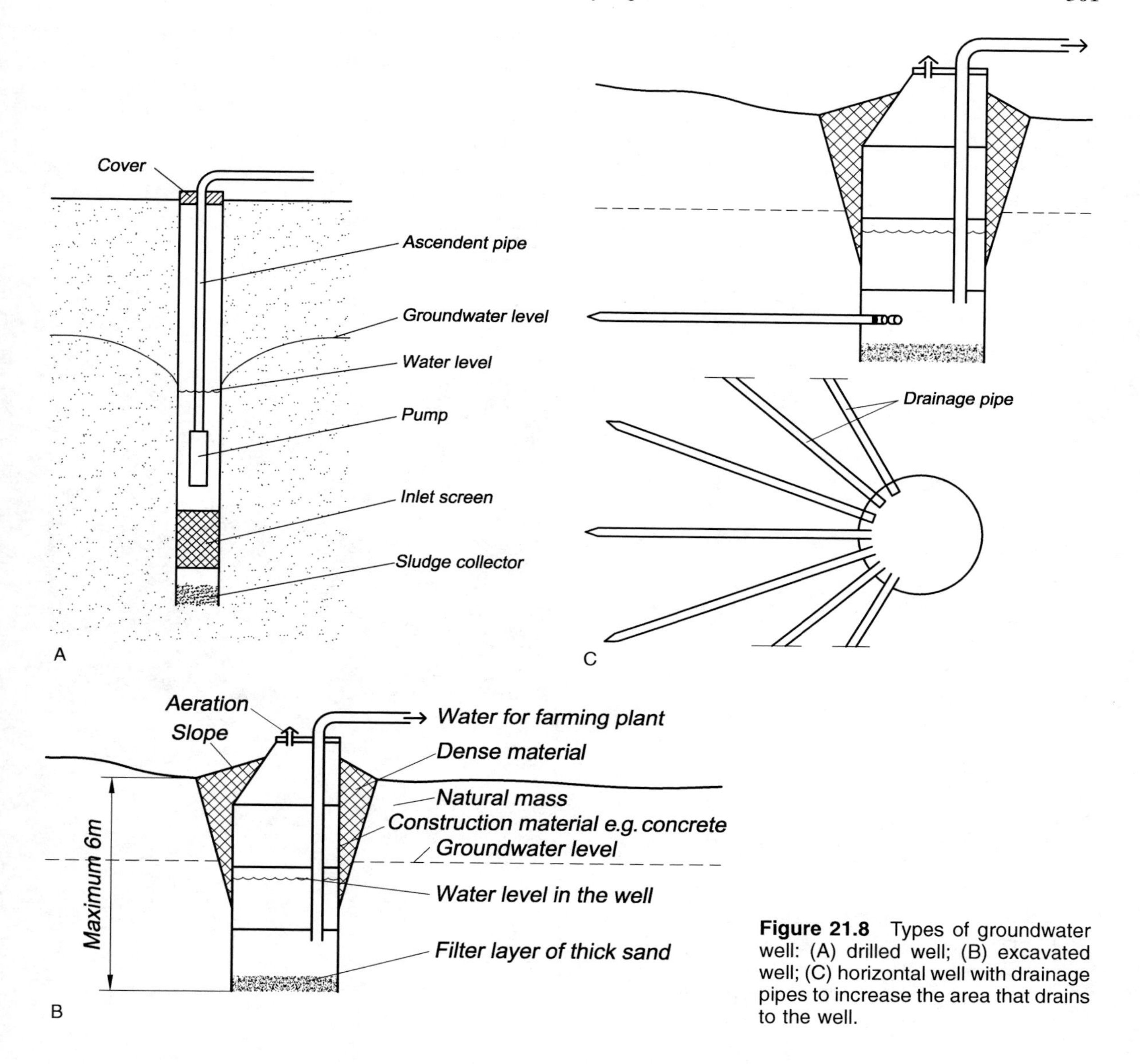

Figure 21.8 Types of groundwater well: (A) drilled well; (B) excavated well; (C) horizontal well with drainage pipes to increase the area that drains to the well.

laid as spokes in a wheel; thus more water is drained into the vertical well, because a larger area is tapped.

Pumping station

If pumps are used to transfer water to the farm a pumping station is needed. The pumps are either dry placed or submerged. Special equipment is necessary to make dry-placed pumps self-priming if the pump is installed at a higher level than the seawater. If such pumping stations are used, special care must be taken to avoid sucking of 'false air'. Dry-placed pumps may be installed in a well where there is pressure from the water, so circumventing this problem; this specially designed well must, in these cases, be excavated and cast. Submerged pumps will always be placed in some type of well (Fig. 21.9); both propeller and centrifugal pumps can be used. Propeller pumps are recommended with large water flows and low lift height (below 10m). If submerged pumps are used, the shore where the pumping station is to be placed must be shielded from the waves. It is also recommended that areas with large waves are avoided when siting pipes to dry-placed pumping stations. There are also experiences with drilling the inlet pipe in rock and siting the entire pumping station inside a mountain.[2] The pumping station can also be placed on the seashore and water is lead into the station through pipes. This requires appropriate ground conditions, sand or gravel for example.

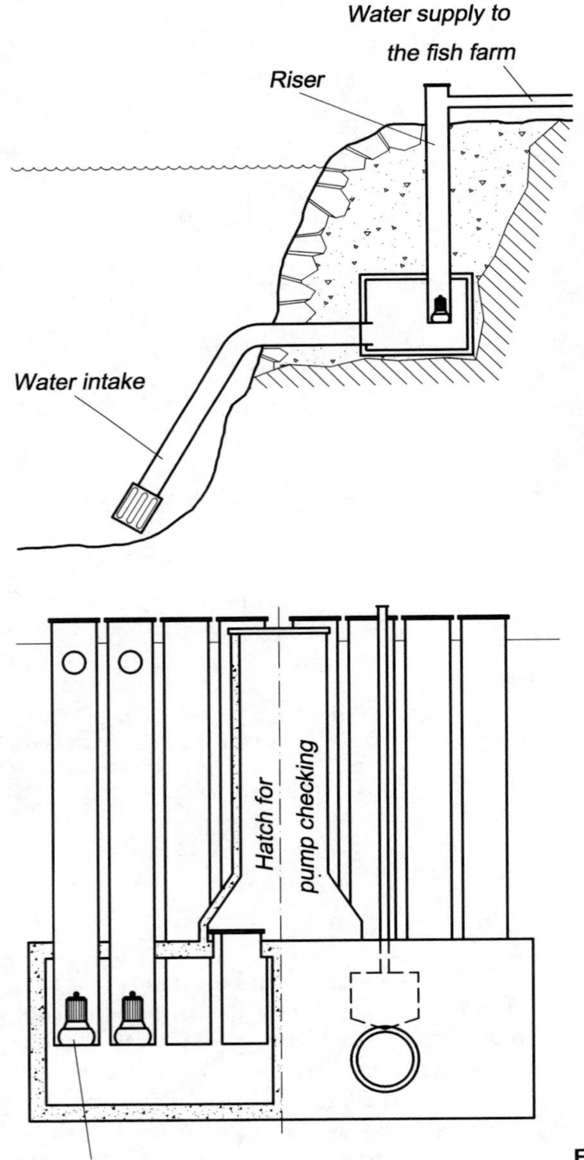

Figure 21.9 Pumping station with submerged pumps for pumping seawater to a land-based fish farm.

For emergency reasons, the water should be supplied to a farm using at least two pumps. It is important that the total efficiency of the pumps is high to reduce the cost per cubic metre of water. If the requirements for water delivery vary, it could be advantageous to use several pumps of different size; however standardization of the pumps so that it is possible to change spare parts and have a common stock of the most needed spares, will be difficult in such cases. There should be at least one stand-by pump in the pumping station. A programme that alternates between the pumps should be used so that all pumps will be run for some period. In this way the stand-by pump is always in use and not found to have seized up with rust when needed.

The pumps are installed in a pumping station.

Figure 21.10 A power generator may be included on the transfer pipeline if there is large height difference between the water source and the fish farm.

Dry-placed pumps are placed in a building on shore. For submerged or dry-placed pumps in a well, the pumping station is placed below the low tide mark in the sea. The pumping station is often made of prefabricated concrete or fibreglass. Submerged pumps should be placed on guide rails so that is easy to take them up for maintenance.

Transfer pipeline

As mentioned earlier, transfer pipes should be as short as possible to avoid unnecessary expense. If only a low head is available (<5m) or if the water is pumped, it is especially important to use piping that is as smooth as possible. Single resistances with high friction coefficients must be avoided; 90° elbows should be substituted with shallower bends, for instance three 30° elbows, or one long smooth elbow. Reduced velocity through the pipelines could also be used to reduce the friction loss, but it must be above 0.5 m/s to avoid settling.

If the water source is at a sufficiently higher altitude than the farm (>30 m) the water velocity in the pipeline could be very high; velocities above 3 m/s should be avoided to prevent breakage in the water column and vacuum effects. Increased velocities will also increase the requirements for anchoring the pipes. In such circumstances a small generator can be installed to utilize the energy in the inlet water (Fig. 21.10); it is important to choose one made of material that is not toxic to the fish. After the turbine, it is important to aerate the water to avoid possible supersaturation with nitrogen gas.

Care must be taken if the pipe is laid over a hill or ridge (high crest) from a lake and is functioning as a siphon (Fig. 21.11) because a vacuum effect will

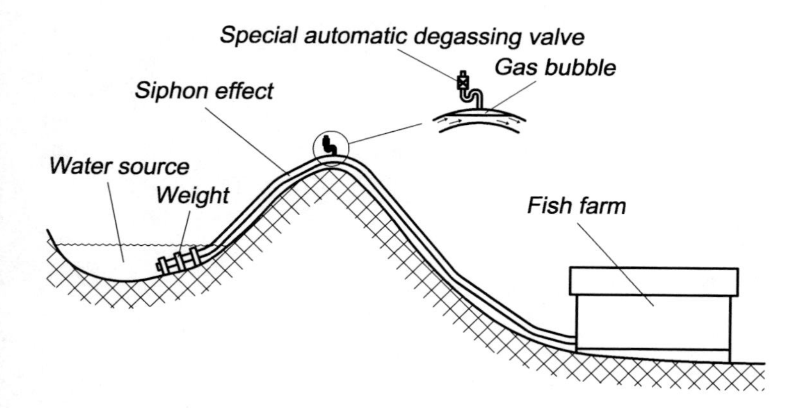

Figure 21.11 On a high crest there may be a vacuum in the inlet pipe.

occur on the top crest and the pipe may collapse. To avoid this, it is important to use a pipe of a sufficiently high pressure class which has a thick wall. Another problem that may occur when the pipeline is not evenly sloped in one direction, is the creation of gas bubbles at the crests. Normally there will always be some gas bubbles in the water and they will collect at the crests; when numerous bubbles collect at this point the effective cross-sectional area of the pipe is reduced. This will again reduce the water flowing through the pipe and in the worst case the water flow can be totally blocked by the gas bubble. The use of a siphon on the inlet pipe is therefore not optimal, and if possible should be avoided. Instead of laying the pipe across the shore of a lake it may be advantageous to excavate a channel out of the lake in which the pipe is placed. It is also important that the pipe is laid with a good and even slope.

If there are possibilities for gas collection at points in the transfer pipeline or pipelines inside the farm, special degassing valves can be used. These can be automatic and discharge gas when necessary. Degassing can also be done manually by installing a valve at the critical point in the pipeline and open it at fixed intervals. It should be remembered that there must be pressure in the system when the valve is opened to force the gas out. The valve can be closed as soon as the water comes out. In a siphon construction a special valve must be used to avoid breaking the siphon.

If fouling occurs in the inlet pipeline, it can be cleaned out by sending plugs through the inlet pipe, which is known as plugging. The cleaning plug can be inserted at the start of the inlet pipe and withdrawn at a point inside the farm (Fig. 21.12). The plug can, for instance, be made of some type of foam rubber. It is important to have enough water pressure to force the plug through the transfer pipeline.

Figure 21.12 A cleaning plug for sending through the pipeline to remove fouling.

21.2.3 Water treatment department

In this department the equipment to control and eventually improve the water quality is installed. It is a very difficult department to design well, especially when much equipment is needed (Fig. 21.13). In several established farms this department looks a mess. When planning, it is important to allow sufficient space. The department will, in all probability, be changed and modified several times. It is advantageous to include several valves in the system so that water flows can easily be stopped and sent in different directions. It must be possible to remove all the individual pieces of equipment without having to shut off the inlet water supply to the farm.

The amount of equipment needed in this department varies with the quality of inlet water and therefore the need to treat it. In this department equipment is typically installed for:

- Aeration
- Disinfection
- Oxygenation
- pH control
- Removal of suspended solids
- Heating and cooling.

Figure 21.13 Good planning of the water treatment department is a challenge because there are several components and it is easy to arrange them poorly.

If using a central re-use system, the equipment for ammonia removal and the re-use pumps may also be placed here.

Before starting to plan a water-treatment department on a new farm, it is always recommended that a flow chart be drawn that includes the different free water surfaces to prevent mistakes (see Chapter 22). It is quite normal to site the water treatment department in two rooms, a machine room and a water treatment room. Equipment for oxygen production together with equipment for heating and cooling can be placed in the machine room. In the water treatment room there are large free water surfaces and therefore high humidity, so proper ventilation is necessary here. Examples of equipment placed in this room include that for aeration, ammonia removal and solids removal. By having two rooms, the expensive mechanical equipment can be placed in a separate room with lower humidity.

It is advantageous to locate some equipment close to the water inlet or where the water transfer pipe to the farms starts. This ensures some exposure time before the inlet water reaches the farm. When using ozone as a disinfectant, or when adding chemicals for changing the pH, it can be done in the inlet and the need for a large retention basin inside the farm avoided.

There are advantages in having a feeder tank as a last step before the water reaches the production system. This will ensure equal pressure in the internal pipelines. At the same time, the pressure will not be too high. High pressure in the internal pipelines will create a lot of noise in the pipes and valves in the production hall. In addition, it may be necessary to use pipes and parts of a higher pressure class, which is more expensive. A feeder tank is also a suitable place to install the alarm sensors, because the level will immediately drop when there is trouble with the water supply (Fig. 21.14); some reaction time is also achieved if the water flow

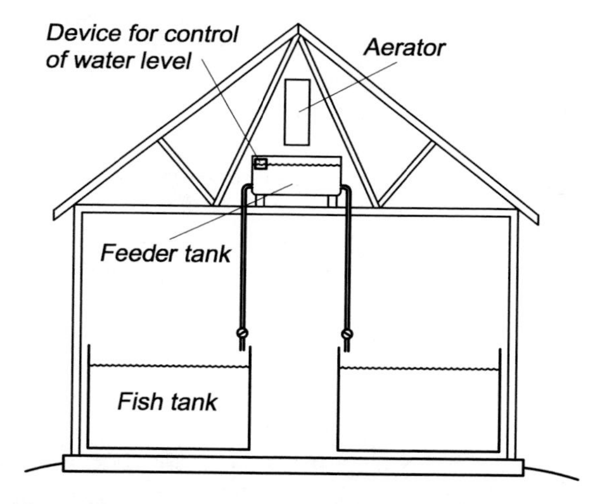

Figure 21.14 If possible, it is advisable to have a feeder tank in the water supply before the production room to ensure equal water pressure in the pipes to the production tanks. The feeder tank is also a suitable place for alarm sensors.

drops, depending on the volume of the header tank. The disadvantages of using a feeder tank are that it is necessary to lift the water to a higher level, if no pressurized water is available. For this reason seawater is sometimes sent directly through channels or pipes into the production room.

21.2.4 Production rooms

The production rooms are normally divided into a hatchery, a first feeding department and an ongrowing department. The hatchery can be separated into an incubation room and a hatching room depending on the species grown. Reasons for division could be to inhibit disease transfer, or to improve the control and the possibilities for different light regimes for different production units. It may also be that the on-growing department is outdoors.

Hatchery

Hatchery designs are based on the species farmed. For some species separate rooms are used for the incubation of fertilized eggs and for the hatching of the eggs, because different production units are required; water temperature and light conditions may also be different. In a hatchery for salmonids, for example, it is common to use trays with 40 cm × 40 cm troughs (Fig. 21.15). Two to four trays are used in a stack, and are commonly 2.1 m or 3.6 m long. The water inlet is at one end of the tray and the outlet is at the other. To control the eggs easy accessibility to all units is important. Because of this it is rather difficult with four tray stacks. If producing eye eggs for salmonids, cylinders may be used; this will, however, require another arrangement. Two cylinders placed at different heights on both sides of the walkway might be used.

The main inlet pipeline to the hatchery comes from the feeder tank. It is recommended that there is a separate supply to the hatchery to provide water at the correct temperature and quality; this can also include removal of smaller particles, use of protein skimmers and disinfection of the water. The flow rate to the hatchery will be quite low. Inlet pipes are either run along the wall or in the roof with valves to the separate trays. The main outlet pipeline is laid in the floor (see below).

To avoid disease transfer when personnel enter the hatchery, they must first pass through a disinfection area. If selling eggs, there should be a room specially for disinfection and packaging of eggs. Personnel movement between these rooms should be avoided. A hatch in the wall that connects the two rooms might be a solution. Between the hatchery and the first feeding department the use of a hatch in the wall to avoid movement of personnel and possible disease transfer is also common (Fig. 21.16).

Figure 21.15 Typical arrangement of trays in a hatchery for salmonid production.

Figure 21.16 Eggs and yolk sac fry can be transported from hatchery to first feeding department through a hatch in the wall without the need for staff to move directly between the departments.

Normally no windows are used in the hatcheries. Sunshine may injure the eggs of some species. Some light sources must also be avoided because they can damage the eggs; for example, blue–violet light tubes. Photomanipulation is also necessary for some species, and having no windows makes this possible. Some species will also require total darkness, or the use of only red light during some periods in the hatchery; for example, halibut.

The hatchery must be easy to clean and disinfect between the hatching seasons. It is important to remember this when designing the roof and walls, and installing the equipment.

If a combined hatching and start feeding is used, a mixed department is built and there is no separate hatchery. In smaller farms this might be a good solution, but of course, this has disadvantages concerning disease control.

First feeding department

The first feeding department is normally a separate room, but can also be combined with the hatchery or be a part of the on-growing department. Tank sizes also vary with the species; this is also the case with the method of first feeding, for instance if live feed is used. Normal tanks sizes are in the range 1–8 m^2 surface area.

It is important to have good accessibility, because first feeding is normally the most difficult stage in the production cycle and where control is most important (Fig. 21.17). The lighting conditions over every tank must therefore ensure good visibility. The foundations of the production results are laid in this department.

Two storey tanks could be used, but this arrangement can inhibit the accessibility and control of the tanks and is a challenge in the planning process (Fig. 21.17). The water can be supplied directly from a general feeder tank or warm water can be supplied from a separate feeder tank. Photomanipulation can be used to improve the first feeding results, so no windows are required in this department. The tanks can be placed along the walls or in the centre, or both with wider buildings. Walkways between the tank rows must be at least 1 m wide, or wider, depending on the systems used for transport of feed and fish.

On-growing department

The on-growing department can be inside or outside. It has become quite common to have parts of it inside, because of possibilities for photomanipulation, increased growth and better disease control, even if this increases the costs. If the on-growing department is outside it is normal to have a bird net above to protect the fish from birds. This may also be combined with a shadow net that shades the fish from the sun and protects some species against sunburn.

The tanks are normally quite large with a diameter of 5–20 m and water height up to 5 m. Because of their size it can be cheaper to have individual superstructures for the tanks. The recommended tank shape is circular, or square with cut corners. Circular tanks optimize material utilization, while square tanks optimize utilization of the farming area.

Depending on the arrangement of the tanks, the plan can be similar to that in the first feeding department. Because the tanks are normally higher, visual control of the fish in them can be a problem. The following solutions can be used (Fig. 21.18):

- Raised walkways for individual tanks
- Raised walkways for several tanks
- Submerged tanks

Raised walkways are simply constructed. If using a building, all installations can theoretically be on the floor and the building will then have a high second hand value. Submerged tanks are best for fish

Figure 21.17 Different layouts in a first feeding department.

farming purposes from the point of view of hygiene, and all the area around the tanks is easy to clean. This requires a concrete floor, located on the upper part of the tanks. On very large tanks (>10 m) a walkway across, like a bridge, can be used to increase the general view of the fish in the tanks.

Since the fish are now larger, the methods chosen for fish handling are more important. More feed is also used, and the feed handling method will therefore also influence the design.

Water supply, inlet pipelines

The water supply can be routed on the walls, in the roof or in the middle of the room. Normally, the water is supplied in pipes, but open channels can also be used although they create more noise and water under pressure cannot be used. A continuous fall from the feeder tank to all the production tanks is required.

It is important that the main inlet pipe is large enough to avoid significant variations in the total flow if the water flow to one tank is increased. Alternatively, the main pipeline can be installed as a ring connected at both ends (Fig. 21.19). This equalizes the pressure at each point in the pipeline, and ensures enough water in the tanks at the end of the pipeline. Use of a ring pipeline will also eliminate the risk of stagnant water staying in the pipeline, when some length of the pipe is not in use. In seawater this can be a particular problem because of the biological processes that occur in the stagnant water.

The water velocity is normally between 1 and 1.5 m/s; flows over 1.5 m/s create high head loss. Lower velocities are advantageous in achieving a larger

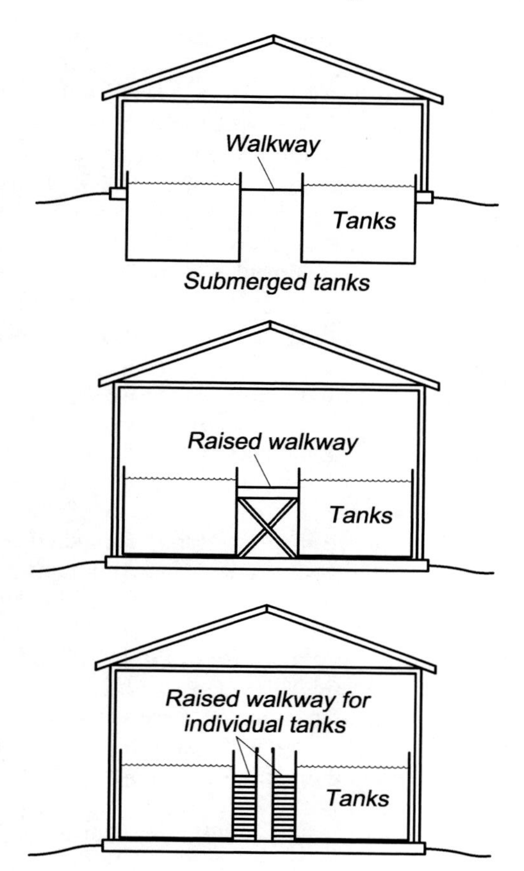

Figure 21.18 Methods of access to higher tanks for inspection, feeding, etc. include use of submerged tanks, a raised walkway to a series of tanks, or raised walkways to individual tanks.

Figure 21.19 If possible, a ring pipeline should be used to equalize the pressure and avoid flow variation in the connected tanks.

reservoir and reducing the variation in flow. However, settling in the pipes may now be a problem and it must be possible to flush the pipes out to remove this and prevent blockages. The valves through which the water flows to the individual tanks are set in the sides of the inlet pipe, not the bottom, to avoid settling in closed or partially open valves. One valve is set in the bottom to flush and clean the main inlet pipe. It is best to have a slope from the inlet reservoir to the valves, and from the valves to the fish tank. The main pipeline can also be close to the floor, or below the floor with vertical pipes up to the single tanks. When using such systems, care must be taken to avoid air locks in the pipelines; degassing valves can be used at critical points in the pipe system. Degassing of the inlet pipe is especially critical when using oxygenated water.

Outlet pipelines

The outlet pipeline starts in the tank. To avoid fouling and blockages the use of 90° bends in this pipeline is not recommended; bends should be as shallow as possible. This will also result in a more gentle treatment of the faeces, so crushing to smaller particles can be avoided. The head loss through the outlet system will also be reduced.

It is easy to get sedimentation in and blockage of the outlet pipes. To prevent this, a slope on the pipes of a least 0.5%/m is recommended. In addition it must be possible to flush and plug the pipes. When designing the system these requirements must be met.

For laying the outlet pipes inside the farm, several systems are employed (Fig. 21.20):

(1) Pipes under the floor
(2) Open channels or culverts in the floor
(3) Pipes in culverts in the floor, either covered with a grating or concrete block
(4) Pipes laid upon the floor.

All the alternatives have advantages and disadvantages. Method (1) is best for avoiding smell from the outlet and getting a surface inside the farm that is easy to clean. A major disadvantage is that lack of access prevents remedial measures being taken when something happens. Later reconstructions of the piping system are also difficult. Open channels create much noise and are a source of unwanted

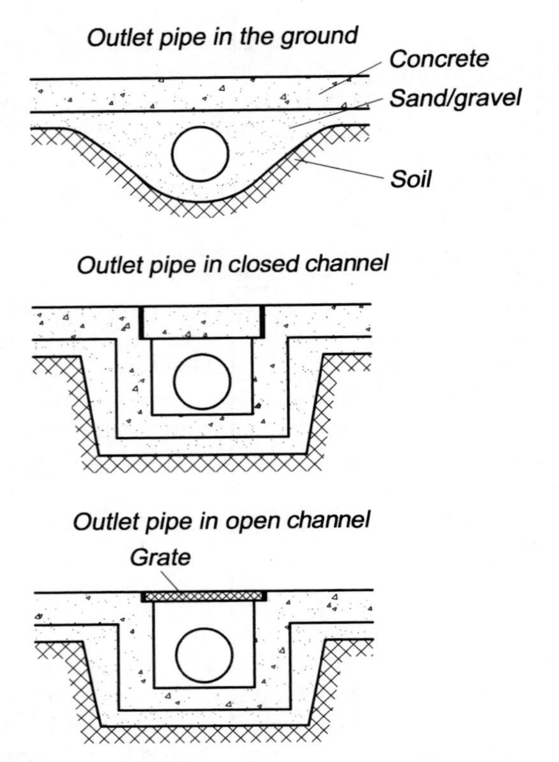

Figure 21.20 The outlet pipes can be laid below the floor, in open or closed channels/culverts or upon the floor.

smell, but are reasonably cheap. Method (4) is not recommended, because it will prevent adequate cleaning of the floor. Its great advantage is, however, that no special arrangement is necessary and that re-use of the buildings for other purposes is easy. Instead of laying pipes directly on the floor, it can be possible to hang them in the tanks, depending on the tanks used, and then use racks to bring the tanks to a correct working height.

Outlet pipes are not always filled with water; two phases can flow, with air above the water. The velocity ought to be above 0.3 m/s to avoid settling of solids; otherwise a good slope and easy flushing and plugging must be possible.

21.2.5 Feed storage

Feed storage depends on the type of feed and how it is packed. If using dry feed, as is most common in intensive aquaculture, the feed can be delivered in bulk and then stored in silos. However, this requires the use of large quantities. Big sacks are used for medium quantities and small sacks for small quantities. Whatever their size, sacks should be stored in buildings, or at least in a sheltered area to protect the contents from animals and birds. The feed sacks must also be protected against direct sunlight to reduce heating and possible destruction of the feed. This is not necessary if insulated buildings are used for storage of dry feed. In the feed storage house concrete floors are typically used because spilled feed is easily cleaned up.

The size of the feed store depends on the feed consumption, types and sizes of feed, and the shelf life. If much feed is bought at the same time, the price for feed and transport is reduced, but this requires a larger feed store. Problems will occur if the feed is stored for too long. The shelf life depends on the composition and the temperature in the feed store, and is given by the feed supplier.

It is important to be aware of the feed handling lines when designing the feed store. If using big sacks, how is the feed going to be transported in and out of the feed store? For instance, if using big sacks, they can be hung up for manual tapping into a wheelbarrow. It is then important to have equipment for lifting the sacks and doors that are wide enough.

21.2.6 Disinfection barrier

To protect the farm against transfer of disease, disinfection is recommended. This includes a disinfection barrier in front of the total farm area and others before entering the different production departments on the farm. A simple method is to utilize a disinfection mat/bath to which a disinfectant is added. The shoes are disinfected when stepping into this bath. It is, however, preferable to use a barrier where the shoes are changed. Having a clean zone, where people only walk in their socks is a possible solution here (Fig. 21.21). Clothes may also be changed in this zone. It is advantageous to use different colours for shoes and working clothes in the different departments. How much sectioning is to be done into separate departments with prior disinfection will always be a problem because it is costly, and more time is needed for changing, especially of clothes.

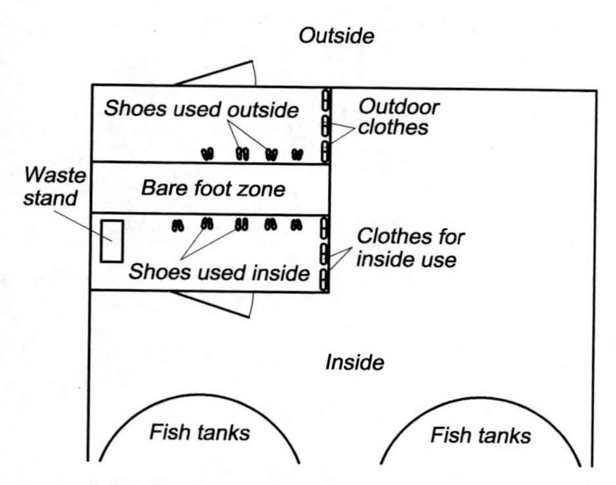

Figure 21.21 A disinfection barrier before the production rooms is recommended.

21.2.7 Other rooms

In a fish farm it is normal to have a small workshop, or at least a place to store the tools, because there is quite a large amount of technical equipment on an intensive fish farm.

Depending on the size of the farm it is normal to have rooms for personnel including a mess room, wardrobe, toilet and bathroom. An office is also necessary on a fish farm; on large farms this could even be an administration building.

21.2.8 Outlet water treatment

If the outlet water is to be treated, this normally only includes equipment for removing particles and storing the sludge. This equipment should be placed as close to the production units as possible to avoid damaging the particles. The water must be treated as gently as possible before it enters this equipment. Normally the effluent water is treated in a separate department or building. It is important to have sufficient slope on the pipes from the production unit to the treatment plant to avoid having to pump because this breaks up the particles. The particle filter is normally of the rotating screen type with a mesh size of 90–100 μm. Outdoor settling ponds may also be used. However, phosphorus can be released from settling ponds.

Contamination of the inlet water by the outlet water must be avoided to reduce the possibilities for disease transfer. Therefore there should be no possibilities for direct movement of personnel between these departments. Neither should there be any possibilities for short circuits between the inlet and outlet pipes. If the inlet water is pumped from a lake or from the sea and the outlet water is sent back to the same source, it is important that short cuts and/or cross contamination are avoided. The inlet and outlet pipes must be spaced far apart both in the vertical and horizontal directions. Transport of the outlet water directly to where the inlet is placed by the main current must also be avoided.

21.2.9 Important equipment

Equipment for feeding and feed handling, and systems for handling fish are important on intensive fish farms because they are used so much. A great deal of equipment is available here (see Chapters 16 and 17). This equipment can to some degree be an integral part of the farm construction, or it can be portable. Therefore the choice of equipment influences the design of the farm, so if a farm is established using a particular system, much reconstruction can be necessary if it is later decided to change that system.

Feed handling

Some types of feeding equipment are commonly used on all intensive land-based farms. This can range from simple feeders on a single production unit, to larger automated central feeding systems or feeding robots. Automation of the feeding system depends on the amount of feed used. It is important that the complete handling process for the feed from delivery to end use is well thought out. The necessary components in the handling line will, of course, depend on the chosen feeding system; if a feeding robot or a complete feeding system is used, the feed must be stored in silos.

If using traditional feeders, the feed must be taken from the store to the fish tanks. Alternatives for performing this operation can be to carry sacks, or to use a trolley or wheelbarrow. Trolleys or wheelbarrows set requirements for the width of the walkways. The feed is normally lifted from floor level up into the hoppers of the feeders manually. The hoppers must therefore be easy to fill. The farm may also be designed with two storeys, a first floor in addition to the ground floor. Feed is stored on the first floor; this is also where hoppers to the tanks on the ground floor are filled, so gravity is utilized and heavy lifting avoided.

Fish handling

Fish handling is an important part of the work in an intensive farm, and equipment for doing this is important, especially when the number of fish and size increase. Reasons for handlings are various and include moving between departments, dividing groups to avoid excessive density, size grading and when delivering fish. Also here it is important to think in complete handling lines, where the fish are taken from the production unit and sent back to the tank. A handling line for moving fish may, for instance, include:

- Crowding in the tank
- Vertical transport out of the tank
- Horizontal transport between tanks.

The first operation, crowding, can be brought about by (1) reducing the water level (it must then be possible to reduce the water level in the tank), (2) using a rotating grid, (3) having removable tank bottom. For the second operation, vertical transport out of the tank, the following methods can be used: (1) net, (2) pumps of various types, such as centrifugal that is lowered into the tank or the handling centre, vacuum (pressure) that also sucks up fish, ejector or airlift that require deep tanks, (3) transport tanks lifted with a forklift truck, (4) fish screw (needs a large area). For horizontal transport the following methods are used: (1) dip net and/or buckets, (2) transport tank, (3) pipelines.

The handling system may also be an integral part of the farm (Fig. 21.22). Examples of such systems are: (1) a centrally placed pumping chamber to where the fish are tapped through in pipelines from all tanks, (2) by having pump adaptors close to the bottom in all tanks and a centrally placed pump, (3) pipelines near the top of all tanks, into which the fish are poured after using a dip net to take them out the tank.

For larger fish the tanks may be equipped with hatches through which the fish either swim voluntarily or are forced to go when the water level is reduced. The first system only requires one channel between the tanks, and there is a combination of voluntary and forced movement by movable grids (Fig. 21.23). The second system is based on channels on two floors and an elevator that lifts the fish between the two levels (Fig. 21.24). This system functions as follows. The water level in the tank is reduced and the fish are forced to leave via the hatch in the tank wall. There is a slope on the

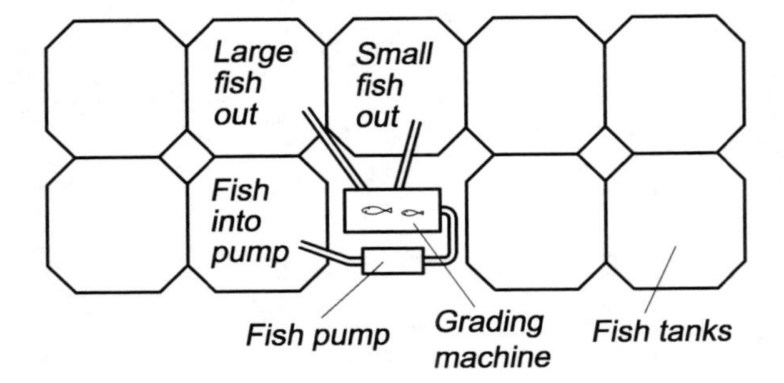

Figure 21.22 Farm in which the handling system is an integral part.

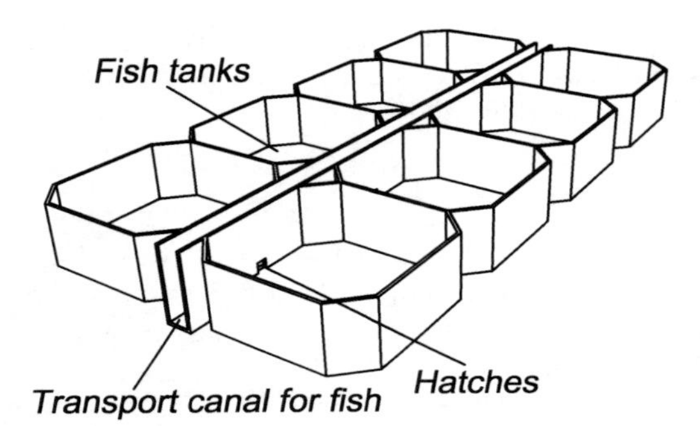

Figure 21.23 Fish farm equipped with channels for transport of fish.

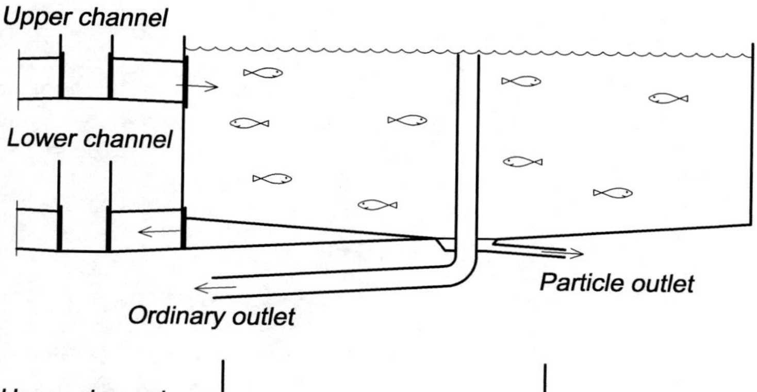

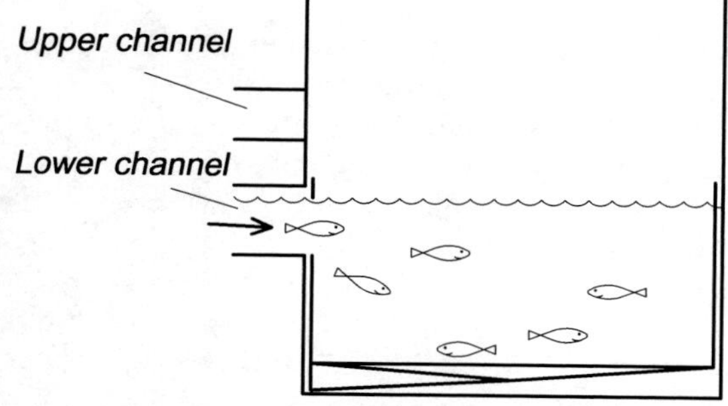

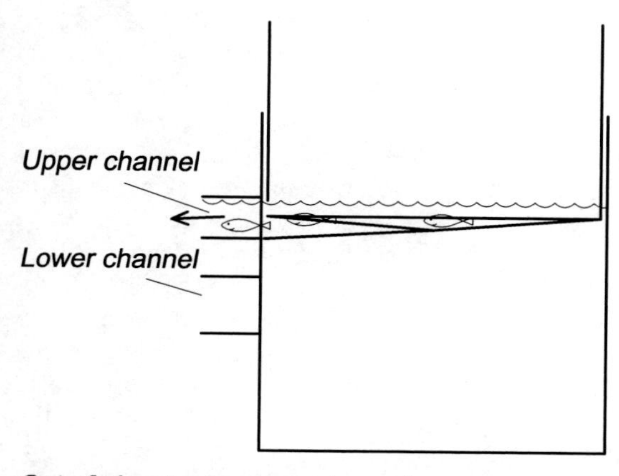

Figure 21.24 Fish farm equipped with channels for voluntary fish movement. Between the two channel levels there is an elevator for vertical transport of fish.

bottom in the lower channel and by continuously reducing the water level the fish are forced into the elevator at the end of the channel system. One tank inside the silo constitutes the elevator. The fish from one tank are collected in the elevator that is actually an ordinary fish tank. By closing the inlet hatch to the elevator and supplying water to the silo the water level will increase and the tank that floats inside will move upwards like an elevator. When the tank reaches the upper channel level the hatch is opened and the fish will start to swim out; by continuing the elevating process they will gradually be forced to leave. After this the water level in the upper channel will be gradually reduced and the fish forced back into their tank. Then fish from a new tank can be moved in a stepwise process.

The solutions for size grading must also be decided before the farm is established. If the farm is small and small amounts of fish are going to be graded, a cradle used directly in the production tanks might be a solution. Otherwise a machine can be used, such as a roller, belt or band type. If using a level grader the high head loss must be remembered. If thinking about the complete system, the grading machine can be placed in a grading centre, or it can be portable and moved to the tank that is to be graded. In the latter case it is important to have sufficient space in the walkways to place the grader.

21.3 On-growing production, sea cage farms

21.3.1 General

An on-growing farm bases its production on buying fry or juvenile fish and producing fish ready for slaughtering. An inexpensive system for on-growing production is to use cages, in lakes, calm rivers or in the sea. Later in this chapter a brief description of the design of a complete cage farm is given, focusing on a sea-based farm.

A total sea cage farm includes the following components (Fig. 21.25):

- The cage farm with fixed equipment
- An operations base
- A boat
- Net handling equipment.

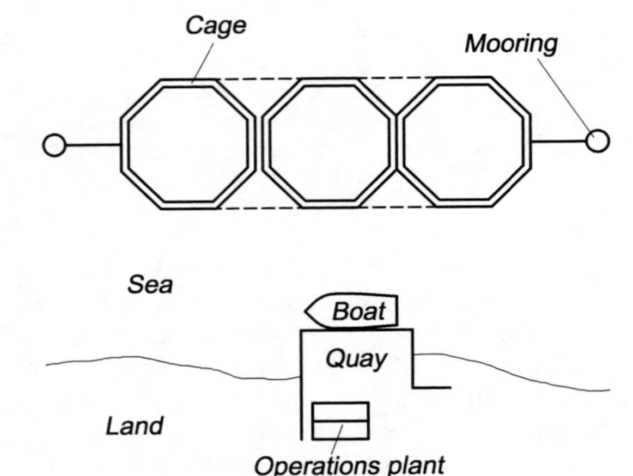

Figure 21.25 Plan of a typical sea cage farm.

The cage farm includes the cages with the mooring system. Equipment fixed to the cage can be feeding equipment, equipment for collection of dead fish and feed loss, and a lighting system.

21.3.2 Site selection

When selecting the sites for sea cage farms several criteria are important. Some species independent criteria used when evaluating a site are as follows:

- Stable water quality (the actual quality requirements will be species dependent)
- Good water exchange, but not too high a velocity (below 1 m/s is recommended)
- Minimum depth under the cages of 5 m
- Good infrastructure
- Temperature above 0°C to avoid icing; otherwise temperature appropriate for the species
- Not close to potential sources of water contamination
- For wave heights above 3 m the equipment costs are rather high.

When having a sea cage farm the use of several sites is recommended. If having species that need more than one year to reach the marked size or when having several inputs of juveniles every year it is advisable to use different sites (Fig. 21.26) to reduce the possibility of disease transfer between the inputs/generations. In addition, is it advisable to let the sites rest for a year after some years in production. This has been shown to improve production on the site. If this production model is used, the sites must not be too close together. The distance depends on the current conditions between the sites, but typically at least 1.5 km is recommended.

21.3.3 The cages and the fixed equipment

Cages and net bags

The number, type and size of the cages depends on the production regime and the farmed species. The exposure of the site to waves is also important. If the site is very exposed, special offshore cages are used. Large cages have become increasingly popular because the production costs can be reduced.[3] In the Norwegian salmon industry, circu-

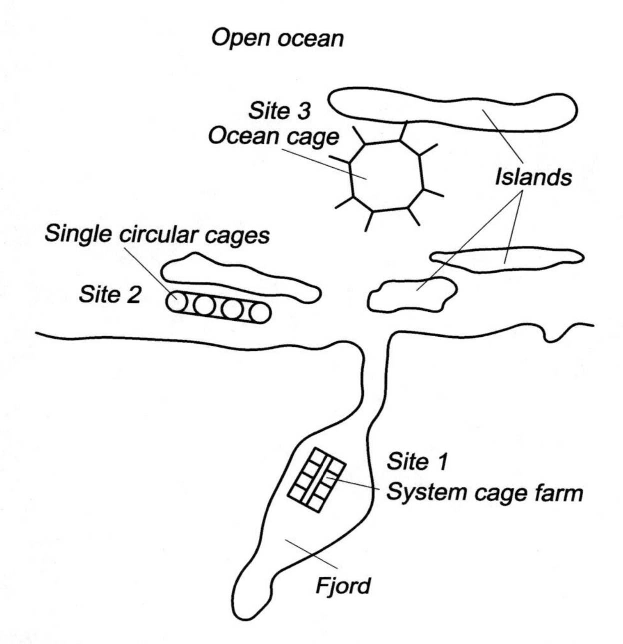

Figure 21.26 The use of several sites is recommended, both to inhibit to transfer of disease and to let the sites rest.

lar cages with a circumference of 90 or 110 m are becoming quite common. However, extra demands are made on the equipment used for handling the nets. Another trend is to utilize more exposed water for farming, which sets extra requirements for the cages, mooring systems and operating boats.

Sea cage farms may also be established close to the shore, so that a walkway can be constructed from the land directly to the cages (Fig. 21.27). In such conditions it is normal to use a system farm, or at least a farm with a floating walkway, to where the cages are moored.

A dead fish collector may be used in the cage: this can either be in the form of a stocking below the ordinary net bag, or a basket inside the cage. The basket can easily be lifted to the surface with a rope, and the dead fish collected.

Lighting system

To increase the growth and reduce early maturation, additional and continuous lights can be used in the sea cages.[4,5] In particular, during the winter season in high latitudes, this has been shown to improve the production results for salmoinds by 20–30%.[6] Thus in almost every sea cage in Norway extra light is used to increase growth and reduce maturation. The light source can stay above or beneath the surface. When having the light above the water surface it must be quite strong because it has to go through the water surface (Fig. 21.28). Light above the surface may be detrimental to the surroundings; for instance, there have been complaints from ships. For example, on a cage 15 m × 15 m four 500 W floodlights have been used; the necessary brightness is around 180 lux, the same as recommended for rooms not in continuous use, such as storerooms.

When planning a light installation, the following factors are important:

- Type of light source
- Placement of light source above or beneath the surface
- Number of lights
- When in the year it shall be used
- For how large a part of the day it shall be used
- Source of energy
- How to get the energy to the cage; where shall electrical cables be laid or shall generators be used?

Today, underwater lights have become much more commonly used, because they are closer to what is to be illuminated. Also, they do not cause the problems associated with non-submerged lights, e.g. navigational confusion.

Feeding system

Several methods are used for feeding fish in cages, ranging from hand feeding where no additional equipment is required to automatic feeding where the additional installations depend on the chosen feeders (see Chapter 16). If traditional feeders are used, they are placed on a platform that could be integrated in the collar. The hoppers can be quite large (>1 m^3) and additional buoyancy is necessary. If central feeding systems are used it is only the pipes with the feed that enter the cages and no additional equipment is necessary. It can be possible to use additional equipment inside the cage such as detectors for uneaten feed and biomass estimation. Such equipment will, however, be portable

Figure 21.27 Cages can be individual, or part of a system with and without a walkway to land.

Figure 21.28 Lighting system used on a cage farm to improve fish growth.

and have no influence on the construction of the cages.

21.3.4 The base station

A base station is necessary for a sea cage farm (Fig. 21.29). This can be land based or sea based (lying on a raft), or it can be a combination of the two alternatives. The size and content of the base station depends on the size of the production unit and the management of the farm, including what services are bought from subcontractors. A base station may contain various pieces of equipment and storage facilities.

It is normal to have at least a wardrobe, toilet and mess room for the workers on the base. In addition there is a small workshop, or at least a place to have some tools.

Every farm also needs a system for taking care of the dead fish; to prevent them becoming an environmental problem, they must not be dropped into the water. A tank to which acid is added can be used, and the fish are ensiled (Fig. 21.30); the produce can, for example, be used to feed fur-bearing animals. The dead fish may also be stored in a cold-storage room, for later collection and utilization.

It is also normal to have a feed store on the base. A central feeding system may also be installed on the base. Feeding barges have become quite popular because they can be moved between sites if the sites are fallowed. Common construction materials are steel or concrete; old ferries have proved quite popular as feeding barges.

The base, whether it is on land or sea, must be of a design that makes it easy for boats to dock. On a sea-based barge this is quite simple; if the base is on land a quay is necessary.

The base may or may not include net storage, net washing equipment and impregnation equipment. However, today it is quite common to employ subcontractors to do all the net handling and storage.

21.3.5 Net handling

Net handling represents a major part of the total workload on a sea cage farm, and requires additional equipment. For many sites fouling is a large problem for nets in the sea and to reduce the degree of fouling the mesh size needs to be as large as possible. This means that it is an advantage to change the nets according to fish growth, so as large a mesh size as possible can be maintained. One or two sizes are typically used per year, but this depends on growth and species; nets can be much more frequently changed on exposed sites.[7] Net exchange is heavy work, especially on larger cages; large cranes are required to handle the nets. This is also one reason for using subcontractors, because they have large equipment specially adapted for the purpose.

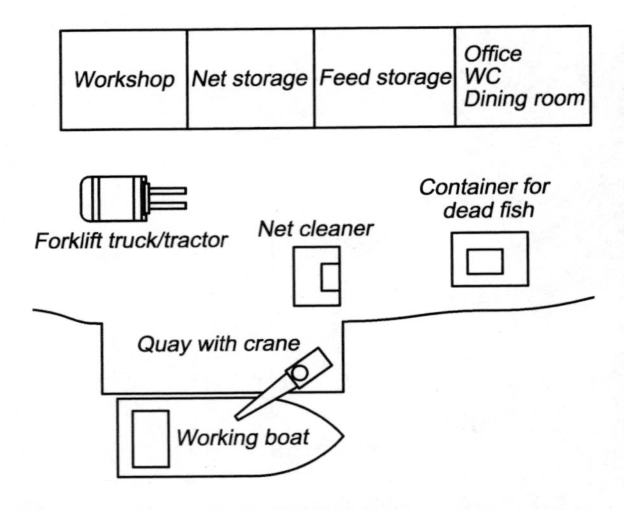

Figure 21.29 Normally a cage farm will be equipped with a base station which is either land or sea based.

Figure 21.30 A tank for ensiling dead fish.

If there is much fouling on the site cleaning or washing of the nets is necessary. This ensures enough water passes through the net panel to supply oxygen and remove waste products. On exposed fouling sites, the nets need to be washed several times a year, up to once a month. Washing may either be done when the net bags are standing in the sea, or they can be removed and taken to shore for washing. Special washing equipment is used by divers to wash nets in the sea. If the nets are taken to shore, large washing machines are used. These machines are similar to a traditional domestic washing machine but have a larger drum into which the nets are loaded (Fig. 21.31). The effluent from the washing machines has a high content of fouling materials. This is discharged at one place through the outlet pipe and the point outlet may be too high. Today there are calls for purification of wastewater from such washing machines; there is a particular problem with discharge of antifouling agents, because normally at least 20% of the antifouling agents remain on the nets at the start of the washing process.

The nets will also need regular repair, because the mesh will break. This is normally done in connection with washing and before the nets are set out again.

It is also usual to treat the net with an antifouling agent to inhibit fouling so that they can stay in the sea for longer before removal for washing. Copper-based antifouling agents are widely used and quite effective. However, there are environmental concerns regarding the use of copper, but no other antifouling agents have so far achieved the same efficiency. Different types of biocides may be used, but there are also environmental concerns

Figure 21.31 Equipment for washing net bags: (A) and (B) when the net is on shore with the use of large washing machines; (C) when the net is in the sea.

about these. Much research is being undertaken to find alternatives. Before setting out the impregnated nets they are dried so that the antifouling stays on the net.

21.3.6 Boat

All sea-based on-growing farms need a boat (Fig. 21.32). In the past, small boats of polyethylene or fibreglass with an outboard motor were often used. Today, larger boats of steel or aluminium with a working deck are becoming increasingly common as a result of the trend towards using larger cages and sites with more feed and heavier equipment. Faster boats are more common than previously, and speeds of up to 15–20 knots are normal. One reason for this is that the distance from the coast is increasing and

Figure 21.32 Different types of boat used on sea cages farms. Today there is a trend towards larger and faster boats.

more exposed water is being used for farming. Catamarans with wide decks and lengths in excess of 10 m are much used today. It is common to have a crane on the boats for handling nets and feed sacks.

References

1. Huguenin, J.E., Colt, J. (2002) *Design and operating guide for aquaculture seawater systems.* Elsevier Science.
2. Lekang, O.I. (1991) *Lukkede produksjonsanlegg for laksefisk i Norge.* ITF-rapport nr. 18, Norwegian University of Life Science (in Norwegian, English summary).
3. Guldberg, B., Kittelsen, A., Rye, M., Åsgård, T. (1993) Improved salmon production in large cage systems. In: *Fish farming technology. Proceedings of the first international confernce on fish farming technology* (eds H., Reinertsen, L.A., Dahle, L. Jørgensen, K. Tvinnereim). A.A. Balkema.
4. Kråkness, R., Hansen, T., Stefansson, S.O., Taranger, G.L. (1991) Continuous light increase growth rate of Atlantic salmon (*Salmo salar* L.) post-smolts in sea cages. *Aquaculture*, 95: 281–287.
5. Hansen, T., Stefanson, S., Tarnager, G.L. (1992) Growth and sexual maturation in Atlantic salmon, *Salmo salar* L., reared in sea cages at two light regimes. *Aquaculture and Fisheries Management*, 23: 275–280.
6. Willougby. S. (1999) *Manual of salmonid farming.* Fishing News Books, Blackwell Publishing.
7. Lucas, J.S., Southgate, P.C. (2003) *Aquaculture, farming aquatic animals and plants.* Fishing News Books, Blackwell Publishing.

22
Planning Aquaculture Facilities

22.1 Introduction

Planning of aquaculture facilities, of whatever type, is a complicated process that requires much knowledge to achieve a good result. It is, for example, more difficult than planning a typical industrial production plant, such as for manufacturing metal parts. Aquaculture facilities involve living individuals. The production result depends, for instance, on the suitability of the tanks, water flow conditions and whether water quality meets the requirements of the individuals, whether fish or shellfish. Planning faults will reduce the performance of the individuals, which can be manifest as reduced growth or more frequent disease problems, for example.

The requirements for planning will vary according to the type of facility. The planning of a farm with one or a few excavated ponds is fairly simple. A rather more complex situation occurs when planning a land-based fish farm for indoor juvenile production. The planning will be even more complex if the farm is to include water re-use technology in addition to flow-through technology. Such complex planning tasks involve several fields of competence: for instance, sanitary, electrical, building and architectural. These are all technological, but because the production involves living organisms it is also necessary to have biological knowledge, for example of the optimal environmental growth condition for the fish. Since so many subjects are involved, planning is not simple to perform and a number of specialists will normally be involved, at least when planning larger farms.

A number of theories and methods have been introduced to optimize the planning process, especially for planning buildings and industrial facilities.[1,2] Despite the biological aspects involved in the planning of aquaculture facilities, some of these basic theories and methods can be employed in addition to the important matters regarding aquaculture.

In this chapter important elements in the planning process of an aquaculture facility are described. This is done by presenting one simplified method for performing the planning process. This is utilized as a tool to avoid missing important elements during the planning process which, because of the complexity of the planning process for intensive fish farms is very easy, at least for personnel who are not highly trained. It is important to remember that this is only one of several methods, and planners will often have developed their own based on experience.

22.2 The planning process

When the planning process starts there is always an initiative to alter the prevailing situation. For example, someone may want to establish a completely new farm, or only be a minor reconstruction of the farm is wanted. Both require a planning process to have been completed before building commences. For the planner, it is important to ensure that they really understand the needs of the applicant to be able to execute the planning process optimally.

The planning process may be separated into the following parts from the choice of site, to when the facility is finished and in production:

(1) Site evaluation and selection
(2) Production plan

(3) Room programme
(4) Necessary analyses, such as function, form, technology, environmental impact and economy
(5) Development of alternative solutions based on the analysis
(6) Evaluation and synthesis of the alternative solutions
(7) Actual design, making the necessary drawings and description, calculation of costs
(8) Drawing up invitations to tender, choice of contractor, starting building
(9) Function test of the plant, with and without fish
(10) Project review.

If there is only one site available, it will only need to be evaluated. It is, however, important to show the limitations of the site. If an extension of an existing plant is wanted, the same will be the case. Independent of this, it is always important to carry out the site evaluation and control to ascertain whether the site really can tolerate the extension and what problems may occur. This may also set additional requirements in the planning process, for instance that there will be a limitation in the water supply and that re-use technology is required.

A production target or a given production plan may also be the starting point of the planning process. If this is the case, a proper site has to be chosen based on these requirements.

22.3 Site selection

To choose a good site is of course of major importance for future production results and possible problems, so proper investigations about site performance must be carried out. In relation to planning this description will give the criteria for the further planning process. If several sites could be used, the description will give the necessary basis for evaluation and selection of the site.

The site chosen will, of course, depend on the type of farm that is being planned – a hatchery or on-growing, land- or sea-based. An extremely important selection criterion when talking about aquaculture facilities is, of course, the amount and quality of available water. There are many stories of land-based freshwater aquaculture facilities suffering from lack of water after some years in production.

For cage farming, the water quality and current are of great interest. The depth and bottom conditions are also important because of the mooring requirements (see Chapter 15). For land-based farms the water quality will also be of major importance, but here the amount of water available is also of great interest. It is important to remember that when a farm is planned it is designed for a given water flow. In almost every case, after a period in production there will be a desire to increase production and therefore the need for water will increase. It is therefore advantageous to include this possibility in the planning process. When checking the possible amounts of water that could be withdrawn from the water source and used in the farm, it is important to find values for the possible water supply for every month all year round. A monitoring programme before establishing a farm must be implemented and surprises resulting from dry seasons must be avoided. Therefore it is important to look up as much historical data as possible regarding the water source. Possibilities for regulation of the water level in lakes, or damming up rivers must also be evaluated.

Good water quality will always be the best, regardless of species farmed. Some species, such as carps, will not have such high requirements for the water quality, while others such as salmonids are more stringent. Water of poor quality can be used on a salmonid farm but requires additional treatment before use, so resulting in increased costs.

Available infrastructure is also important when selecting a site. To have easy access to electricity, good roads and telephone lines reduces the costs of establishing the farm.

22.4 Production plan

In the production plan an estimate of the future production is given, for instance how much fish and of what size is going to be in the farm at different times of the year, normally every month at least, to ensure a given production (Table 22.1). This includes the requirements for oxygen and water, eventually requirements for heating or cooling of water, and necessary units for storing of the fish.

Table 22.1 Example of a production plan for a land-based farm.

	Month											
	Jan	Feb	Mar	Apr	May	Jun	Jul	Aug	Sep	Oct	Nov	Dec
Water temperature (°C)	1	1	1	2	8	10	12	12	8	6	4	2
Egg water temperature (°C)											8	8
Temperature 0+ (°C)	8	8.0/12.0	12	12	12	12	12	12	8	6	4	2
Temperature 1+ (°C)	1	1	1	2	8	10						
Weight 0+ (g)	0.2	0.18	0.27	0.69	1.72	4.31	9.87	22.5	48.5	69.3	88.7	97.3
Weight 1+ (g)	103	107	110	114	117	158						
Number of eggs											683 971	674 419
Number of 0+ fish	657 343	629 583	590 031	554 732	550 631	546 561	542 521	538 510	534 529	530 578	526 656	522 762
Number of 1+ fish	520 581	518 898	511 254	507 475	503 725	500 000						
Total number of fish	1 177 624	1 143 790	1 101 285	1 062 189	1 050 285	1 046 561	542 521	538 510	534 529	530 578	526 656	522 762
Biomass eggs (l)											1 371	1 371
Biomass 0+ fish (kg)	131	113	158	383	949	2 355	5 353	12 135	25 897	36 766	46 730	50 887
Biomass 1+ fish (kg)	53 739	55 020	56 332	57 675	58 991	78 924						
Total fish biomass (kg)	53 870	55 133	56 490	58 050	59 940	81 279	5 353	12 135	25 897	36 766	46 730	50 887
Water requirement eggs (l/min)											120	120
Water requirement 0+ fish (l/min)	221	364	507	1 228	2 230	4 239	8 029	16 989	15 020	13 604	11 213	7 633
Water requirement 1+ fish (l/min)	6 986	7 153	7 323	8 651	29 496	58 404						
Total water for water (l/min)	7 207	7 517	7 830	9 879	31 726	62 643	8 029	16 989	15 020	13 604	11 213	7 633
Hatching trays (No. of)											69	69
Tank volume 0+ (m^3)	69 trays	63 m^2	59 m^2	38	95	236	535	607	1 295	1 838	1 557	1 696
Tank volume 1+ (m^3)	1 791	1 834	1 878	1 923	1 966	2 631						
Total tank volume (m^3)	1 791	1 834	1 878	1 961	2 061	2 867	535	607	1 295	1 838	1 557	1 696

Table 22.2 A room programme for a small fish production plant.

Area	Size
Rearing section	
broodstock	2 tanks, >8 m^2
hatchery	14 hatching trays
first feeding	12 m^2
on-growing	800 m^3
Water treatment	50 m^2
Various	
feed storage	15 m^2
workshop/storage	20 m^2
office/mess room	15 m^2
wardrobe	7 m^2
WC/shower	5 m^2
disinfection barrier	4 m^2

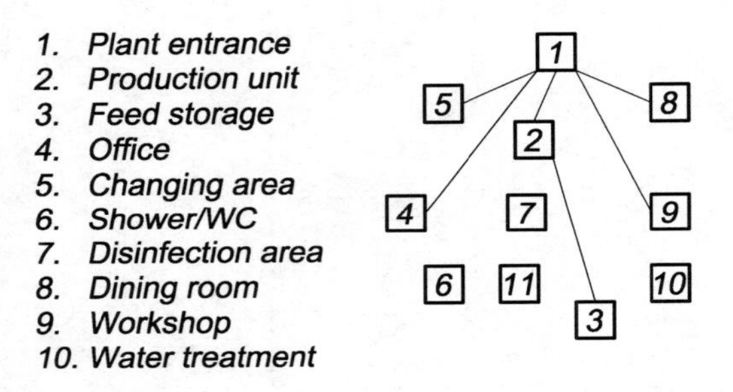

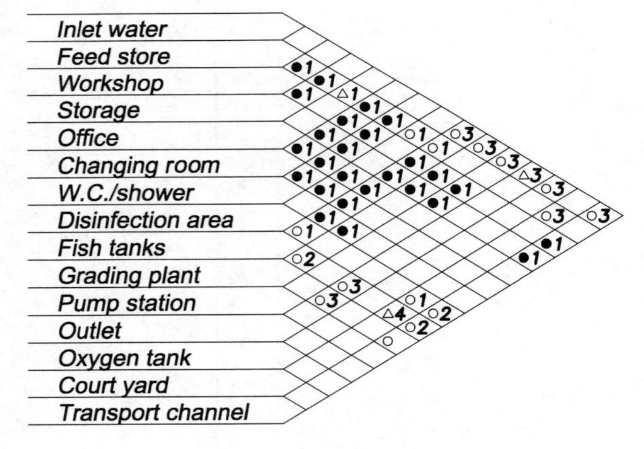

Importance
- ○ *Essential*
- ● *Important*
- □ *No interest*
- △ *Not wanted*

Cause
1. *Personnel traffic*
2. *Fish transport*
3. *Material transport*
4. *Infection, pollution*

Figure 22.1 Different ways to prepare a connection analysis.

A well prepared and detailed production plan is the basis for the rest of the planning process. It is important to allow sufficient time when developing the production plan; mistakes here will be transferred to mistakes in the planned facility.

22.5 Room programme

The real planning process can now start, and the first task is to get a survey of the main installations, including the buildings with necessary rooms and heavy fixed installations. This is done through a room programme (Table 22.2). Here a first estimation of the need for size and area is also given. A

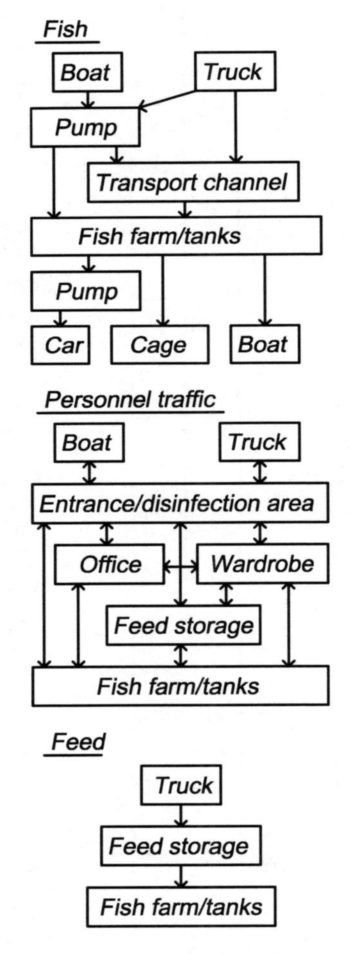

Figure 22.2 Drawing process diagrams can be a tool in the planning process.

Grading of fish

Method	1. Out of tank	2. To grader	3. Up on grader	4. Grading	5. From grader	6. Up into tank
A	Dip net	Dip net	Dip net	Roller	Dip net	Dip net
B	Tap	Pipe	Pump	Several floor	Pump	Tap
C	Tap	Pipe	Pipe	Belt	Pipe, slope	Tap

A - Small investments, much manual work
B - Larger investments
C - Depending on slope

Figure 22.3 Various methods for fish handling illustrated in an alternatives chart.

Figure 22.4 Blasting for establishing aquaculture facilities in the coastal area may result in unsightly scars in the landscape.

given production will, for instance, need a tank volume based on the requirements of the fish. If the production of fish is known, some assumptions about the space for feed storage can be made. Some rooms, such as a changing room, bathroom, mess room and office may also be necessary. The aim of the room programme is to obtain some idea about the size of the different rooms and installations which can be used in the further analyses, not to make a complete list of what the facility is going to look like when it is finished. By summing all these components an estimate of the total size of the plant is obtained.

22.6 Necessary analyses

Part of the aim with analysis is to remember to think through the different possible solutions. Advantages and disadvantages of the different solutions are to be discussed, which is really helpful in the planning process. Whilst analysis can be performed on many topics, it is important to perform the main analyses. This part must not be confused with the description of the chosen solution; it must be an analysis.

One necessary analysis is that concerning area connections; which areas in the plant are or are not to have connections. This can be illustrated with an example: a farm is fenced in and the only entrance is through a disinfection barrier where the shoes are disinfected in a bath; there should be no possibilities for direct entrance to the plant in other ways. During the planning process an area connection analysis will identify such relations.

One method of performing the analysis is to spread out the rooms and areas from the room pro-

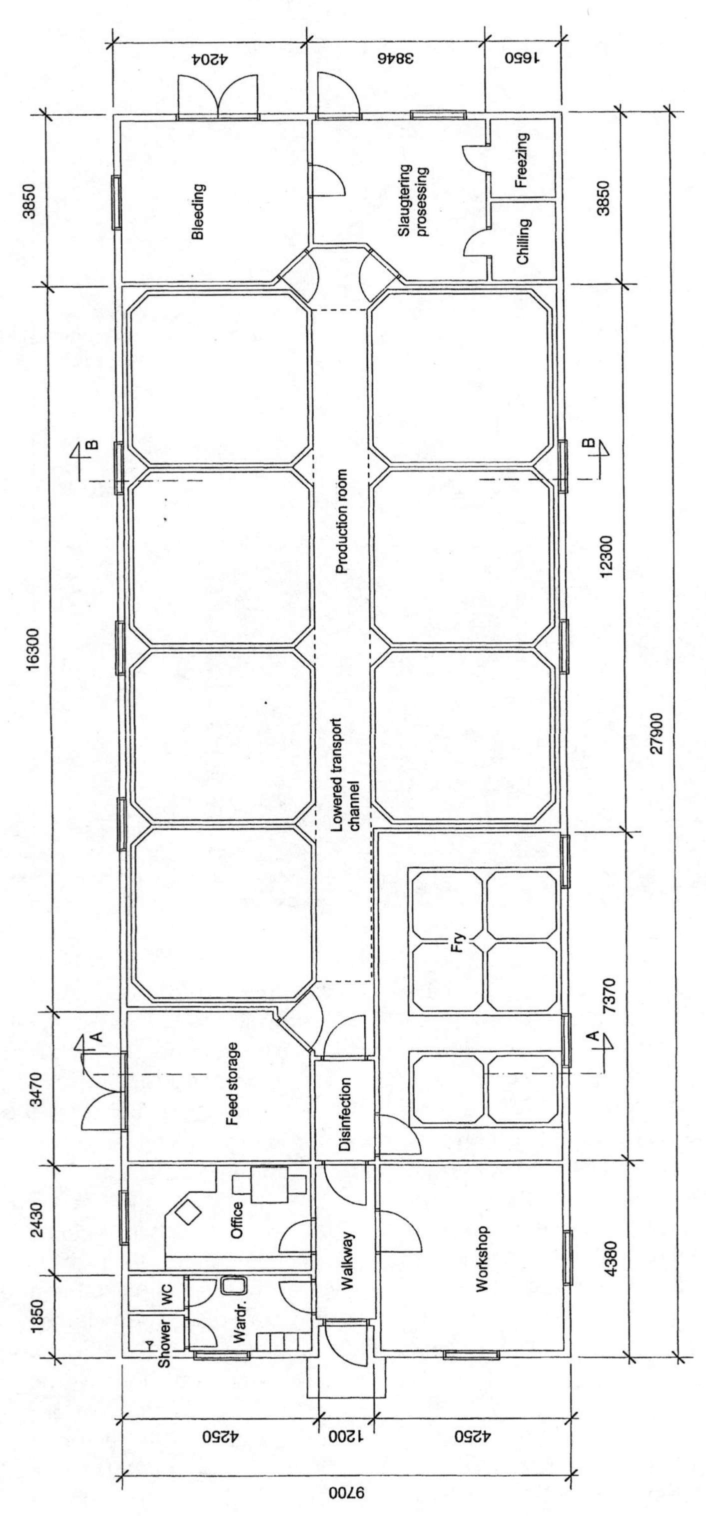

Figure 22.5 Alternative plans should be developed before choosing the layout of the fish farm.

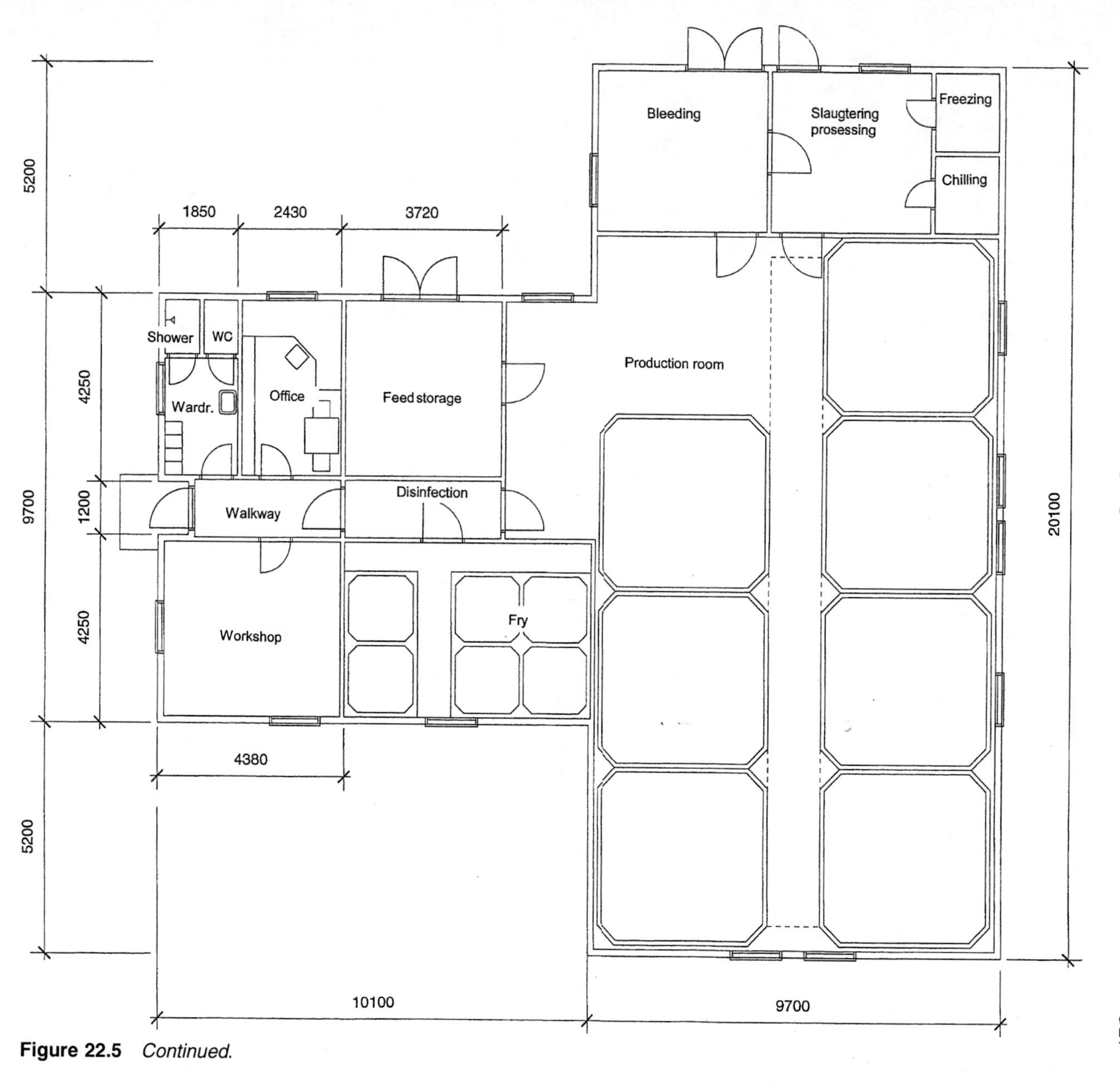

Figure 22.5 *Continued.*

gramme like pieces of a puzzle and draw lines between the areas were connections are wanted (Fig. 22.1). The same can also be done with an arrow diagram, where connections and the reason for connections are illustrated (Fig. 22.1).

To remember the different process that must take place process diagrams can used as a tool (Fig. 22.2). The technical analysis includes a survey of ways of solving technical problems with their advantages and disadvantages. For instance, if the water is to be aerated, what types of aerators are to be used and what are the advantages and disadvantages of the different types; another analysis can be whether or not to use oxygen. Analysis of different materials includes the advantages and disadvantages of each. Process diagrams and alternatives charts are also examples of assistance tools; for instance, alternatives charts are helpful for showing various handling methods (Fig. 22.3).

Form and situation analysis includes where in the terrain the farm can be located, with advantages and disadvantages; for example, should it be in the ground or on top. Aesthetic considerations must also be included.

Analysis of environmental impact is becoming increasingly important for aquaculture facilities. How to reduce the discharge is an important analysis. To establish aquaculture facilities near beach zones may result in large impacts in the landscape, caused, for instance, by blasting operations that create large 'scars' in the landscape (Fig. 22.4). The need for proper analysis is necessary in such cases.

An area function analysis of the different areas is also commonly included where the requirements and their function are discussed. Taking the feed storage as an example, this could include the following analysis: will there be possibilities for expansion or not; will there be possibilities for draining the floor or not; are there any special requirements for the surface of the floor or not?

22.7 Drawing up alternative solutions

Based on the analyses, the development and planning of alternative solutions may start. This includes simple sketches of the different possible options. The reason for stressing development of alternative options to meet planning requirements is that this functions as a tool to develop optimal solutions. The plans can, with advantage, be as different as possible from each other. This stresses the variability, which is important in improving creativity. At least two or three alternatives should be developed (Fig. 22.5); these can be discussed with the owner of the farm, to involve them in the planning process and to make sure that the developed solution meets their requirements. The water levels are extremely important, and when planning land-based fish farms it is important to have control of the free water levels; therefore it is helpful to prepare diagrams showing this.

22.8 Evaluation of and choosing between the alternative solutions

The next step in the planning process is to evaluate the alternatives and choose from among the developed solutions. On this basis the chosen plan can be further developed. Hopefully the analysis and consideration of different solutions have improved the plan compared to first proposals. All the developed solutions will of course have advantages and disadvantages, and these must be weighed when developing the final plan which will often be a mix of the alternatives.

22.9 Finishing plans, detailed planning

After choosing a solution, this can be further developed, by preparing more detailed plans and drawings of constructions and/or buildings. Here more detailed design of the necessary components is also included and, based on this, a more detailed calculation of the costs. This is labour-intensive. It may also be a two-step planning process with a pilot planning project followed by the detailed planning process.

The next step is usually to draw up invitations to tender with the necessary descriptions. When a tender is accepted, the building process can start. During the building process is it important to check progress regularly.

22.10 Function test of the plant

After finishing building the plant or part of the plant, a period of function testing is necessary, starting with single components and ending with the entire farm. First is it performed without fish in the plant and when everything is functioning the testing can be continued with fish in the system. Sufficient time must be taken at this important stage, and when establishing advanced facilities can take up to several months. This is an important stage that is often underestimated. To put the fish into the facilities too early may end in disaster if something fails. If contractors build the entire project, the owner of the farm must not take it over before operational testing of components and the whole farm has been carried out with satisfactory results.

22.11 Project review

It is important to undertake a post-hoc review of the building process and of the chosen options. The major object of doing this is to optimise the process in later planning, and to create future optimal solutions. Post-hoc project reviews are mainly for the benefit of the planner.

References

1. Muther, R. (2000) *Systematic planning of industrial facilities (SPIF)*. Management of Industrial Research Publication, Kansas City.
2. Svardal, S. (1994) *Planning of rural buildings. Theory and method*. Lecture notes. Norwegian University of Life Sciences (in Norwegian).

Index

Printed in the USA/Agawam, MA
December 1, 2010

555391.010